THE MIS SOLUTION

STUDENT RESOURCES

- Interactive eBook
- Graded Quizzes
- New Practice Quiz Generator
- New Trackable Activities
- Flashcards

- Videos
- Games
- Case Studies
- Review Cards

Students sign in at **www.cengagebrain.com**

INSTRUCTOR RESOURCES

- All Student Resources
- Engagement Tracker
- LMS Integration
- Instructor's Manual
- Test Bank
- PowerPoint® Slides
- Instructor Prep Cards
- Discussion Questions
- Online Supplemental Chapters

Instructors log in at **www.cengage.com/login**

△ Print

MIS5 delivers all the key terms and all the content for the **Management Information Systems** course through a visually-engaging and easy to reference print experience.

△ CourseMate

CourseMate provides access to the full **MIS5** narrative, alongside a rich assortment of quizzing, flashcards, and interactive resources for convenient reading and studying.

CENGAGE
Learning®

MIS5
Hossein Bidgoli

Vice President, General Manager:
  Neil Marquardt

Product Director: Steve Joos

Product Manager: Laura Redden

Content Developer: Adam Goetz

Product Assistant: Mandira Jacob

Marketing Manager: Eric La Scola

Content Project Manager: Darrell E. Frye

Media Developer: Tricia Hempel

Manufacturing Planner: Ron Montgomery

Production Service: Integra Software
  Services Pvt. Ltd

Sr. Art Director: Stacy Jenkins Shirley

Internal and Cover Designer: Craig Ramsdell

Cover Image: © AzmanJaka/iStock Photo

Intellectual Property
  Analyst: Sara Crane
  Project Manager: Kathryn Kucharek

Internal Ads:

  © iStockphoto.com/A-Digit | © Cengage
    Learning 2011

  © Go Media | © Cengage Learning 2011

  © iStockphoto.com/A-Digit | © Cengage
    Learning 2011

Common Art: © iStockphoto.com/hrstklnkr

To so many fine memories of my mother, Ashraf,
my father, Mohammad, and my brother, Mohsen,
for their uncompromising belief in the power of
education.
–Hossein Bidgoli

For product information and technology assistance, contact us at
Cengage Learning Customer & Sales Support, 1-800-354-9706

For permission to use material from this text or product,
submit all requests online at **www.cengage.com/permissions**
Further permissions questions can be emailed to
permissionrequest@cengage.com

Library of Congress Control Number: 2014945778

Student Edition ISBN: 978-1-285-83646-1

Student Edition with CourseMate PAC ISBN: 978-1-285-83645-4

Cengage Learning
20 Channel Center Street
Boston, MA 02210
USA

Cengage Learning is a leading provider of customized learning solutions with office locations around the globe, including Singapore, the United Kingdom, Australia, Mexico, Brazil, and Japan. Locate your local office at:
www.cengage.com/global

Cengage Learning products are represented in Canada by
Nelson Education, Ltd.

To learn more about Cengage Learning Solutions, visit **www.cengage.com**

Purchase any of our products at your local college store or at our preferred online store **www.cengagebrain.com**

Printed in the United States of America

Print Number: 02   Print Year: 2015

© Joshua Resnick/Shutterstock.com

CONTENTS

3 Database Systems, Data Warehouses, and Data Marts 44

4 Personal, Legal, Ethical, and Organizational Issues of Information Systems 66

Part 2

Data Communication, the Internet, E-Commerce, and Global Information Systems

7 The Internet, Intranets, and Extranets 136

Part 3
IS Development, Enterprise Systems, MSS, and Emerging Trends

© agsandrew/Shutterstock.com

10 Building Successful Information Systems 200

11 Enterprise Systems 224

12 Management Support Systems 242

1 | Information Systems: An Overview

LEARNING OUTCOMES

After studying this chapter, you should be able to:

- **1-1** Discuss common applications of computers and information systems.
- **1-2** Explain the differences between computer literacy and information literacy.
- **1-3** Define transaction-processing systems.
- **1-4** Define management information systems.
- **1-5** Describe the four major components of an information system.
- **1-6** Discuss the differences between data and information.
- **1-7** Explain the importance and applications of information systems in functional areas of a business.
- **1-8** Discuss how information technologies are used to gain a competitive advantage.
- **1-9** Explain the Five Forces Model and strategies for gaining a competitive advantage.
- **1-10** Review the IT job market.
- **1-11** Summarize the future outlook of information systems.

After you finish this chapter, go to **PAGE 19** for the **STUDY TOOLS**

© Monkey Business Images/Shutterstock.com

> **Organizations use computers and information systems to reduce costs and gain a competitive advantage in the marketplace.**

This chapter starts with an overview of common uses for computers and information systems, explains the difference between computer literacy and information literacy, and then reviews transaction-processing systems as one of the earliest applications of information systems. Next, the chapter discusses the components of a management information system (MIS), including data, databases, processes, and information, and then delves into how information systems relate to information technologies. This chapter also covers the roles and applications of information systems and explains the Five Forces

Model used to develop strategies for gaining a competitive advantage. Finally, the chapter reviews the IT job market and touches on the future of information systems.

1-1 COMPUTERS AND INFORMATION SYSTEMS IN DAILY LIFE

Organizations use computers and information systems to reduce costs and gain a competitive advantage in the marketplace. Throughout this book, you will study many information system applications. For now, you will look at some common applications used in daily life.

Computers and information systems are all around you. As a student, you use computers and office suite software and might take online classes. Computers are often used to grade your exam answers and generate detailed reports comparing the performance of each student in your class. Computers and information systems also calculate grades and GPAs and can deliver this information to you.

Computers and information systems are commonly used in grocery and retail stores as well. For example, a point-of-sale (POS) system speeds up service by reading the universal product codes (UPCs) on items in your shopping cart (see Exhibit 1.1). This same system also manages store inventory, and some information systems can even reorder stock automatically. Banks, too, use

desktop in seconds. The Internet is also used for social purposes. With social networking sites—such as Facebook, Twitter, Google+, LinkedIn, and Foursquare—you can connect with friends, family, and colleagues online and meet people with similar interests and hobbies. Twitter (*www.twitter.com*), for example, is a social networking and short-message service. Users can send and receive brief text updates, called tweets. These posts are displayed on one's profile page, and other users can sign up to have them delivered to their in-boxes. As an example, the author of this text-book sends daily tweets that consist of links to current articles about information systems applications, new developments, breaking news, IT jobs, and case examples. You can read these tweets in Twitter, Facebook, or LinkedIn.

Organizations also use social networking sites to give customers up-to-date information and how-to support via videos. These sites can reduce organizations' costs by providing an inexpensive medium for targeting a large customer base.

computers and information systems for generating your monthly statement, running ATM machines, and for many other banking activities.

Many workers are now telecommuters who perform their jobs at home, and others often use their PDAs (personal digital assistants) to conduct business while on the go. The most common PDA is a smartphone (such as an iPhone, Galaxy, Droid, or a Blackberry). A typical PDA includes a calendar, address book, and task-listing programs; more advanced PDAs often allow for wireless connection to the Internet and have built-in MP3 players. Smartphones are mobile phones with advanced capabilities, much like a mini PC. They include e-mail and Web-browsing features, and most have a built-in keyboard or an external USB keyboard (see Exhibit 1.2). Increasingly, tablet computers, such as iPads, are being used as PDAs. These tablets come with apps (small programs) for common applications, and they can improve the user's efficiency.

The Internet is used for all kinds of activities, from shopping to learning to working. Search engines and broadband communication bring information to your

Exhibit 1.2
Examples of smartphones

© iStockphoto.com/SKrow

© iStockphoto.com/pressureUA

A New Era of Marketing: YouTube

Companies use newspapers, magazines, TV shows, and search engines to promote their products, services, and brands. YouTube is a popular video-sharing service that can be used as a marketing tool. The videos on YouTube are very well indexed and organized. They are categorized and sorted by "channels." The channels range from film and animation to sports, short movies, and video blogging. Individual YouTube users have used this marketing tool to share videos and stories. One of the popular applications is watching how-to videos for repairing cars, home appliances, and so forth. Corporations can also take advantage of this popular platform. YouTube represents a great opportunity for marketers to reach consumers who are searching for information about a brand or related products and services. It can also be used as a direct marketing tool. The following are examples of corporations that are using YouTube to promote their products and services:

Quiksilver—This manufacturer of apparel and accessories, including the Roxy brand, frequently posts new videos of its products, continually renewing its Web presence.

Ford Models—Since 2006, it has uploaded over 554 videos promoting its brand.

University of Phoenix Online—This site has hundreds of video testimonials, reviews, and documentaries that promote the university's degree programs.

The Home Depot—Here you will find free content, including practical knowledge and money-saving tips for home improvements.

Nikefootball—Nike maintains several distinct YouTube channels that cater to specific audiences. Consumers can find content that is relevant to their needs without having to sift through everything.[1,2]

In addition, people use video-sharing sites to watch news, sporting events, and entertainment videos. One of the most popular sites is YouTube (*www.youtube.com*). You can upload and share video clips via Web sites, mobile devices, blogs, and e-mail. Users upload most of the content on YouTube, although media corporations such as CBS, BBC, Sony Music Group, the Sundance Channel, and others also provide content. Anyone can watch videos on YouTube, but you must register to upload videos. (This book has a YouTube channel on which you can watch many practical videos related to information systems.) Businesses are increasingly using YouTube to promote their products and services. See the information box above, which highlights a few such companies.

So what do all these examples mean to you? Computers and information technology will help the knowledge workers of the future perform more effectively and productively, no matter what profession they choose. In addition, these workers will be able to connect to the rest of the world to share information, knowledge, videos, ideas, and almost anything else that can be digitized. Throughout this book, these opportunities, as well as the power of computers and information systems, are explored.

© StockLite/Shutterstock.com

Computers and information technology will help the knowledge workers of the future perform more effectively and productively, no matter what profession they choose.

As you read, keep in mind that the terms *information systems* and *information technologies* are used interchangeably. Information systems are broader in scope than information technologies, but the two overlap in many areas. Both are used to help organizations be more competitive and to improve their overall efficiency and effectiveness. Information technologies offer many advantages for improving decision making but involve some challenges, too, such as security and privacy issues.

Computer literacy is skill in using productivity software, such as word processors, spreadsheets, database management systems, and presentation software, as well as having a basic knowledge of hardware and software, the Internet, and collaboration tools and technologies.

Information literacy is understanding the role of information in generating and using business intelligence.

The information box below describes one of the potential challenges.

 1-2 **COMPUTER LITERACY AND INFORMATION LITERACY**

In the 21st century, knowledge workers need two types of knowledge to be competitive in the workplace: computer literacy and information literacy. **Computer literacy** is skill in using productivity software, such as word processors, spreadsheets, database management systems, and presentation software, as well as having a basic knowledge of hardware and software, the Internet, and collaboration tools and technologies. **Information literacy**, on the other hand, is understanding the role of information in generating and using business intelligence.

SOCIAL NETWORKING AND THE VULNERABILITY OF PERSONAL INFORMATION

The popularity of social networking sites such as Facebook, Twitter, Google+, and Foursquare is on the rise. As of March 31, 2013, Facebook had more than 1.11 billion registered users, and the number is increasing on a daily basis.[3] But so is the potential risk. According to an InfoWorld study published on May 4, 2010, over half of all users of social networks in this country are putting themselves at risk by posting information that could be misused by cybercriminals. Many social networkers post their full birth dates, their home addresses, photos of themselves and their families, and the times when they will be away from home. This information could be used by cybercriminals for malicious purposes. According to the report, 9 percent of the 2,000 people who participated in the study had experienced some kind of computer-related trouble, such as malware infections, scams, identity theft, or harassment. To reduce risk and improve the privacy of your personal information, the study offers several tips[4]:

- Always use the privacy controls offered by the social networking sites.
- Use long passwords (8 characters or longer) that mix uppercase and lowercase letters with numbers and symbols.
- Do not post a phone number or a full address.
- Do not post children's names, even in photo tags or captions.
- Do not be specific when posting information about vacations or business trips.

Business intelligence (BI) is more than just information. It provides historical, current, and predictive views of business operations and environments and gives organizations a competitive advantage in the marketplace. (BI is discussed in more detail in Chapter 3.) To summarize, knowledge workers should know the following:

- Internal and external sources of data
- How data is collected
- Why data is collected
- What type of data should be collected
- How data is converted to information and eventually to business intelligence
- How data should be indexed and updated
- How data and information should be used to gain a competitive advantage

 ## 1-3 THE BEGINNING: TRANSACTION-PROCESSING SYSTEMS

For the past 60 years, **transaction-processing systems (TPSs)** have been applied to structured tasks such as record keeping, simple clerical operations, and inventory control. Payroll, for example, was one of the first applications to be automated. TPSs focus on data collection and processing, and they have provided enormous reductions in costs.

Computers are most beneficial in transaction-processing operations. These operations are repetitive, such as printing numerous checks, or involve enormous volumes of data, such as inventory control in a multinational textile company. When these systems are automated, human involvement is minimal. For example, in an automated payroll system, there is little need for managerial judgment in the task of printing and sending checks, which reduces personnel costs.

 ## 1-4 MANAGEMENT INFORMATION SYSTEMS

A **management information system (MIS)** is an organized integration of hardware and software technologies, data, processes, and human elements designed to produce timely, integrated, relevant, accurate, and useful information for decision-making purposes.

The hardware components, which are discussed in more detail in Chapter 2, include input, output, and memory devices and vary depending on the application and the organization. MIS software, also covered in Chapter 2, can include commercial programs, software developed in-house, or both. The application or organization determines the type of software used. Processes are usually methods for performing a task in an MIS application. The human element includes users, programmers, systems analysts, and other technical personnel. This book emphasizes users of MISs.

In designing an MIS, the first task is to clearly define the system's objectives. Second, data must be collected and analyzed. Finally, information must be provided in a useful format for decision-making purposes.

Many MIS applications are used in both the private and public sectors. For example, an MIS for inventory control provides data (such as how much of each product is on hand), what items have been ordered, and what items are back-ordered. Another MIS might forecast sales volume for the next fiscal period. This type of system uses recent historical data and mathematical or statistical models to generate the most accurate forecast, and sales managers can use this information for planning purposes. In the public sector, an MIS for a police department, for example, could provide information such as crime statistics, crime forecasts, and allocation of police units. Management can examine these statistics to spot increases and decreases in crime rates or types of crimes and analyze this data to determine future deployment of law enforcement personnel.

As you will see in this book, many organizations use information systems to gain a competitive advantage. The information box on Domino's Pizza describes one example of this. (*Note*: MISs are often referred to as just *information systems*, and these terms are used interchangeably in this book.)

Business intelligence (BI) provides historical, current, and predictive views of business operations and environments and gives organizations a competitive advantage in the marketplace.

Transaction-processing systems (TPSs) focus on data collection and processing; the major reason for using them is cost reduction.

A **management information system (MIS)** is an organized integration of hardware and software technologies, data, processes, and human elements designed to produce timely, integrated, relevant, accurate, and useful information for decision-making purposes.

INFORMATION TECHNOLOGIES AT DOMINO'S PIZZA

In 1960, Domino's Pizza opened its first store. Today, there are nearly 11,000 stores, half of them outside the United States. In 2007, Domino's started online and mobile ordering. Today, customers can order online at *www.dominos.com* or they can use apps for the iPhone, Android, or Kindle Fire.[5] This allows them to customize their pizzas with any combination of ingredients, enhancing their sense of participation while also saving Domino's the labor costs associated with phone orders. After placing the order, the customer can track it all the way to when it is sent out for delivery, keeping an eye on an estimated delivery time.

In 2012, for the first time, Domino's surpassed $1 billion in annual sales through its Web site, proving that electronic sales will continue to play a large role in the company's success.[6]

At Domino's, online ordering seamlessly accomplishes multiple objectives without the customer even taking notice. First, it creates the feeling among customers that they are an active part of the pizza-making process. Second, it results in greater efficiency at the various stores because employees do not have to spend as much time taking orders. They merely need to prepare the orders, which appear in an instant order queue, with all the customers' specifications.

Domino's now has the ability to store its online orders in its database. This data can then be used for many purposes, including target marketing and deciding which pizzas to offer in the future. The company is also actively using social media, including Facebook and Twitter, to promote its products and gather customers' opinions.

1-5 MAJOR COMPONENTS OF AN INFORMATION SYSTEM

In addition to hardware, software, and human elements, an information system includes four major components, which are discussed in the following sections: data, a database, a process, and information (see Exhibit 1.3).[7]

1-5a Data

The **data** component of an information system is considered the input to the system. The information that users need affects the type of data that is collected and used. Generally, there are two sources of data: external and internal. An information system should collect data from both sources, although organizational objectives and the type of application also determine what sources to use. Internal data includes sales records, personnel records, and so forth. The following list shows some examples of external data sources:

- Customers, competitors, and suppliers
- Government agencies and financial institutions
- Labor and population statistics
- Economic conditions

Typically, data has a time orientation, too. For example, past data is collected for performance reports, and current data is collected for operational reports. In addition, future data is predicted for budgets or cash flow reports. Data can also be collected in different forms, such as aggregated (e.g., subtotals for categories of information) or disaggregated (e.g., itemized lists). An organization might want disaggregated data to analyze sales by product, territory, or salesperson. Aggregated data can be useful for reporting overall performance during a particular sales quarter, for example, but it limits the ability of decision makers to focus on specific factors.

If an organization has defined its strategic goals, objectives, and critical success factors, then structuring the data component to define what type of data is collected and in what form is usually easy. On the other hand, if there are conflicting goals and objectives or the company is not aware of critical success factors, many problems in data collection can occur, which affects an information system's reliability and effectiveness.

Data consists of raw facts and is a component of an information system.

Exhibit 1.3
Major components of an information system

Data	Database	Process	Information

© Cengage Learning®

If an organization has defined its strategic goals, objectives, and critical success factors, then structuring the data component to define what type of data is collected and in what form is usually easy.

1-5b Database

A **database**, the heart of an information system, is a collection of all relevant data organized in a series of integrated files. (You learn more about databases in Chapter 3.) A comprehensive database is essential for the success of any information system. To create, organize, and manage databases, a database management system (DBMS) is used, such as Microsoft Access or FileMaker Pro for home or small-office use. In a large organization, a DBMS such as Oracle or IBM DB2 might be used.

Databases are also important for reducing personnel time needed to gather, process, and interpret data manually. With a computerized database and a DBMS, data can be treated as a common resource that is easy to access and use.

1-5c Process

The purpose of an information system's **process** component is generating the most useful type of information for making decisions. This component generally includes transaction-processing reports and models for decision analysis that can be built into the system or accessed from external sources.

An information system can include a wide range of models to support all levels of decision making. Users should be able to query an information system and generate a variety of reports. In addition, an information system should be able to grow with the organization so users can redefine and restructure models and incorporate new information into their analyses.

1-5d Information

Although they might seem the same, data and information are different. Data consists of raw facts and by itself is difficult to use for making decisions. **Information**—the output of an information system—consists of facts that have been analyzed by the process component and are therefore more useful to the MIS user. For example,

XYZ Company's total sales last month were $5,000,000. This number is data, because it does not tell you how the company performed. Did it meet the sales goal? Did sales increase or decrease from the previous month? How did the company perform against its top competitors? These questions and more can be answered by the information that an information system provides.

The quality of information is determined by its usefulness to users, and its usefulness determines the success of an information system. Information is useful if it enables decision makers to make the right decision in a timely manner. To be useful, information must have the following qualities:

- Timeliness
- Integration with other data and information
- Consistency and accuracy
- Relevance

If information lacks any of these qualities, the results are incorrect decisions, misallocation of resources, and overlooked windows of opportunity. If the system cannot give users a minimum level of confidence in its reliability, it will not be used or users might dismiss the reports it generates. Information must provide either a base for users to explore different options or insight into tasks.

Another factor affecting the usefulness of information is the information system's user interface. Because this interface must be flexible and easy to use, most information systems make use of graphical user interfaces (GUIs), with features such as menus

A **database** is a collection of all relevant data organized in a series of integrated files.

The **process** component of an information system generates the most useful type of information for decision making, including transaction-processing reports and models for decision analysis.

Information consists of facts that have been analyzed by the process component and is an output of an information system.

© Blend Images/Jupiter Images

built-in query capability that can generate the following information:

- How many students are in each major?
- Which major is the fastest growing?
- What is the average age of the student body?
- Among the international students, which country is home to the highest number of students?
- What is the ratio of male to female students in each major?

Many other types of analysis can be done. A forecasting model (part of the process component) could be used to generate the estimated number of students for 2018, for instance. In addition, predictions could be made or improved, based on information this system provides. For example, knowing which major is the fastest growing can help with decisions on hiring faculty, and knowing the estimated number of students for 2018 can help with planning facilities.

Example 2 Teletech, an international textile company, uses a database to store data on products, suppliers, sales personnel, costs, and so forth. The process component of the information system conducts analysis on the data to provide the following information about the preceding month:

- Which salesperson generated the highest sales?
- Which product generated the highest sales? The lowest sales?
- Which region generated the highest sales?

Again, forecasting models can be used to generate predictions for the next sales period, and these predictions can be broken down by product, region, and salesperson. Based on this information, many decisions could be made, such as allocating the advertising budget to different products and regions.

 1-6 ## USING INFORMATION SYSTEMS AND INFORMATION TECHNOLOGIES

Information systems are designed to collect data, process the collected data, and deliver timely, relevant, and useful information that can be used for making decisions. To achieve this goal, an information system might use many different **information technologies**. For example, organizations often use the Internet as a worldwide network to communicate with one another. Computer

and buttons. To be useful, information systems should also produce information in different formats, including graphics (e.g., pie charts and bar graphs), tables, and exception reports, which highlight information that is outside a specified range. Supplying information in a variety of formats increases the likelihood of users understanding and being able to use the information. Note that, in addition to the formal information that an information system generates, users need to be able to make use of informal information—such as rumors, unconfirmed reports, and stories—when solving problems.

The ultimate goal of an information system is to generate business intelligence (BI), described earlier in this chapter. As you will learn throughout this book, many different tools, techniques, and types of information system technologies are used to generate BI.

1-5e Examples of Information Systems

To better understand the four main components of an information system, take a look at the following two examples.

Example 1 A state university stores all student data in a database. The collected data includes each student's first name, last name, age, gender, major, nationality, and so forth. The process component of the information system performs all sorts of analysis on this data. For example, the university's DBMS has a

---

Information technologies support information systems and use the Internet, computer networks, database systems, POS systems, and radio-frequency-identification (RFID) tags.

INFORMATION TECHNOLOGIES AT THE HOME DEPOT

The Home Depot revolutionized the do-it-yourself home-improvement industry in the United States. Its stores use a POS system for both fast customer service and improved inventory management as well as a wireless network for efficient in-store communication.[8] The Home Depot has a Web site to communicate with customers and increase sales with online orders. It also uses RFID tags to better manage inventory and improve the efficiency of its supply chain network.

The Home Depot maintains a high-speed network connecting its stores throughout the United States and Canada, and it uses a data-warehousing application to analyze variables affecting its success—customers, competitors, products, and so forth.[9] The information system gives The Home Depot a competitive advantage by gathering, analyzing, and using information to better serve customers and plan for customers' needs.

In 2010, The Home Depot launched a transition to Fujitsu U-Scan self-checkout software in its United States and Canadian retail stores. The software offers retailers the flexibility to quickly make changes to their POS systems, and offers savings in labor costs.[10]

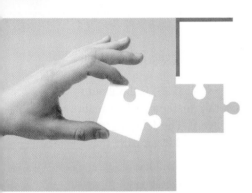

> Information systems are designed to collect data, process the collected data, and deliver timely, relevant, and useful information that can be used for making decisions.

networks (wired and wireless), database systems, POS systems, and radio-frequency-identification (RFID) tags are just a few examples of information technologies used to support information systems. The information box on The Home Depot gives you an idea of how companies use information technologies to stay competitive.

1-6a The Importance of Information Systems

Information is the second most important resource (after the human element) in any organization. Timely, relevant, and accurate information is a critical tool for enhancing a company's competitive position in the marketplace and managing the four Ms of resources: manpower, machinery, materials, and money.

To manage these resources, different types of information systems have been developed. Although all have the major components shown in Exhibit 1.3, they vary in the kind of data they collect and the analyses they perform. This section discusses some major types of information systems, focusing on the types of data and analysis used in each.

A personnel information system (PIS) or human resource information system

(HRIS) is designed to provide information that helps decision makers in personnel carry out their tasks more effectively. Web technologies have played a major role in improving the efficiency and effectiveness of HR departments. For example, intranets are often used to provide basic HR functions, such as employees checking how much vacation time they have left or looking up how much they have in their 401(k) plans. Intranets reduce personnel costs and speed up responses to common employee requests. As

© Miodrag Gajic/Shutterstock.com

discussed in Chapter 7, an intranet is a network within an organization that uses Internet protocols and technologies for collecting, storing, and disseminating useful information that supports business activities such as sales, customer service, human resources, and marketing. The main difference between an intranet and the Internet is that intranets are private and the Internet is public. A PIS/HRIS supports the following decisions, among others:

- Choosing the best job candidate
- Scheduling and assigning employees
- Predicting the organization's future personnel needs
- Providing reports and statistics on employee demographics
- Allocating human and financial resources

The information box above highlights some real life applications of HRIS.

A logistics information system (LIS) is designed to reduce the cost of transporting materials while maintaining safe and reliable delivery. The following are a few examples of decisions supported by an LIS:

- Improving routing and delivery schedules
- Selecting the best modes of transportation
- Improving transportation budgeting
- Improving shipment planning

The information box featuring UPS shows uses of information systems and information technologies, particularly logistics information systems.

A manufacturing information system (MFIS) is used to manage manufacturing resources so companies can reduce manufacturing costs, increase product quality, and make better inventory decisions. MFISs can perform many types of analysis with a high degree of timeliness and accuracy. For example, managers could use an MFIS to assess the effect on final product costs of a 7 percent increase in raw materials or to determine how many assembly-line workers are needed to produce 200 automobiles in the next 3 weeks. Here are some decisions that an MFIS supports:

- Ordering decisions
- Product cost calculations
- Space utilization
- The bid evaluation process used with vendors and suppliers
- Analysis of price changes and discounts

The goal of a financial information system (FIS) is to provide information to financial executives in a timely manner. Here are some decisions an FIS is used to support:

- Improving budget allocation
- Minimizing capital investment risks
- Monitoring cost trends
- Managing cash flows
- Determining portfolio structures

In addition, marketing information systems (MKISs) are used to improve marketing decisions. An effective MKIS should provide timely, accurate, and integrated information about the marketing mix-4Ps: price, promotion, place, and product. Here are some decisions that an MKIS supports:

- Analyzing market share, sales, and sales personnel
- Sales forecasting
- Price and cost analysis of items sold

1-6b Using Information Technologies for a Competitive Advantage

Michael Porter, a professor at Harvard Business School, identified three strategies for successfully competing in the marketplace[17]:

- Overall cost leadership
- Differentiation
- Focus

Information systems can help organizations reduce the cost of products and services and, if designed correctly, they can assist with differentiation and focus strategies. Throughout this book, you will see many examples of the cost savings that organizations have achieved with information systems and technologies. For example, Walmart has been using overall cost leadership strategies successfully (see the information box below).

INFORMATION TECHNOLOGIES AT WALMART

Walmart (*http://walmartstores.com*), the largest retailer in the world, built the Walmart Satellite Network, which is the largest private satellite communication system in the United States. The network links branch stores with the home office in Bentonville, Arkansas, by using two-way voice and data and one-way video communication. In addition to the POS systems used for many years, Walmart uses the following information technologies to gain a competitive advantage:

- Telecommunications is used to link stores with the central computer system and then to suppliers' computers. This system creates a seamless connection among all parties.
- Network technologies are used to manage inventory and implement a just-in-time inventory system. As a result, products and services can be offered at the lowest possible prices.
- Walmart uses an extranet, called RetailLink, to communicate with suppliers. Suppliers are able to review product sales records in all stores and track current sales figures and inventory levels.[18] (Extranets are discussed in Chapter 7.)
- Electronic data interchange (EDI), discussed in Chapter 11, is used to streamline the order-invoice-payment cycle, reduce paperwork, and improve accuracy.
- Walmart is a major user of RFID technologies, which have improved its supply chain and inventory management systems.

Information technologies can help bottom-line and top-line strategies. The focus of a bottom-line strategy is improving efficiency by reducing overall costs. A top-line strategy focuses on generating new revenue by offering new products and services to customers or increasing revenue by selling existing products and services to new customers. For example, e-commerce businesses are adapting business models to reduce distribution costs dramatically. A good example is antivirus vendors using the Internet to distribute software. For a subscription fee of around $30, you can download the software and receive updates for a year. Without the Internet for easy, inexpensive distribution, vendors could not afford to offer software at such a low price.

As discussed in Chapter 11, many organizations use enterprise systems, such as supply chain management (SCM), customer relationship management (CRM), enterprise resource planning (ERP), knowledge management (KM), and collaboration software, to reduce costs and improve customer service. The goal of these systems is to use information technologies to create the most efficient, effective link between suppliers and consumers. A successful CRM program, for example, helps improve customer service and create a long-term relationship between an organization and its customers.

For differentiation strategies, organizations try to make their products and services different from their competitors. Apple has been successful with this strategy by designing its computers to look very different than PCs and focusing on its computers' ease of use. As another example, Amazon.com has differentiated its Web site by using certain information technologies, such as personalization technologies (covered in more detail in Chapter 11) to recommend products to customers based on their previous purchases. Amazon.com also uses the one-click system for fast checkout. With this system, customers can enter credit card numbers and addresses once and, in subsequent visits, simply

Information technologies can help bottom-line and top-line strategies.

click once to make a purchase, without having to enter information again.

With focus strategies, organizations concentrate on a specific market segment to achieve a cost or differentiation advantage. Apple has also used this strategy to target iPhones to consumer users rather than business users. Similarly, Macintosh computers are heavily marketed to creative professionals such as designers, photographers, and writers. As another example, Abercrombie & Fitch targets high-end clothing to low-income customers, such as teenagers and young adults, while Nordstrom targets its high-end clothing to high-income customers. Information technologies could assist these companies in reaching their target market segments more cost effectively.

Remember that focus and differentiation strategies work only up to a certain point. Customers are often willing to pay more for a unique product or service or one with a specific focus. However, cost still plays a major role. If a product or service becomes too expensive, customers might not be willing to purchase it.

1-6c Porter's Five Forces Model: Understanding the Business Environment

Harvard Business School's Michael Porter created a comprehensive framework called the **Five Forces Model** for analyzing an organization, its position in the marketplace, and how information systems could be used to make the organization more competitive.[19] The five forces, shown in Exhibit 1.4, are:

- Buyer power
- Supplier power
- Threat of substitute products or services
- Threat of new entrants
- Rivalry among existing competitors

Buyer power is high when customers have many choices and low when they have few choices. Typically, organizations try to limit buyers' choices by offering services that make it difficult for customers to switch, which is essentially using a differentiation strategy. For example, Dell Computer was among the first to offer computer customization options to customers, and other computer manufacturers followed suit. Grocery

Michael Porter's **Five Forces Model** analyzes an organization, its position in the marketplace, and how information systems could be used to make it more competitive. The five forces include buyer power, supplier power, threat of substitute products or services, threat of new entrants, and rivalry among existing competitors.

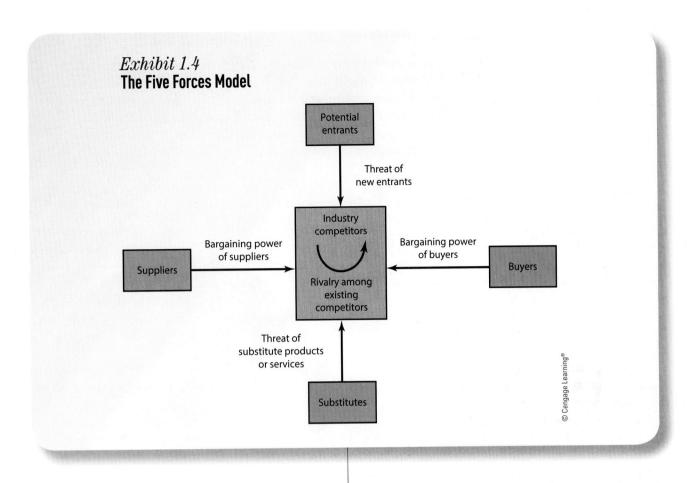

Exhibit 1.4
The Five Forces Model

- Potential entrants
- Threat of new entrants
- Industry competitors
- Rivalry among existing competitors
- Bargaining power of suppliers
- Suppliers
- Bargaining power of buyers
- Buyers
- Threat of substitute products or services
- Substitutes

© Cengage Learning®

stores, such as Sam's Club, offer club cards that encourage customers to shop by giving them big discounts, an example of overall cost leadership strategies. Similarly, airlines and hotels offer free mileage and points when customers use their services. Information systems can make managing these strategies easier and more cost effective.

Organizations use these strategies to increase customer loyalty, thus combating the threat of new entrants or substitute products. However, certain information technology tools, such as the Internet, have evened the playing field by giving customers more access to all sorts of data, such as the various

prices for products and services. This increases buyers' bargaining power and decreases supplier power, which is discussed next.

Supplier power is high when customers have fewer options and low when customers have more options. Organizations might use information systems to make their products and services cheaper or to offer more services in order to distinguish themselves from competitors (another use of a differentiation strategy). Netflix, for example, uses information technologies to offer products and services, which increases its power in the marketplace. (For examples of how these focus strategies get used, see the information box on Netflix.)

Certain information technology tools, such as the Internet, have evened the playing field by giving customers more access to all sorts of data, such as being able to compare prices.

INFORMATION TECHNOLOGIES AT NETFLIX

Using a wide variety of devices, over 37 million Netflix subscribers in the United States and around the world connect to the company's Web site and watch unlimited movies and TV episodes.[20] The users can also receive DVDs through the mail. The Internet, personalization technologies, and effective inventory management techniques have revolutionized rental entertainment at Netflix. Any user with an Internet connection can connect to the Web site and watch movies and TV episodes streamed and delivered instantly to a user's device. Netflix is currently using Amazon Web Services (AWS), which has provided the company with a high degree of availability and speed.[21]

Using data-mining and collaborative filtering technologies, Netflix's personalization system actively searches information to discover relationships and patterns and make recommendations based on a user's past movie-viewing history and questions that were answered on the Web site. Based on these techniques, Netflix has created CineMatch, an algorithm that recommends other movies the customer might enjoy.[22]

The inventory system used by Netflix is fully automated, and computers and robots play a major role in managing CDs and DVDs by scanning them as soon as they arrive, notifying the customer that the shipment has been received, and quickly making the CDs and DVDs available for other customers. (Netflix owns over 89 million discs.) A machine called a "stuffer" processes DVDs—puts the discs into the envelopes, folds and closes the envelope, and applies the sticker to the center of the envelope.[23]

Organizations have tools other than information systems and technologies to increase their power. For example, drug companies obtain patents for their products to reduce competition.

The threat of customers choosing substitute products or services is high when many alternatives to an organization's products and services are available. Some organizations add services—such as Amazon.com's personalized recommendations—to make their positions in the marketplace more distinctive. Other organizations use fees to discourage customers from switching to a competitor, such as when cell phone companies add charges for switching to another provider before the customer contract is up.

The threat of new entrants into the marketplace is low when duplicating a company's product or service is difficult. Organizations often use focus strategies to ensure that this threat remains low. For example, developing a search engine that could compete successfully with Google would be difficult. In addition, organizations use information technologies to increase customer loyalty, as mentioned previously, which reduces the threat of new entrants. For example, banks offer free bill paying to attract customers and keep them from switching to another bank; setting up a bill-paying service at another bank takes time that most customers do not want to spend. Similarly, after customizing their home pages with options offered by sites such as Yahoo! and Google, many users do not want to repeat this process at a new site.

Rivalry among existing competitors is high when many competitors occupy the same marketplace position; it is low when there are few competitors. For example, online brokerage firms operate in a highly competitive environment, so they use information technologies to make their services unique.

1-7 THE IT JOB MARKET

During the past decade, the IT job market has been one of the fastest growing segments in the economy, and it continues to be so. Even during the economic downturn, certain segments of the IT job market, such as Web design, infrastructure, and computer and network security, have shown growth compared to the rest of the job market. Currently, cloud computing-related jobs (discussed in Chapter 14) are in high demand.[24] Broadly speaking, IT jobs fall into the following categories:

- Operations and help desk
- Programming
- Systems design
- Web design and Web hosting
- Network design and maintenance
- Database design and maintenance
- Robotics and artificial intelligence

The educational backgrounds for an IT position can include an AA, BA, BS, MS, or MBA in information systems and related fields. The salaries vary based on educational background, experience, and the job's location. They range from $52,000 for a programmer to over $180,000 for a CIO.

Popular jobs in the information systems field are described in the following sections.

1-7a CTO/CIO

The top information systems job belongs to either the chief technology officer (CTO) or the chief information officer (CIO). This person oversees long-range planning and keeps an eye on new developments in the field that can affect a company's success. Some organizations also have a chief privacy officer (CPO). This executive position includes responsibility for managing the risks and business impacts of privacy laws and policies.

1-7b Manager of Information Systems Services

This person is responsible for managing all the hardware, software, and personnel within the information systems department.

1-7c Systems Analyst

This person is responsible for the design and implementation of information systems. In addition to computer knowledge and an information systems background, this position requires a thorough understanding of business systems and functional areas within a business organization.

1-7d Network Administrator

This person oversees a company's internal and external network systems, designing and implementing network systems that deliver correct information to the right

decision maker in a timely manner. Providing network and cybersecurity is part of this position's responsibility.

1-7e Database Administrator

A database administrator (DBA) is responsible for database design and implementation. Additionally, a database administrator should have knowledge and understanding of data warehouses and data-mining tools.

1-7f Computer Programmer

A computer programmer writes computer programs or software segments that allow the information system to perform a specific task. There are many computer languages available, and each one requires a specific knowledge suitable for a specific application. Because of the popularity of smartphones and mobile devices, many programmers are now developing apps for iOS and Android devices. There is a huge demand for these applications and for the programmers who develop these apps.

1-7g Webmaster

A webmaster designs and maintains the organization's Web site. Because of the popularity of e-commerce applications, webmasters have been in high demand.

In recent years, new IT-related jobs have been created that appear to also be in high demand.[25, 26] These jobs include:

- Social media/online-community manager
- Social media architect
- Telework manager or coordinator
- Search engine optimization specialist
- Business architect
- Data scientist
- Mobile technology expert
- Enterprise mobile developer
- Cloud architect

1-8 OUTLOOK FOR THE FUTURE

By examining various factors related to designing, implementing, and using information systems, the following predictions can be made:

- Hardware and software costs will continue to decline, so processing information will be less expensive. These cost savings should make information systems affordable for any organization, regardless of its size and financial status.

- Artificial intelligence and related technologies will continue to improve and expand, which will have an impact on information systems. For example, further development in natural language processing should make information systems easier to use. Also, robots will play a major role in the workforce of the future.

- The computer literacy of typical information system users will improve, as computer basics are taught more in elementary schools.

- Networking technology will improve, so connecting computers will be easier, and sending information from one location to another will be faster. Compatibility issues between networks will become more manageable, and integrating voice, data, and images on the same transmission medium will improve communication quality and information delivery.

- Personal computers and tablets will continue to improve in power and quality, so most information system software will be able to run on them without problems. This trend should make information systems more affordable, easier to maintain, and more appealing to organizations.

- Internet growth will continue, which will put small and large organizations on the same footing, regardless of their financial status. Internet growth will also make e-collaboration easier, despite geographical distances.

- Computer criminals will become more sophisticated, and protecting personal information will become more difficult.

The Industry Connection highlights Microsoft and its products and services.

Industry Connection: Microsoft Corporation[27]

Microsoft, founded in 1975, is the world's largest software company and is involved in all aspects of desktop computing. It is best known for the Disk Operating System (DOS), Windows operating systems, and office software suites such as Office. Here are some of the products and services Microsoft offers:

Windows: The most popular operating system for PCs and PC-compatible computers

Windows XP, Windows Vista, Windows 7, and Windows 8: Four widely used OSs for PCs

Windows Server 2003, Server 2008, Server 2010, Server 2012: Four widely used server operating systems used in network environments

Office: The most widely used office suite; includes Word, Excel, Access, and PowerPoint

Office 365: An online service, similar to Google Apps, that lets users collaborate on documents, spreadsheets, and e-mail using a combination of subscription desktop software and Web apps; includes Word, Excel, Exchange, and PowerPoint

Internet Explorer: A popular Web browser

Expression Web (replacing FrontPage): An HTML editor and Web design program for developing Web pages and other HTML applications

MSN: An Internet portal combining Web services and free Web-based e-mail (Hotmail)

SharePoint Server: Groupware for facilitating information sharing and content management

SQL Server 2008, 2010, 2012, and 2014: Widely used database management systems

Xbox: A video game system

Visual Studio: An integrated development environment (IDE) that can be used to program applications in a number of different languages (such as C++, Java, Visual Basic, and C#); used for console or GUI applications as well as Web applications

Windows Live ID: A single sign-on service for multiple Web sites

Skype: A software application that allows users to communicate using voice, videos, and data over the Internet

Surface: A tablet designed to compete with the iPad

STUDY TOOLS 1

LOCATED AT BACK OF THE TEXTBOOK
☐ Rip out Chapter Review Card

LOCATED AT WWW.CENGAGE.COM/LOGIN
☐ Review Key Terms Flash Cards (Print or Online)
☐ Download audio and visual summaries to review

☐ Complete both Practice Quizzes to prepare for tests
☐ Play "Beat the Clock" and "Quizbowl" to master concepts
☐ Complete "Crossword Puzzle" to review key terms
☐ Watch video on "Method" for a real company example

KEY TERMS

business intelligence (BI) 7
computer literacy 6
data 8
database 9
Five Forces Model 14
information 9

information literacy 6
information technologies 10
management information system (MIS) 7
process 9
transaction-processing systems (TPSs) 7

REVIEWS AND DISCUSSIONS

1. What are some of the goals and objectives of transaction-processing systems?

2. What are three examples of information technologies used in retail business?

3. What are the four main components of an MIS?

4. What are Michael Porter's three strategies for competing in the marketplace? Give an example of a company that uses each strategy.

5. What is Michael Porter's Five Forces Model? How can businesses use this model to improve their competiveness in the marketplace?

6. What are two IT jobs that are in high demand?

7. What are three applications of an HRIS?

8. Which information technologies are being used by Netflix to improve its competiveness?

PROJECTS

1. Identify three applications of information systems at the college or the university that you are attending. Write a one-page paper that describes these three applications, and provide an example of the type of decisions that are being improved by each application.

2. Grocery chains have been using information technologies for several decades. After reading the information

presented in this chapter and other sources, write a one-page paper that describes three such technologies.

3. RFID tags are being increasingly used by companies such as Macy's, Walmart, and Home Depot. Identify an additional company that uses RFIDs and write a one-page paper that describes the company's specific application of RFIDs.

4. After reading the information presented in this chapter and other sources, write a one-page paper that describes the ways two different companies use Michael Porter's three strategies. How are information systems assisting these companies in implementing each strategy?

5. After reading the information presented in this chapter and other sources, write a one-page paper that supports the claim that, in the future, computer criminals will become more sophisticated and that protecting personal information will become more difficult.

6. Banks are promoting online banking to a broad range of customers. After reading the information presented in this chapter and other sources, write a one-page paper that lists three advantages and three disadvantages of online banking. Why are some customers reluctant to use online banking?

ARE YOU READY TO MOVE ON?

1. Transaction-processing systems (TPSs) have been applied to structured tasks such as record keeping, simple clerical operations, and inventory control. True or False?

2. The data component of an information system is considered the output of the system. True or False?

3. Michael Porter's strategies for successfully competing in the marketplace include overall cost leadership, differentiation, and focus. True or False?

4. Which of the following is not included in Michael Porter's Five Forces Model?
   a. Buyer power
   b. Supplier power
   c. Threat of new entrants
   d. Information overload

5. Which of the following is not considered part of an IT job title?
   a. Telemarketer
   b. Cloud architect
   c. Systems analyst
   d. Programmer

6. Which of the following is not true regarding the future of information systems?
   a. Hardware and software costs will continue to decline.
   b. Computer criminals will become more sophisticated.
   c. Artificial intelligence and related technologies will continue to improve.
   d. Internet growth will slow down.

CASE STUDY 1-1

Using Information Technologies at Federal Express

Federal Express (FedEX), founded in 1971, handles an average of 3 million package-tracking requests every day (http://about.fedex.designcdt.com). To stay ahead in a highly competitive industry, the company focuses on customer service by maintaining a comprehensive Web site, FedEx.com, where it assists customers and reduces costs. For example, every request for information that is handled at the Web site rather than by the call center saves an estimated $1.87. Federal Express has reported that customer calls have decreased by 83,000 per day since 2000, which saves the company $57.56 million per year. And because each package-tracking request costs Federal Express 3 cents, costs have been reduced from more than $1.36 billion per year to $21.6 million per year by customers using the Web site instead of the call center.

Another technology that improves customer service is Ship Manager, an application installed on customers' sites so users can weigh packages, determine shipping charges, and print shipping labels. Customers can also link their invoicing, billing, accounting, and inventory systems to Ship Manager.[28]

However, Federal Express still spends almost $326 million per year on its call center in order to reduce customers' frustration when the Web site is down or when customers have difficulty using it. The company uses customer relationship management software called Clarify in its call centers to make customer service representatives' jobs easier and more efficient and to speed up response time.[29]

© Denise Kappa/Shutterstock.com

Answer the following questions:

1. Is technology by itself enough to ensure high-quality customer service?
2. What are Federal Express's estimated annual savings from using information technology?
3. What are a couple of examples of information technologies used by Federal Express?

CASE STUDY 1-2

Mobile Technology: A Key Player for Future Shopping

Faced with strong competition by online stores, retailers are looking for new ways to improve customer service and lower operating costs. They have found mobile technology to be the key for achieving this goal. Scan-as-you-go mobile devices are a logical next step after the self-checkout used by many retailers. Retail experts predict the new mobile-based retail devices could eventually bring about the end of traditional cash register systems. The mobile checkout stations pioneered at Apple stores appear to be the future. The goal is to speed up and improve customer service and to keep consumers in stores and spending.

Ahold USA's Stop & Shop retail stores use a mobile device called Scan It that hangs on the handle of the shopping cart and allows customers to shop and scan as they go through the aisles. If there is a coupon for an item, the device quickly gives the customers a credit and the total is recalculated. The device is smart enough to alert the customer if there is a coupon for a complementary item, such as coffee creamer if the customer has purchased coffee. Shoppers

© lofoto/Shutterstock.com

who use the Scan It device spend about 10 percent more than average.

The clothing retailer Nordstrom is also using mobile devices, which it issues to its sales associates on the floor so they can scan items on the spot and let customers pay without going through the cash registers. The Home Depot uses a device called First Phones as an inventory tracker. If the item is out of stock, First Phones quickly notifies the customer whether a nearby store has it, then holds the item for the customer for pick up. Starbucks is taking steps toward a digital-wallet model that allows customers to pay using their smartphones.[30]

Answer the following questions:

1. According to this case study, what is an upcoming key technology that will be used in retail stores to improve customer service?
2. What is the name of the device used by Ahold USA's Stop & Shop retail stores?
3. What will be the role of smartphones in the future of shopping?

2 | Computers: The Machines Behind Computing

LEARNING OUTCOMES

After studying this chapter, you should be able to:

2-1 Define a computer system, and describe its components.

2-2 Discuss the history of computer hardware and software.

2-3 Explain the factors distinguishing the computing power of computers.

2-4 Summarize computer operations.

2-5 Discuss the types of input, output, and memory devices.

2-6 Explain how computers are classified.

2-7 Describe the two major types of software.

2-8 List the generations of computer languages.

After you finish this chapter, go to **PAGE 40** for the **STUDY TOOLS**

If airplanes had developed as computers have developed, today you would be able to go around the globe in less than 20 minutes for just 50 cents.

In this chapter, you will learn about the major components of a computer and what factors distinguish computing power. You will review a brief history of computer hardware and software and learn an overview of computer operations. You will also go into more detail on specific computer components: input, output, and memory devices. You will learn how computers are classified, based on size, speed, and sophistication, and about the two major types of software—system software and application software—and the five generations of computer languages.

2-1 DEFINING A COMPUTER

If airplanes had developed as computers have developed, today you would be able to go around the globe in less than 20 minutes for just 50 cents. Computers have gone through drastic changes in a short time. For example, a computer that weighed more than 18 tons 60 years ago has been replaced by one that now weighs less than 2 pounds. Today's computer is 100 times more powerful and costs less than 1 percent of the 60-year-old computer.

As you learned in Chapter 1, you use computers every day for a multitude of purposes. You even use them indirectly when you use appliances with embedded computers, such as TVs and microwaves. Computers have become so ubiquitous, in fact, that a cashless and checkless society is likely just around the corner. Similarly,

computers might eliminate the need for business travel. Even now, executives seldom need to leave their offices for meetings in other locations because of technologies such as computer conferencing and telepresence systems.

Computers are used in a wide variety of tasks, including report distribution in businesses, rocket guidance control in the NASA space program, and DNA analysis in medical research. This book could not have been published in such a timely manner without the use of computers. The text was typed and revised with word-processing software, and composition software was used to typeset the pages. Printing, warehousing, inventory control, and shipping were accomplished with the help of computers.

So what is a computer? Many definitions are possible, but in this book a **computer** is defined as a machine that accepts data as input, processes data without human intervention by using stored instructions, and outputs information. The instructions, also called a *program*, are step-by-step directions for performing a specific task, written in a language the computer can understand. Remember that a computer only processes data (raw facts); it cannot change or correct the data that is entered. If data is erroneous, the information the computer provides is also erroneous. This rule is sometimes called GIGO: garbage in, garbage out.

To write a computer program, first you must know what needs to be done, and then you must plan a method to achieve this goal, including selecting the right language for the task. Many computer languages are available; the language you select depends on the problem being solved and the type of computer you are using. Regardless of the language, a program is also referred to as the *source code*. This source code must be translated into object code—consisting of binary 0s and 1s. Binary

© Yellowj/Shutterstock.com

code—a set of instructions used to control the computer—uses 0s and 1s, which the computer understands as on or off signals. You will learn more about the binary system and computer languages later in this chapter.

2-1a Components of a Computer System

A computer system consists of hardware and software. Hardware components are physical devices, such as keyboards, monitors, and processing units. The software component consists of programs written in computer languages.

Exhibit 2.1 shows the building blocks of a computer. Input devices, such as keyboards, are used to send

A **computer** is a machine that accepts data as input, processes data without human intervention by using stored instructions, and outputs information.

Exhibit 2.1
The building blocks of a computer

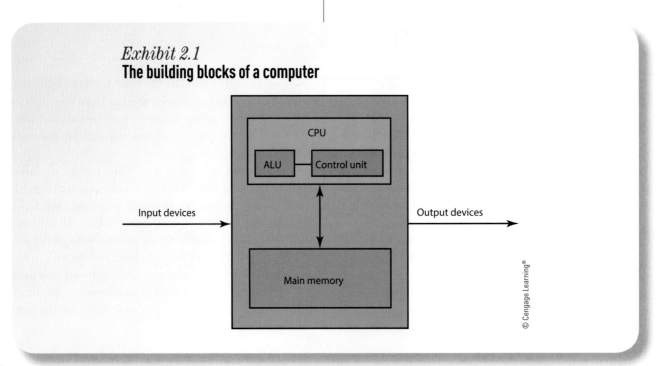

© Cengage Learning®

data and information to the computer. Output devices, such as monitors and printers, display the output a computer generates.

Main (primary) memory is where computers store data and instructions, similar to a human brain. The **central processing unit (CPU)** is the heart of a computer. It is divided into two components: the **arithmetic logic unit (ALU)** and the **control unit**. The ALU performs arithmetic operations (+, −, *, /) as well as comparison or relational operations (<, >, =); the latter are used to compare numbers. The control unit tells the computer what to do, such as instructing the computer which device to read or send output to.

Some computers have a single processor; other computers, called *multiprocessors*, contain multiple processors. Multiprocessing is the use of two or more CPUs in a single computer system. Generally, a multiprocessor computer performs better than a single-processor computer in the same way that a team would perform better than an individual on a large, time-consuming project. Some computers use a dual-core processor, which is essentially two processors in one, to improve processing power. Dual-core processors are common in new PCs and Apple computers. In recent years multicore processors have been introduced. A quad-core processor contains four cores, a hexa-core processor contains six cores, and an octa-core processor contains eight cores. Simply put, these new chips are making computers faster than their predecessors.

Another component that affects computer performance is a **bus**, which is the link between devices connected to the computer. A bus can be parallel or serial, internal (local) or external. An internal bus enables communication between internal components, such as a video card and memory; an external bus is capable of communicating with external components, such as a USB device.

Other factors that affect computer performance include the processor size and the operating system (OS). In recent years, 32-bit and 64-bit processors and OSs have created a lot of interest. A 32-bit processor can use 2^{32} bytes (4 GB) of RAM; and, in theory, a 64-bit processor can use 2^{64} bytes (16 EB, or exabytes)

of RAM. So a computer with a 64-bit processor can perform calculations with larger numbers and be more efficient with smaller numbers; it also has better overall performance than a 32-bit system. However, to take advantage of this higher performance, you must also have a 64-bit OS.

Exhibit 2.2 shows additional components of a computer system. A **disk drive** is a peripheral device for recording, storing, and retrieving information. A **CPU case** (also known as a computer chassis or tower) is the enclosure containing the computer's main components.

The **central processing unit (CPU)** is the heart of a computer. It is divided into two components: the arithmetic logic unit (ALU) and the control unit.

The **arithmetic logic unit (ALU)** performs arithmetic operations (+, −, *, /) as well as comparison or relational operations (<, >, =); the latter are used to compare numbers.

The **control unit** tells the computer what to do, such as instructing the computer which device to read or send output to.

A **bus** is a link between devices connected to the computer. It can be parallel or serial, internal (local) or external.

A **disk drive** is a peripheral device for recording, storing, and retrieving information.

A **CPU case** is also known as a computer chassis or tower. It is the enclosure containing the computer's main components.

Exhibit 2.2
Components of a computer system

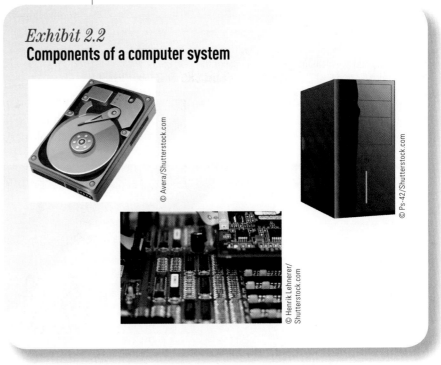

© Avera/Shutterstock.com

© Henrik Lehnerer/Shutterstock.com

© Ps-42/Shutterstock.com

A **motherboard** is the main circuit board containing connectors for attaching additional boards. In addition, it usually contains the CPU, Basic Input/Output System (BIOS), memory, storage, interfaces, serial and parallel ports, expansion slots, and all the controllers for standard peripheral devices, such as the display monitor, disk drive, and keyboard. A serial port is a communication interface through which information is transferred one bit at a time; a parallel port is an interface between a computer and a printer that enables the computer to transfer multiple bits of information to the printer simultaneously.

2-2 THE HISTORY OF COMPUTER HARDWARE AND SOFTWARE

Major developments in hardware have taken place over the past 60 years. To make these developments more clear, computers are often categorized into "generations" that mark technological breakthroughs. Beginning in the 1940s, first-generation computers used vacuum tube technology. They were bulky and unreliable, generated excessive heat, and were difficult to program. Second-generation computers used transistors and were faster, more reliable, and easier to program and maintain. Third-generation computers operated on integrated circuits, which enabled computers to be even

A **motherboard** is the main circuit board containing connectors for attaching additional boards. It usually contains the CPU, Basic Input/Output System (BIOS), memory, storage, interfaces, serial and parallel ports, expansion slots, and all the controllers for standard peripheral devices, such as the display monitor, disk drive, and keyboard.

smaller, faster, more reliable, and more sophisticated. Remote data entry and telecommunications were introduced during this generation. Fourth-generation computers continued several trends that further improved speed and ease of use: miniaturization, very-large-scale integration (VLSI) circuits, widespread use of personal computers, and optical discs (discs written or encoded and read using a laser optical device). The current fifth-generation computers include parallel processing (computers containing hundreds or thousands of CPUs for rapid data processing), gallium arsenide chips that run at higher speeds and consume less power than silicon chips, and optical technologies. Table 2.1 summarizes these hardware generations.

Because silicon cannot emit light and has speed limitations, computer designers have concentrated on technology using gallium arsenide, in which electrons move almost five times faster than in silicon. Devices made with this synthetic compound can emit light, withstand higher temperatures, and survive much higher doses of radiation than silicon devices. The major problems with gallium arsenide are difficulties in mass production. This material is softer and more fragile than silicon, so it breaks more easily during slicing and polishing. Because of the high costs and difficulty of production, the military is currently the major user of this technology. However, research continues to eliminate some shortcomings of this technology.

In October 2012, IBM announced that it will start using carbon nanotubes (CNTs) instead of silicon in its computer chips. CNTs are developed using nanotechnology, which is discussed in Chapter 14. It is one of the methods that should keep chip sizes shrinking after the current silicon-based technology has reached its limit.[1]

The field of optical technologies involves the applications and properties of light, including its interactions with lasers, fiber optics, telescopes, and so forth. These technologies offer faster processing speed, parallelism

TABLE 2.1 HARDWARE GENERATIONS

Generation	Date	Major technologies	Example
First	1946–1956	Vacuum tube	ENIAC
Second	1957–1963	Transistors	IBM 7094, 1401
Third	1964–1970	Integrated circuits, remote data entry, telecommunications	IBM 360, 370
Fourth	1971–1992	Miniaturization, VLSI, personal computers, optical discs	Cray XMP, Cray II
Fifth	1993–present	Parallel processing, gallium arsenide chips, optical technologies	IBM System zEnterprise EC12

© Cengage Learning®

(several thousand light beams can pass through an ordinary device), and interconnection; much denser arrays of interconnections are possible because light rays do not affect each other. Optical computing is in its infancy, and more research is needed to produce a full-featured optical computer. Nevertheless, storage devices using this technology are revolutionizing the computer field by enabling massive amounts of data to be stored in very small spaces.

Computer languages and software have also developed through five generations. They are discussed in more detail in the "Computer Languages" section, but Table 2.2 summarizes these generations.

The information box titled "IBM Watson Wins Jeopardy" describes a supercomputer that has several features of a fifth-generation computer as well as artificial intelligence capabilities.

2-3 THE POWER OF COMPUTERS

Computers draw their power from three factors that far exceed human capacities: speed, accuracy, and storage and retrieval capabilities. These are discussed in the following sections.

> Computers draw their power from three factors that far exceed human capacities: speed, accuracy, and storage and retrieval capabilities.

TABLE 2.2 COMPUTER LANGUAGE TRENDS

Generation	Major attribute
First	Machine language
Second	Assembly language
Third	High-level language
Fourth	Fourth-generation language
Fifth	Natural language processing (NLP)

© Cengage Learning®

2-3a Speed

Computers process data with amazing speed. They are capable of responding to requests faster than humans can, which improves efficiency. Today's high-speed computers make it possible for knowledge workers to perform tasks much faster than with the slower computers of the past. Typically, computer speed is measured as the number of instructions performed during the following fractions of a second:

- Millisecond: 1/1,000 of a second
- Microsecond: 1/1,000,000 of a second
- Nanosecond: 1/1,000,000,000 of a second
- Picosecond: 1/1,000,000,000,000 of a second

2-3b Accuracy

Unlike humans, computers do not make mistakes. To understand computer accuracy more clearly, take a look at these two numbers:

- 4.00000000000000000000000001
- 4.00000000000000000000000002

To humans, these two numbers are so close that they are usually considered equal. To a computer, however, these two numbers are completely different. This degree of accuracy is critical in many computer applications. On a space mission, for example, computers are essential for calculating reentry times and locations for space shuttles. A small degree of inaccuracy could lead the space shuttle to land in Canada instead of the United States.

2-3c Storage and Retrieval

Storage means saving data in computer memory, and retrieval means accessing data from memory. Computers can store vast quantities of data and locate a specific item quickly, which makes knowledge workers more efficient in performing their jobs.

In computers, data is stored in bits. A bit is a single value of 0 or 1, and 8 bits equal 1 byte. A byte is the size of a character. For example, the word *computer* consists of 8 characters or 8 bytes (64 bits). Table 2.3 shows storage measurements.

TABLE 2.3 STORAGE MEASUREMENTS (APPROXIMATIONS)

1 bit	A single value of 0 or 1
8 bits	1 byte or character
2^{10} bytes	1000 bytes, or 1 kilobyte (KB)
2^{20} bytes	1,000,000 bytes, or 1 megabyte (MB)
2^{30} bytes	1,000,000,000 bytes, or 1 gigabyte (GB)
2^{40} bytes	1,000,000,000,000 bytes, or 1 terabyte (TB)
2^{50} bytes	1,000,000,000,000,000 bytes, or 1 petabyte (PB)
2^{60} bytes	1,000,000,000,000,000,000 bytes, or 1 exabyte (EB)

© Cengage Learning®

Every character, number, or symbol on the keyboard is represented as a binary number in computer memory. A binary system consists of 0s and 1s, with a 1 representing "on" and a 0 representing "off," similar to a light switch.

Computers and communication systems use data codes to represent and transfer data between computers and network systems. The most common data code for text files, PC applications, and the Internet is American Standard Code for Information Interchange (ASCII), developed by the American National Standards Institute. In an ASCII file, each alphabetic, numeric, or special character is represented with a 7-bit binary number (a string of 0s or 1s). Up to 128 (2^7) characters can be defined. There are two additional data codes used by many operating systems: Unicode and Extended ASCII. Unicode is capable of representing 256 (2^8) characters, and Extended ASCII is an 8-bit code that also allows representation of 256 characters.

Before the ASCII format, IBM's Extended Binary Coded Decimal Interchange Code (EBCDIC) was popular. In an EBCDIC file, each alphabetic, numeric, or special character is represented with an 8-bit binary number.

2-4 COMPUTER OPERATIONS

Computers can perform three basic tasks: arithmetic operations, logical operations, and storage and retrieval operations. All other tasks are performed using one or a combination of these operations. For example, playing a computer game could require a combination of all three operations. During a game, your computer may perform calculations in order to make a decision (such as whether to move from point A to point B), it may compare two numbers, and it may perform storage and retrieval functions for going forward with the process.

Computers can add, subtract, multiply, divide, and raise numbers to a power (exponentiation), as shown in these examples:

A + B (addition)	$5 + 7 = 12$
A − B (subtraction)	$5 - 2 = 3$
A * B (multiplication)	$5 * 2 = 10$
A / B (division)	$5 / 2 = 2.5$
A ^ B (exponentiation)	$5 \wedge 2 = 25$

Computers can perform comparison operations by comparing two numbers. For example, a computer can compare x to y and determine which number is larger.

Computers can store massive amounts of data in very small spaces and locate a particular item quickly. For example, you can store the text of more than one million books in a memory device about the size of your fist. Later in this chapter, you will learn about different storage media, such as magnetic disks and tape.

You can store the text of more than one million books in a memory device about the size of your fist.

2-5 INPUT, OUTPUT, AND MEMORY DEVICES

To use a computer and process data, three major components are needed: input, output, and memory devices. These are discussed in the following sections.

2-5a Input Devices

Input devices send data and information to the computer. They are constantly being improved to make data input easier. Examples of input devices include:

- *Keyboard*—This is the most widely used input device. Originally, keyboards were designed to resemble typewriters, but several modifications have been made to improve their ease of use. For example, most keyboards include control keys, arrow keys, function keys, and other special keys. In addition, some keyboards, such as the split keyboard, have been developed for better ergonomics. You can perform most computer input tasks with keyboards, but for some tasks a scanner or mouse is faster and more accurate.

- *Mouse*—This pointing device moves the cursor on the screen, allowing fast, precise cursor positioning. With programs that use graphical interfaces, such as Microsoft Windows or Mac OS, the mouse has become the input device of choice.

- *Touch screen*—This is a combination of input devices, usually working with menus. Some touch screens rely on light detection to determine which menu item has been selected; others are pressure sensitive. Touch screens are often easier to use than keyboards, but they might not be as accurate because selections can be misread or mistouched. You probably saw touch screens used extensively on television during the 2012 presidential election to quickly show electoral maps and analyze election data in different ways. The information box on the next page discusses touchless computing, which may become popular in near future.

- *Light pen*—This is a light-sensitive stylus connected to the monitor with a cable. When it is placed on an on-screen location, the data in that spot is sent to the computer. The data can be characters, lines, or blocks. A light pen is easy to use, inexpensive, and accurate, and it is particularly useful for engineers and graphic designers who need to make modifications to technical drawings.

- *Trackball*—This is kept in a stationary location, but it can be rolled on its axis to control the on-screen cursor. Trackballs occupy less space than a mouse, so they are ideal for notebook computers. However, positioning with a trackball is sometimes less precise than with a mouse.

- *Data tablet*—This consists of a small pad and a pen. Menus are displayed on the tablet, and you make selections with the pen. Data tablets are used most widely in computer-aided design and manufacturing applications.

- *Barcode reader*—This is an optical scanner that uses lasers to read codes in bar form. These devices are fast and accurate and have many applications in inventory, data entry, and tracking systems. They are used mostly with UPC systems in retail stores.

- *Optical character reader (OCR)*—This works on the same principle as a barcode reader but reads text instead. OCRs must be able to recognize many

input devices send data and information to the computer. Examples include a keyboard and mouse.

TOUCHLESS COMPUTING: THE NEW PARADIGM IN USER SYSTEM INTERFACE

Imagine using your tablet, PC, or Macintosh without needing a mouse, keyboard, or touchpad. Like Tom Cruise in *Minority Report*, you may soon be able to interact with your device without touching it, using only gestures. In fact, touchless computing enables you to control what is on the screen with the movements of your finger. The goal of touchless computing is to provide an environment similar to the real world, in which you are able to manipulate objects using only your fingers.[5]

This technology should be particularly useful for social networking sites such as Facebook, in situations where a user wants to navigate through different types of data in multiple settings. A San Francisco-based startup company named Leap Motion has designed a device called the Leap that is the size of a thumb drive. The Leap enables you to control your computer with high degree of accuracy by waving your hands around. This technology could also radically change our interaction with our TV sets. You will be able to navigate through channels and programs without touching the remote or the TV set.[6]

Engineers who use this technology will be able to manipulate 3D models on the screen by mere gesture. The *New York Times* has developed a 3D news reader and video reader based on this technology.[7]

special characters and distinguish between uppercase and lowercase letters, so using one is more difficult than using a barcode reader. Nevertheless, OCRs have been used successfully in many applications and are improving steadily. The United States Postal Service uses OCRs to sort mail.

- *Magnetic ink character recognition (MICR) system*—This reads characters printed with magnetic ink and is used primarily by banks for reading the information at the bottom of checks.

An **output device** is capable of representing information from a computer. The form of this output might be visual, audio, or digital; examples include printers, display monitors, and plotters.

- *Optical mark recognition (OMR) system*—This is sometimes called a "mark sensing" system because it reads marks on paper. OMRs are often used to grade multiple-choice and true/false tests.

2-5b Output Devices

Many **output devices** are available for both mainframes and personal computers. Output displayed on a screen is called "soft copy." The most common output devices for soft copy are cathode ray tube (CRT), plasma display, and liquid crystal display (LCD).

The other type of output is "hard copy," for which the most common output device is a printer. Inkjet and laser printers are standard printers used today. Inkjet printers produce characters by projecting onto paper electrically charged droplets of ink that create an image. High-quality inkjet printers use multicolor ink cartridges for near-photo quality output and are often used to print digital photographs. Inkjet printers are suitable for home users who have limited text and photo printing needs. When selecting a printer, consider cost (initial and maintenance), quality, speed, space that it occupies, and networking facilities.

Laser printers use laser-based technology that creates electrical charges on a rotating drum to attract toner. The toner is fused to paper using a heat process that creates high-quality output. Laser printers are better suited to larger office environments with high-volume and high-quality printing requirements. Other output devices include plotters for converting computer output to graphics and voice synthesizers for converting computer output to voice. Voice synthesis has become common. Cash registers at grocery stores use it to repeat item prices. When you call directory assistance, the voice you hear is probably computer generated. Voice output is also being used for marketing. A computer can dial a long list of phone numbers and deliver a message. If a number is busy, the computer makes a note and dials it later. Although the value of this method has been questioned, it is constructively used in some political campaigns to deliver messages about voting.

2-5c Memory Devices

Two types of memory are common to any computer: main memory and secondary memory. **Main memory** stores data and information and is usually volatile, meaning its contents are lost when electrical power is turned off. **Secondary memory**, which is nonvolatile, holds data when the computer is off or during the course of a program's operation. It also serves as archival storage. Main memory plays a major role in a computer's performance; to some extent, the more memory a computer has, the faster and more efficient its input/output (I/O) operations are. Graphics cards, also called video adapters, enhance computer performance, too. High-end graphics cards are important for graphic designers, who need fast rendering of 3D images. Many video games also require high-end graphics cards for the best display.

Main Memory Devices

The most common type of main memory is a semiconductor memory chip made of silicon. A semiconductor memory device can be volatile or nonvolatile. Volatile memory is called **random access memory (RAM)**, although you could think of it as "read-write memory." In other words, data can be read from and written to RAM. Some examples of the type of information stored in RAM include open files, the Clipboard's contents, running programs, and so forth.

A special type of RAM, called **cache RAM**, resides on the processor. Because memory access from main RAM storage generally takes several clock cycles (a few nanoseconds), cache RAM stores recently accessed memory so the processor is not waiting for the memory transfer.

Nonvolatile memory is called **read-only memory (ROM)**; data cannot be written to ROM. The type of data usually stored in ROM includes BIOS information and the computer system's clock. There are two other types of ROM. Programmable read-only memory (PROM) is a type of ROM chip that can be programmed with a special device. However, after it has been programmed, the contents cannot be erased. Erasable programmable read-only memory (EPROM) is similar to PROM, but its contents can be erased and reprogrammed.

Secondary Memory Devices

Secondary memory devices are nonvolatile and used for storing large volumes of data for long periods. As mentioned earlier, they can also hold data when the computer is off or during the course of a program's operation. There are three main types: magnetic disks, magnetic tape, and optical discs. Large enterprises also use storage area networks and network-attached storage (discussed in the next section) for storing massive amounts of data in a network environment.

A **magnetic disk**, made of Mylar or metal, is used for random-access processing. In other words, data can be accessed in any order, regardless of its order on the surface. Magnetic disks are much faster but more expensive than tape devices.

Magnetic tape, made of a plastic material, resembles a cassette tape and stores data sequentially. Records can be stored in a block or separately, with a gap between each record or block, called the interrecord gap (IRG). Magnetic tape is sometimes used for storing backups, although other media are more common now. Exhibit 2.3 shows both types of magnetic memory devices.

main memory stores data and information and is usually volatile; its contents are lost when electrical power is turned off. It plays a major role in a computer's performance.

secondary memory, which is nonvolatile, holds data when the computer is off or during the course of a program's operation. It also serves as archival storage.

random access memory (RAM) is volatile memory, in which data can be read from and written to; it is also called read-write memory.

cache RAM resides on the processor. Because memory access from main RAM storage takes several clock cycles (a few nanoseconds), cache RAM stores recently accessed memory so the processor is not waiting for the memory transfer.

read-only memory (ROM) is nonvolatile; data cannot be written to ROM.

A **magnetic disk**, made of Mylar or metal, is used for random-access processing. In other words, data can be accessed in any order, regardless of its order on the surface.

magnetic tape, made of a plastic material, resembles a cassette tape and stores data sequentially.

Exhibit 2.3
Magnetic memory devices

© Claudio Divizia/Shutterstock.com

© Avera/Shutterstock.com

Optical discs use laser beams to access and store data. Optical technology can store vast amounts of data and is durable. Three common types of optical storage are CD-ROMs, WORM discs, and DVDs.

Compact disc read-only memory (CD-ROM), as the name implies, is a read-only medium. CD-ROMs are easy to duplicate and distribute and are widely used in large permanent databases, such as for libraries, real estate firms, and financial institutions. They are sometimes used for multimedia applications and to distribute software products. However, because of its larger capacity—a minimum of 4.7 GB—digital versatile disc read-only memory (DVD-ROM) is used more often now, particularly for software distribution.

A write once, read many (WORM) disc is also a permanent device. Information can be recorded once and cannot be altered. A major drawback is that a WORM disc cannot be duplicated. It is used mainly to store information that must be kept permanently but not altered—for example, annual reports and information for nuclear power plants, airports, and railroads. SanDisk recently announced a Secure Digital (SD) card, a type of WORM disc, that can store data for 100 years but can be written on only once.[8]

Unlike with CD-ROMs and WORM discs, information stored on an erasable optical disc can be erased and altered repeatedly. These discs are used when high-volume storage and updating are essential.

Other secondary memory devices include hard disks, USB flash drives, and memory cards (see Exhibit 2.4). Hard disks come in a variety of sizes and can be internal or external, and their costs have been decreasing steadily. Memory sticks have become popular because of their small size, high storage capacity, and decreasing cost. Flash memory is nonvolatile memory that can be electronically erased and reprogrammed. It is used mostly in memory cards and USB flash drives for storing and transferring data between computers and other devices. Table 2.4 compares the capacity of common storage devices.

A **redundant array of independent disks (RAID)** system is a collection of disk drives used for fault tolerance and improved performance, and is typically found in large network systems. Data can be stored in multiple places to improve the system's reliability. In other words, if one disk in the array fails, data is not lost. In some RAID configurations, sequences of data can be read from multiple disks simultaneously, which improves performance.

Cloud storage (discussed in Chapter 14) has become a popular option for many organizations and individuals in recent years. Used for online storage and backup, it involves multiple virtual servers that are usually

---

Optical discs use laser beams to access and store data. Examples include CD-ROMs, WORM discs, and DVDs.

A **redundant array of independent disks (RAID)** system is a collection of disk drives used for fault tolerance and improved performance, and is typically found in large network systems.

Used for online storage and backup, **cloud storage** involves multiple virtual servers that are usually hosted by third parties. Customers buy or lease storage space from third parties based on their current or future needs.

TABLE 2.4 CAPACITY OF SECONDARY MEMORY DEVICES	
Device	**Storage capacity**
Memory stick	16 GB
Hard disk	2 TB
CD-ROM, CD-R, CD-RW	800 MB
DVD-ROM, DVD-R, DVD-RW	4.7 GB or more
Blu-Ray (latest generation optical disc)	Up to 25 GB on a single-layer disc and 50 GB on a dual-layer disc

© Cengage Learning®

Exhibit 2.4
Examples of memory devices

© Timurpix/Shutterstock.com
© Martin Petransky/Shutterstock.com
© Yojik/Shutterstock.com

hosted by third parties. Examples include JustCloud, ZipCloud, MyPCBackup, and Dropbox. Cloud storage may reduce the demand for some of the traditional storage devices or even replace some of them.[9]

Large corporations and enterprises that need a lot of storage space use a server farm or server cluster. This is a collection of hundreds or thousands of computers that require a large amount of power to run and keep cool.

Storage Area Networks and Network-Attached Storage

A **storage area network (SAN)** is a dedicated high-speed network consisting of both hardware and software used to connect and manage shared storage devices, such as disk arrays, tape libraries, and optical storage devices. A SAN network makes storage devices available to all servers on a local area network (LAN) or wide area network (WAN). (LANs and WANs will be discussed in Chapter 6.) Because a SAN is a dedicated network, servers can access storage devices more quickly and do not have to use their processing power to connect to these devices. Typically, SANs are used only in large enterprises because of their cost and installation complexity.

SANs speed up data access performance, and despite their cost, they are more economical than having storage devices attached to each server. A SAN's capacity can be extended easily, even to hundreds of terabytes.

Network-attached storage (NAS), on the other hand, is essentially a network-connected computer dedicated to providing file-based data storage services to other network devices. Software on the NAS handles features such as data storage, file access, and file and storage management.

When choosing between a SAN and a NAS system, consider the following factors:

- Hybrid solutions (combining a SAN and a NAS) might be available.
- A SAN offers only storage; a NAS system offers both storage and file services.

- NAS is popular for Web servers and e-mail servers because it lowers management costs and helps make these servers more fault tolerant. It is also becoming a useful solution for providing large amounts of heterogeneous data (text, documents, voice, images, movies, and so forth) for consumer applications.

- The biggest issue with NAS is that, as the number of users increases, its performance deteriorates. However, it can be expanded easily by adding more servers or upgrading the CPU.

2-6 CLASSES OF COMPUTERS

Usually, computers are classified based on cost, amount of memory, speed, and sophistication. Using these criteria, computers are classified as subnotebooks, notebooks, personal computers, minicomputers, mainframes, or supercomputers. Supercomputers are the most powerful; they also have the highest storage capabilities and the highest price.

Mainframe computers are usually compatible with the IBM System/360 line introduced in 1965. As mentioned in the Industry Connection later in this chapter, IBM System zEnterprise EC12 is the latest example in this class. Systems that are not based on System/360 are referred to as "servers" (discussed in the next section) or supercomputers. Supercomputers are more expensive, much bigger, faster, and have more memory than personal computers, minicomputers, and mainframes.

A **storage area network (SAN)** is a dedicated high-speed network consisting of both hardware and software used to connect and manage shared storage devices, such as disk arrays, tape libraries, and optical storage devices.

Network-attached storage (NAS) is essentially a network-connected computer dedicated to providing file-based data storage services to other network devices.

A notebook computer today has more power than a mainframe of the 1970s, and all indications suggest that this trend will continue.

POPULAR iPAD BUSINESS APPLICATIONS

The iPad is a tablet computer designed and developed by Apple. iPad users can browse the Web, read and send e-mail, share photos, watch HD videos, listen to music, play games, read e-books, and much more by using a multitouch user interface. Currently, iPads are used by many business professionals, including the following:[10,11]

Health care workers—To access medical applications and for bedside care

Sales agents and service workers—To perform on-the-road sales presentations and to display product information

Insurance agents—To display quotes

Real estate agents—To provide remote, interactive, visual home tours for interested home buyers

Legal professionals—To access legal documents and conduct a NexisLexis search from a car, office, or courtroom

Teachers and students—To access Windows applications and resources

Financial professionals—To access Windows trading applications, dashboards, documents, real-time quotes, Bloomberg Anywhere, and portfolio analysis tools

Corporate campus workers—To access corporate data so employees can collaborate with their colleagues while moving from office to office

Remote and mobile workers—To access Windows business applications, desktops, and data while on the road

Applications for computers include anything from doing homework (subnotebook, notebook, and personal computer) to launching space shuttles (supercomputer). Because all computers are steadily increasing in speed and sophistication, delineating different classes of computers is more difficult now. For example, a notebook computer today has more power than a mainframe of the 1970s, and all indications suggest that this trend will continue. The information box above highlights some of the business applications available for the iPad.

Experts believe we are entering the era of wearable, ubiquitous computing. See the information box below for more on this topic.

UBIQUITOUS COMPUTING

Experts describe the current state of computing as ubiquitous computing or pervasive computing. It is also called the third wave. The first wave was identified by mainframe computers, the second wave by personal computers, and the third wave by small computers embedded into many devices used daily—cellphones, cameras, watches, and so forth. Because people usually carry these devices around, the term "wearable" has been coined to describe them. Wearable computers are also used in medical monitoring systems, and they can be helpful when people need to use computers—to enter text, for example—while standing or walking around. There are many platforms and several players in this market. The following are three popular examples[12]:

- *Google Glass*—Displays information in a hands-free format; can communicate with the user and the Web in a natural language
- *Nike+ FuelBand*—Uses a sensor that runners can slip into their shoes to track performance
- *Jawbone Up*—Tracks one's movements around the clock

A new type of wearable computing using ingestible sensors and implantable chips may soon become common. These devices can be used to collect information about what is going on inside a person's body. They can be used for treating chronic illnesses and could assist the medical community in coming up with more suitable drugs faster. At the same time, these devices may create some legal issues. For instance, they could reveal that someone has a particular illness, which could result in a higher insurance fee.[13]

2-6a Server Platforms: An Overview

A **server** is a computer and all the software for managing network resources and offering services to a network. Many different server platforms are available for performing specific tasks, including the following:

- *Application servers* store computer software, which users can access from their workstations.
- *Database servers* store and manage vast amounts of data for access from users' computers.
- *Disk servers* contain large-capacity hard drives and enable users to store files and applications for later retrieval.
- *Fax servers* contain software and hardware components that enable users to send and receive faxes.
- *File servers* contain large-capacity hard drives for storing and retrieving data files.
- *Mail servers* are configured for sending, receiving, and storing e-mails.
- *Print servers* enable users to send print jobs to network printers.
- *Remote access servers (RAS)* allow off-site users to connect to network resources, such as network file storage, printers, and databases.
- *Web servers* store Web pages for access over the Internet.

2-7 WHAT IS SOFTWARE?

Software is all the programs that run a computer system. It can be classified broadly as system software and application software. For example, Microsoft Windows is the OS for most PCs and belongs to the system software group. This type of software works in the background and takes care of housekeeping tasks, such as deleting files that are no longer needed. Application software is used to perform specialized tasks. Microsoft Excel, for example, is used for spreadsheet analyses and number-crunching tasks. The information box below highlights Google Docs, a Web-based software application.

2-7a Operating System Software

An **operating system (OS)** is a set of programs for controlling and managing computer hardware and software. It provides an interface between a computer and the user and increases computer efficiency by helping users share computer resources and by performing repetitive tasks for users. A typical OS consists of control programs and supervisor programs.

Control programs manage computer hardware and resources by performing the following functions:

- *Job management*—Control and prioritize tasks performed by the CPU.
- *Resource allocation*—Manage computer resources, such as storage and memory. In a network, control programs are also used for tasks, such as assigning a print job to a particular printer.

A **server** is a computer and all the software for managing network resources and offering services to a network.

An **operating system (OS)** is a set of programs for controlling and managing computer hardware and software. It provides an interface between a computer and the user and increases computer efficiency by helping users share computer resources and by performing repetitive tasks for users.

GOOGLE DOCS: APPLICATIONS AND CHALLENGES

Google Docs is a free Web-based application for creating word-processor documents, spreadsheets, presentations, and forms. You can use it to create and edit documents online while collaborating in real time with other users, and you can send files you create to others via the Internet or e-mail. You can save files in a variety of formats: .doc, .xls, .rtf, .csv, .ppt, and so forth. By default, files are saved to Google's servers, using the Google cloud computing platform. Cloud computing is covered in more detail in Chapter 14; basically, it makes data and applications more portable so you can work with a program from anywhere. Another popular feature of Google Docs is collaboration. Multiple users can share and edit files at the same time. With spreadsheets, users can even be notified of changes by e-mail.[14] Google Drive is used to store documents, photos, music, and videos in one place. It syncs with your mobile devices and your computer so if you make a change using one device, the change will automatically show up on your other devices.[15]

Cloud computing and Google Docs do present some challenges and security risks. On March 10, 2009, Google Docs revealed some private documents to unauthorized users because of a security flaw. According to Google, the flaw has been corrected.[16]

- *Data management*—Control data integrity by generating checksums to verify that data has not been corrupted or changed. When the OS writes data to storage, it generates a value (the checksum) along with the data. The next time this data is retrieved, the checksum is recalculated and compared with the original checksum. If they match, the integrity is intact. If they do not, the data has been corrupted somehow.
- *Communication*—Control the transfer of data among parts of a computer system, such as communication between the CPU and I/O devices.

The supervisor program, also known as the kernel, is responsible for controlling all other programs in the OS, such as compilers, interpreters, assemblers, and utilities for performing special tasks.

In addition to single-tasking and multitasking OSs, time-shared OSs allow several users to use computer resources simultaneously. OSs are also available in a variety of platforms for both mainframes and personal computers. Microsoft Windows, Mac OS, and Linux are examples of personal computer OSs, and mainframe OSs include UNIX and OpenVMS as well as some versions of Linux.

Two new operating systems for smartphones and other handheld devices have attracted much attention in recent years: iOS and Android. The iOS and its various versions run on iPad, iPhone, and iPod Touch. (See the information box below on iOS.) The Android operating system by Google runs on non-Apple smartphones and handheld devices, such as HTC Inspire, Samsung Galaxy, and LG Thrill.

2-7b Application Software

A personal computer can perform a variety of tasks by using **application software**, which can be commercial software or software developed in house. In-house software is usually more expensive than commercial software but is more customized and often fits the users' needs better. For almost any task you can imagine, a software package is available. The following sections give you an overview of common categories of commercial application software for personal computers. In addition to these, many other categories of software are available, such as information management software, Web-authoring software, and photo and graphics software.

> **application software** can be commercial software or software developed in house and is used to perform a variety of tasks on a personal computer.

iOS: The Brain Behind Apple Devices

iOS is the operating system that enables all apps to run on an iPhone, iPad, and iPod Touch. The latest version is iOS7.1 (operating system version 7.1). Some iOS7 features are new, and others have been further enhanced from earlier versions. Features include the following[17]:

Maps—Provides visual and spoken turn-by-turn navigation and real-time traffic updates

Siri—Provides voice interface for many ordinary tasks including information on restaurants, movies, and sports

Facebook integration—Allows direct interaction from your Apple device using voice or typed instructions without leaving your app

AirDrop—Allows you to share photos, videos, and contacts from any app with a Share button

Control Center—Gives you quick access to frequently used apps

Shared photo stream—Allows sharing just the photos you want, with the people you choose

Passbook—Allows you to add passes to Passbook through apps, e-mails, and Web sites from participating airlines, theaters, stores, and many more

FaceTime—Allows you to see the person that you are talking to on your iPhone or iPad

Phone—Allows you to decline an incoming call, instantly reply with a text message, or set a callback

Mail—Allows you set up a VIP list so you will never miss an important message from important people in your list

Safari—In addition to navigation, now allows you to save Web pages as well as links

Accessibility—Makes it easier to use Apple devices for people with vision, hearing, learning, and mobility disabilities

Camera with Panorama—Allows you to shoot up to 240 degrees horizontally as well as vertically

Word-Processing Software

You are probably most familiar with word-processing software used to generate documents. Typically, this includes editing features, such as deleting, inserting, and copying text. Advanced word-processing software often includes sophisticated graphics and data management features. Word-processing software saves time, particularly for repetitive tasks, such as sending the same letter to hundreds of customers. Most word-processing software offers spell checkers and grammar checkers. Some popular word-processing programs are Microsoft Word, Corel WordPerfect, and OpenOffice.

Spreadsheet Software

A spreadsheet is a table of rows and columns, and spreadsheet software is capable of performing numerous tasks with the information in a spreadsheet. You can even prepare a budget and perform a "what-if" analysis on the data. For example, you could calculate the effect on other budget items of reducing your income by 10 percent, or you could see the effect of a 2 percent reduction in your mortgage interest rate. Common spreadsheet software includes Microsoft Excel, IBM's Lotus 1-2-3, and Corel Quattro Pro. Appendices A, B, and C, which can be accessed at the Cengage CourseMate Web site (*www.cengage.com/cengagecourse/coursemate*), explain many of Excel's features.

Database Software

Database software is designed to perform operations such as creating, deleting, modifying, searching, sorting, and joining data. A database is essentially a collection of tables consisting of rows and columns. Database software makes accessing and working with data faster and more efficient. For example, manually searching a database containing thousands of records would be almost impossible. With database software, users can search information quickly and even tailor searches to meet specific criteria, such as finding all accounting students younger than 20 who have GPAs higher than 3.6. You will learn more about databases in Chapter 3. Popular database software for personal computers includes Microsoft Access, FileMaker Pro, and Alpha Software's Alpha Five. High-end database software used in large enterprises includes Oracle, IBM DB2, and Microsoft SQL Server.

Presentation Software

Presentation software is used to create and deliver slide shows. Microsoft PowerPoint is the most commonly used presentation software; other examples include Adobe Persuasion and Corel Presentations. You can include many types of content in slide shows, such as bulleted and numbered lists, charts, and graphs. You can also embed graphics as well as sound and movie clips.

Presentation software also offers several options for running slide shows, such as altering the time interval between slides. In addition, you can usually convert presentations into other formats, including Web pages and photo albums with music and narration. Another option in some presentation software is capturing what is on the computer screen and then combining several screen captures into a video for demonstrating a process, which can be useful in educational settings or employee training seminars, for example.

Graphics Software

Graphics software is designed to present data in a graphical format, such as line graphs and pie charts. These formats are useful for illustrating trends and patterns in data and for showing correlations. Graphics are created with integrated packages, such as Excel, Lotus 1-2-3, and Quattro Pro, or dedicated graphics packages, such as Adobe Illustrator and IBM Freelance. Exhibit 2.5 shows the types of graphs you can create in Microsoft Excel.

Desktop Publishing Software

Desktop publishing software is used to produce professional-quality documents without expensive hardware and software. This software works on a "what-you-see-is-what-you-get" (WYSIWYG, pronounced "wizzy-wig")

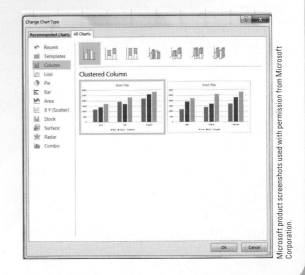

Exhibit 2.5
Types of graphs in Microsoft Excel

Microsoft product screenshots used with permission from Microsoft Corporation.

concept, so the high-quality screen display gives you a good idea of what you will see in the printed output. Desktop publishing software is used for creating newsletters, brochures, training manuals, transparencies, posters, and even books. Many desktop publishing packages are available; three popular ones are Adobe InDesign, QuarkXPress, and Microsoft Office Publisher.

Financial Planning and Accounting Software

Financial planning software, which is more powerful than spreadsheet software, is capable of performing many types of analysis on large amounts of data. These analyses include present value, future value, rate of return, cash flow, depreciation, retirement planning, and budgeting. A widely used financial planning package is Intuit Quicken. Using this package, you can plan and analyze all kinds of financial scenarios. For example, you can calculate how much your $2,000 IRA will be worth at 5 percent interest in 30 years, or you can determine how to save $150,000 in 18 years toward your child's college education, using a fixed interest rate.

In addition to spreadsheet software, dedicated accounting software is available for performing many sophisticated accounting tasks, such as general ledgers, accounts receivable, accounts payable, payroll, balance sheets, and income statements. Some popular accounting software packages include Intuit QuickBooks, a small-business accounting software, and Sage Software's Peachtree.

Project Management Software

A project, such as designing a Web site or setting up an order entry system, consists of a set of related tasks. The goal of project management software is to help project managers keep time and budget under control

by solving scheduling problems, planning and setting goals, and highlighting potential bottlenecks. You can use such software to study the cost, time, and resource impact of schedule changes. There are several project management software packages on the market, including Microsoft Project and Micro Planning International's Micro Planner.

Computer-Aided Design Software

Computer-aided design (CAD) software is used for drafting and design and has replaced traditional tools, such as T-squares, triangles, paper, and pencils. It is used extensively in architecture and engineering firms, but because of major price reductions and increases in PC power, small companies and home users can now afford this software. Widely used CAD software includes Autodesk AutoCAD, Cadkey, and VersaCAD.

2-8 COMPUTER LANGUAGES

As mentioned earlier, computer languages have developed through four generations, and the fifth generation is currently being developed. The first generation of computer languages, **machine language**, consists of a series of 0s and 1s representing data or instructions. Machine language depends on the machine, so code written for one type of computer does not work on another type of computer. Writing a machine-language program is time consuming and painstaking.

Assembly language, the second generation of computer languages, is a higher-level language than machine language but is also machine dependent. It uses a series of short codes, or mnemonics, to represent data or instructions. For example, ADD and SUBTRACT are typical commands in assembly language. Writing programs in assembly language is easier than in machine language.

Third-generation computer languages are machine independent and are called **high-level languages**. Three of the most widely used languages are C++, Java, and VB.NET. These languages are used mostly for Web development and Internet applications. High-level languages are more like English, so they are easier to learn and code. In addition, they are self-documenting, meaning that you can usually understand the programs without needing additional documentation.

Fourth-generation languages (4GLs) are the easiest computer languages to use. The commands are powerful and easy to learn, even for people with little

machine language, the first generation of computer languages, consists of a series of 0s and 1s representing data or instructions. It is dependent on the machine, so code written for one type of computer does not work on another type of computer.

Assembly language, the second generation of computer languages, is a higher-level language than machine language but is also machine dependent. It uses a series of short codes, or mnemonics, to represent data or instructions.

high-level languages are machine independent and part of the third-generation of computer languages. Many languages are available, and each is designed for a specific purpose.

Fourth-generation languages (4GLs) use macro codes that can take the place of several lines of programming. The commands are powerful and easy to learn, even for people with little computer training.

computer training. Sometimes, 4GLs are called nonprocedural languages, which means you do not need to follow a rigorous command syntax to use them. Instead, 4GLs use macro codes that can take the place of several lines of programming. For example, in a 4GL you might issue the PLOT command, a macro code that takes the place of 100 or more lines of high-level programming code. One simple command does the job for you. SQL (structured query language), which will be discussed in Chapter 3, is an example of a 4GL.

Fifth-generation languages (5GLs) use some of the artificial intelligence technologies (discussed in Chapter 13), such as knowledge-based systems, natural language processing (NLP), visual programming, and a graphical approach to programming. Codes are automatically generated and designed to make the computer solve a given problem without a programmer or with minimum programming effort. These languages are designed to facilitate natural conversations between you and the computer. Imagine that you could ask your computer, "What product generated the most sales last year?" Your computer, equipped

> Fourth-generation languages (4GLs) are the easiest computer languages to use.

with a voice synthesizer, could respond, "Product X." Dragon NaturallySpeaking Solutions is an example of NLP. Research continues in this field because of the promising results so far. Programming languages used for Internet programming and Web development include ActiveX, C++, Java, JavaScript, Perl, Python, PHP, Visual Basic, and Extensible Stylesheet Language (XSL). The most important Web development languages are Hypertext Markup Language (HTML) and Extensible Markup Language (XML). Both languages are markup languages, not full-featured programming languages.

The Industry Connection highlights IBM and its most popular product areas.

> **Fifth-generation languages (5GLs)** use some of the artificial intelligence technologies, such as knowledge-based systems, natural language processing (NLP), visual programming, and a graphical approach to programming. These languages are designed to facilitate natural conversations between you and the computer.

Industry Connection: International Business Machines (IBM)[18]

IBM, the largest computer company in the world, is active in almost every aspect of computing, including hardware, software, services, and collaboration tools such as groupware and e-collaboration. IBM has also been a leader in developing mainframe computers; its latest mainframe is the IBM System zEnterprise EC12. IBM's most popular product areas include the following:

Software: IBM offers software suites for all types of computers. Lotus, for example, includes features for e-mail, calendaring, and collaborative applications as well as business productivity software, much like Microsoft Office does. Tivoli, another software suite, has many features for asset management, security management, backup and restore procedures, and optimization of storage systems and data management. IBM also offers several different types of software for business analytics.

Storage: IBM's storage devices include disk and tape systems, storage area networks, network-attached storage, hard disks, and microdrives. A microdrive is a 1-inch hard disk designed to fit in a CompactFlash Type II slot.

Servers: IBM offers a variety of servers, including UNIX and Linux servers, Intel-based servers, AMD-based servers, and more.

IBM is also active in e-commerce software, hardware, and security services, such as encryption technologies, firewalls, antivirus solutions, and more.

STUDY TOOLS 2

LOCATED AT BACK OF THE TEXTBOOK

☐ Rip out Chapter Review Card

LOCATED AT WWW.CENGAGE.COM/LOGIN

☐ Review Key Terms Flash Cards (Print or Online)

☐ Download audio and visual summaries to review

☐ Complete both Practice Quizzes to prepare for tests

☐ Play "Beat the Clock" and "Quizbowl" to master concepts

☐ Complete "Crossword Puzzle" to review key terms

☐ Watch video on "Method" for a real company example

KEY TERMS

application software 36
arithmetic logic unit (ALU) 25
assembly language 38
bus 25
cache RAM 31
central processing unit (CPU) 25
cloud storage 32
computer 24
control unit 25
CPU case 25
disk drive 25
fifth-generation languages (5GLs) 39
fourth-generation languages (4GLs) 38
high-level languages 38
input devices 29

machine language 38
magnetic disk 31
magnetic tape 31
main memory 31
motherboard 26
network-attached storage (NAS) 33
operating system (OS) 35
optical discs 32
output devices 30
random access memory (RAM) 31
read-only memory (ROM) 31
redundant array of independent disks (RAID) 32
secondary memory 31
server 35
storage area network (SAN) 33

REVIEWS AND DISCUSSIONS

1. Define a "computer."

2. What are the distinguishing features of third-generation computers?

3. How was IBM Watson able to beat two former champions of the television game show "Jeopardy"?

4. What are two classes of software? What is an example of each?

5. What are three factors that define the power of computers?

6. What are some of the advantages of touchless computing?

7. Which languages are mostly used in a Web-development environment?

8. What are server farms?

PROJECTS

1. A local law firm needs your advice. It currently has 20 PCs being used by its attorneys and staff, and it is debating whether to use Google Apps or Office 365 as its productivity tools. Write a two-page paper that summarizes your recommendation. What are two advantages and two disadvantages of each choice? Which alternative is more suitable for a law firm with 20 PCs? Which alternative is less expensive?

2. After reading the information presented in this chapter and other sources, write a one-page paper that describes two advantages and two disadvantages of SAN and NAS. Name two vendors of each technology. Which one is less expensive?

3. A classmate of yours is not sure whether to buy a PC or a tablet. He needs the device for schoolwork (mostly Microsoft Office), for Web access, and as an e-reader. After reading the information presented in this chapter and other sources, write a one-page paper that summarizes your recommendation to this fellow student. Also mention two choices that you consider as top of the line for each alternative.

4. IBM Watson has created a lot of excitement in the computing field. After reading the information presented in this chapter and other sources, write a one-page paper that summarizes three commercial applications of this platform. What are two advantages of using Watson compared to using humans in the medical field? What are two disadvantages?

5. Android and iOS are two major operating systems for smartphones and other mobile devices. After reading the information presented in this chapter and other sources, write a two-page paper that summarizes the key features of each OS. What are two advantages and two disadvantages of each?

6. Dragon NaturallySpeaking Solutions is an example of an NLP. After reading the information presented in this chapter and other sources, write a two-page paper that describes eight commercial applications for this platform. Which businesses will benefit the most from this platform? What are two drawbacks of using this platform?

ARE YOU READY TO MOVE ON?

1. RAM is volatile. True or False?

2. SANs are used only in large enterprises because of their cost and installation complexity. True or False?

3. Microsoft Excel is an example of system software. True or False?

4. Fourth-generation computers are indentified by all of the following technologies except:

   a. Miniaturization

   b. Vacuum tubes

   c. Optical discs

   d. VLSI

5. Which of the following is not an example of a secondary memory device?

   a. EPROM

   b. Magnetic tape

   c. Magnetic disk

   d. Optical discs

6. All of the following are examples of application software except:

   a. MS Word

   b. MS Excel

   c. Autodesk AutoCAD

   d. Android

CASE STUDY 2-1

Become Your Own Banker

Using one of the many financial apps (most of them free) that are available for your tablet or smartphone these days, you can become your own banker. In fact, trips to the bank or even to an ATM could soon become a thing of the past. The new apps let you do all the functions that you could do on a bank's Web site and more. For example, to deposit a check into your banking account, you can simply take a photograph of the front and back of the check, after which the app will ask you for the amount you want to deposit. Then you type the account number and then press "OK" to complete the deposit. With the Mint app (*mint.com*) from Intuit, you can even perform all sorts of analytics on your account. For example, you can establish a budget and keep track of the expenses in various categories, such as food, gas, and groceries. Another app that offers similar services is HelloWallet from HelloWallet.com. Still other apps help you track and report work-related travel expenses. An example is an iPhone app called QuickShot. It allows you to take photos of your receipts, and the photos are automatically stored in an account on Dropbox.com, an Internet file-storage service.[19] Banks, brokerage firms, and other financial institutions have developed their own apps that simplify their customers' financial interactions with the institution.

© bloomua/Shutterstock.com

Answer the following questions:

1. Using your smartphone and a financial app, how do you deposit a check into your checking account?
2. Name some financial apps introduced in this case.
3. What are some examples of analytics that can be done using financial apps?
4. What are some other examples of financial apps? What are their applications?

CASE STUDY 2-2

Laptop Versus Tablet: Which One to Buy?

According to IHS iSuppli, a market research firm, annual sales of Internet-enabled devices, such as Apple's iPad, will surpass 750 million units by 2015. This figure significantly exceeds the projected PC sales of 479.1 million.[20] Revolutionized by the Apple iPad, tablet PCs are increasingly being used for a wide variety of tasks, even though they have not yet replaced laptops. There are many other tablets on the market as well, of course—Surface (Microsoft), Galaxy Tab (Samsung), Kindle Fire (Amazon), Optimus Pad (LG), Nexus 7 (Google), and XOOM (Motorola), to name a few.

Tablet computers work best as media players or e-book readers. Having used a standard keyboard on a laptop makes it difficult for some users to switch to touch screens; it is also difficult to switch from a standard mouse and keyboard. However, some tablets can be connected to external keyboards. In most cases, a tablet PC is more expensive than a laptop.

iPad users can employ a vast number of applications, many of which are free, by downloading them from the App Store. Non-Apple tablets use Android operating systems, and there are apps available for these tablets as well.

Size, battery life, availability of productivity tools, hardware/software support, and connectivity are the key factors to consider when choosing between a tablet and a laptop, although the line between the two is becoming

© Zentilia/Shutterstock.com

more and more blurred. A tablet is smaller, lighter, easier to carry, and faster to boot. An iPad's battery life is about 10 hours, whereas a typical laptop has a battery life of about three hours. Laptops are used more as productivity tools for business functions, but with additional accessories, tablets can also perform those tasks. As for connectivity and systems support, there are plenty of peripheral devices for laptops, whereas the hard disk capacity of tablets is much smaller than for laptops. Also, there are more operating systems to choose from on a laptop than on a tablet.

Before deciding which way to go, you must clearly define your needs. If you need a device with lots of hardware and software support for business applications, programming, and gaming, then you should choose a laptop. If you are looking for an easy-to-use smart device with many apps and easy connectivity, then a tablet would be a more suitable choice.[21]

Answer the following questions:

1. What should you do before deciding between a laptop or a tablet?
2. What are some of the key features of each device?
3. What are three important characteristics of each device that might influence your decision?
4. What are three examples of tablets on the market?

WHY CHOOSE?

Every 4LTR Press solution comes complete with a visually engaging textbook in addition to an interactive eBook. Go to CourseMate for (**MIS5**) to begin using the eBook. Access at **www.cengagebrain.com**

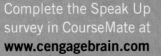

3 | Database Systems, Data Warehouses, and Data Marts

LEARNING OUTCOMES

After studying this chapter, you should be able to:

3-1 Define a database and a database management system.

3-2 Explain logical database design and the relational database model.

3-3 Define the components of a database management system.

3-4 Summarize recent trends in database design and use.

3-5 Explain the components and functions of a data warehouse.

3-6 Describe the functions of a data mart.

3-7 Define business analytics and describe its role in the decision-making process.

3-8 Explain big data and its business applications.

After you finish this chapter, go to **PAGE 63** for the **STUDY TOOLS**

In a database system, all files are integrated, meaning information can be linked.

This chapter gives you an overview of databases and database management systems and their importance in information systems. You learn about the types of data in a database, methods for accessing files, and physical and logical views of information. You also review the most common data model, the relational model, and the major components of a database management system. This chapter also discusses recent trends in database use, including data-driven Web sites, distributed databases, and object-oriented databases. Next, the chapter provides an overview of data warehouses and data marts and their roles in

generating business intelligence, then reviews business analytics and its role in gaining a competitive advantage. The chapter concludes with an overview of big data and its growing business applications.

3-1 DATABASES

A **database** is a collection of related data that is stored in a central location or in multiple locations. You can think of it as being similar to a filing cabinet, in which data is organized and stored in drawers and file folders. As you can imagine, however, retrieving data from a database is much faster.

Although a database can consist of only a single file, it is usually a group of files. A university database, for example, might have files for students, staff, faculty, and courses. In a database, a file is a group of related records, and a record is a group of related fields. This structure is called a **data hierarchy**, as shown in Exhibit 3.1. In the university database example, fields consist of Social

> A **database** is a collection of related data that is stored in a central location or in multiple locations.
>
> A **data hierarchy** is the structure and organization of data, which involves fields, records, and files.

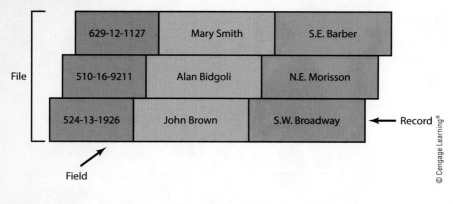

Exhibit 3.1
Data hierarchy

629-12-1127	Mary Smith	S.E. Barber
510-16-9211	Alan Bidgoli	N.E. Morisson
524-13-1926	John Brown	S.W. Broadway

File

← Record

Field

© Cengage Learning®

© lightpoet/Shutterstock.com

Security number, student name, and address. All the fields storing information for Mary Smith, for instance, constitute a record, and all three records in Exhibit 3.1 make up the student file.

In a database system, all files are integrated, meaning information can be linked. For example, you can retrieve the names of all students enrolled in Professor Thomas's MIS 480 course from the courses file, or look up Professor Thomas's record to find out other courses he is teaching in a particular semester.

A database is a critical component of information systems because any type of analysis that is done is based on data available in the database. To make using databases more efficient, a **database management system (DBMS)** is used. A DBMS is software for creating, storing, maintaining, and accessing database files. Its major components are discussed in the "Components of a DBMS" section, later in this chapter. If you are familiar with Microsoft Office software, you know that you use Word to create a document and Excel to create a spreadsheet. You can also use Access to create and modify a database, although it does not have as many features as other DBMSs.

For now, take a look at Exhibit 3.2, which shows how users, the DBMS, and the database interact. The user issues a request, and the DBMS searches the database and returns the information to the user.

In the past, data was stored in a series of files that were called "flat files" because they were not arranged in a hierarchy, and there was no relationship among these files. The problem with this flat file organization was that the same data could be stored in more than one file, creating data redundancy. For example, in a customer database, a customer's name might be stored in more than one table. This duplication takes up unnecessary storage space and can make retrieving data inefficient. Updating a flat file system can be time consuming and inefficient. Data might not be updated in all files consistently, resulting in conflicting reports generated from these files.

In summary, a database has the following advantages over a flat file system:

- More information can be generated from the same data.
- Complex requests can be handled more easily.
- Data redundancy is eliminated or minimized.
- Programs and data are independent, so more than one program can use the same data.
- Data management is improved.
- A variety of relationships among data can be easily maintained.
- More sophisticated security measures can be used.
- Storage space is reduced.

A **database management system (DBMS)** is software for creating, storing, maintaining, and accessing database files. A DBMS makes using databases more efficient.

Exhibit 3.2
Interaction between the user, DBMS, and database

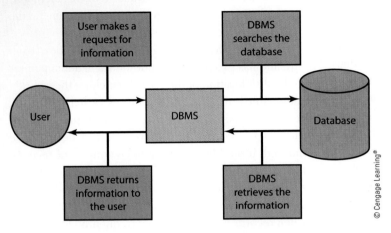

User makes a request for information

DBMS searches the database

User → DBMS → Database

DBMS returns information to the user

DBMS retrieves the information

© Cengage Learning®

3-1a Types of Data in a Database

As you learned in Chapter 1, to generate business intelligence (BI), the database component of an information system needs access to two types of data: internal and external. Internal data is collected from within an organization and can include transaction records, sales records, personnel records, and so forth. An organization might use internal data on customers' past purchases to generate BI about future buying patterns, for example. Internal data is usually stored in the organization's internal databases and can be used by functional information systems.

Collecting and using external data can be more challenging. External data comes from a variety of sources and is often stored in a data warehouse (discussed later in the chapter). The following are examples of sources for external data:

- Competitors, customers, and suppliers
- Distribution networks
- Economic indicators (e.g., the consumer price index)
- Government regulations
- Labor and population statistics
- Tax records

In the section called "Data Warehouses," found later in this chapter, you will learn how information from external data sources is used to conduct analyses and generate reports for BI. The information box below discusses how BI is used in other fields—in this case, law enforcement.

© Phase4Studios/Shutterstock.com

BI IN ACTION: LAW ENFORCEMENT

Business intelligence (BI) is used in law enforcement as well as in the business world. In Richmond, Virginia, data entered into the information system includes crime reports from the past 5 years, records of 911 phone calls, details about weather patterns, and information about special events. The system generates BI reports that help pinpoint crime patterns and are useful for personnel deployment, among other purposes. The system has increased public safety, reduced 911 calls, and helped management make better use of Richmond's 750 officers.

Recently, the department refined its reports by separating violent crimes into robberies, rapes, and homicides to help them discover patterns for certain types of crime. For example, the department discovered that Hispanic workers were often robbed on paydays. By entering workers' paydays into the system and looking at robbery patterns, law enforcement officers were able to identify days and locations on which these incidents were likely to occur. Moving additional officers into those areas on paydays has reduced the number of robberies.[1]

3-1b Methods for Accessing Files

In a database, files are accessed by using a sequential, random, or indexed sequential method. In a **sequential access file structure**, records in files are organized and processed in numerical or sequential order, typically the order in which they were entered. Records are organized based on what is known as a "primary key," discussed later in this chapter, such as Social Security numbers or account numbers. For example, to access record number 10, records 1 through 9 must be read first. This type of access method is effective when a large number of records are processed less frequently, perhaps on a quarterly or yearly basis. Because access speed usually is not critical, these records are typically stored on magnetic tape. Normally, a sequential file structure is used for backup and archive files because they rarely need updating.

In a **random access file structure**, records can be accessed in any order, regardless of their physical locations in storage media. This method of access is fast and very effective when a small number of records needs to be processed daily or weekly. To achieve this speed, these records are often stored on magnetic disks. Disks are random access devices, whereas tapes are sequential access devices. (Consider how much quicker it is to skip a song on a CD as opposed to a tape cassette.)

With the **indexed sequential access method (ISAM)**, records can be accessed sequentially or randomly, depending on the number being accessed. For a small number, random access is used, and for a large number, sequential access is used. This file structure is similar to a book index that lists page numbers where you can find certain topics. The advantage of this method is that both types of access can be used, depending on the situation and the user's needs.[2]

ISAM, as the name suggests, uses an index structure and has two parts: the indexed value and a pointer to the disk location of the record matching the indexed value. Retrieving a record requires at least two disk accesses, once for the index structure and once for the actual record. Because every record needs to be indexed, if the file contains a huge number of records, the index is also quite large. Therefore, an index is more useful when the number of records is small. Access speed with this method is fast, so it is recommended when records must be accessed frequently. This advice, however, was more applicable when processors were slow and memory and storage were expensive. Given the speed and low storage cost of today's computers, the number of records might not be as important, meaning more records could be accessed and processed with this method.

3-2 LOGICAL DATABASE DESIGN

Before designing a database, you need to know the two ways information is viewed in a database. The **physical view** involves how data is stored on and retrieved from storage media, such as hard disks, magnetic tapes, or CDs. For each database, there is only one physical view of data. The **logical view** involves how information appears to users and how it can be organized and retrieved. There can be more than one logical view of data, depending on the user. For example, marketing executives might want to see data organized by top-selling products in a specific region; the finance officer might need to see data organized by cost of raw materials for each product.

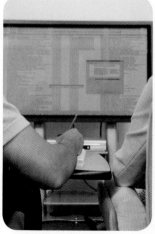

Image Source/Getty Images

The first step in database design is defining a **data model**, which determines how data is created, represented, organized, and maintained. A data model usually includes these three components:

- *Data structure*—Describes how data is organized and the relationship among records
- *Operations*—Describes methods, calculations, and so forth that can be performed on data, such as updating and querying data
- *Integrity rules*—Defines the boundaries of a database, such as maximum and minimum values allowed for a field, constraints (limits on what type of data can be stored in a field), and access methods

Many data models are used. The most common, the relational model, is described in the next section, and you learn about the object-oriented model later in the chapter, in the "Object-Oriented Databases" section. Two other common data models are hierarchical and network, although they are not used as much now.

In a **hierarchical model**, shown in Exhibit 3.3, the relationships among records form a treelike structure (hierarchy). Records are called nodes, and relationships among records are called branches. The node at the top is called the root, and every other node (called a child) has a parent. Nodes with the same parents are called twins or siblings. In Exhibit 3.3, the root node is a supplier, which provides three product lines, each considered a sibling of the other two. Each product line has categories of products, and each category has specific products. (The categories and specific products are also siblings of each other.) For example, Supplier A supplies three product lines: soap, shampoo, and toothpaste. The toothpaste product line has two categories (P6 and P7): whitening and cavity-fighting toothpaste. The P7 category has three specific cavity-fighting products: S, T, and U.

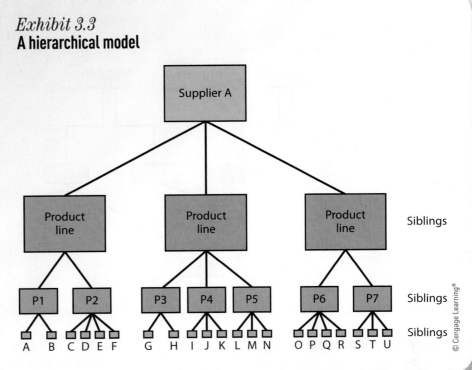

Exhibit 3.3
A hierarchical model

© Cengage Learning®

The **network model** is similar to the hierarchical model, but records are organized differently, as shown in Exhibit 3.4. This model links invoice number, customer number, and method of payment. For example, invoice 111 belongs to customer 2000, who paid with cash. Unlike the hierarchical model, each record in the network model can have multiple parent and child records. For example, in Exhibit 3.4, a customer can have several invoices, and an invoice can be paid by more than one method.

A **data model** determines how data is created, represented, organized, and maintained. It usually contains data structure, operations, and integrity rules.

In a **hierarchical model**, the relationships between records form a treelike structure (hierarchy). Records are called nodes, and relationships between records are called branches. The node at the top is called the root, and every other node (called a child) has a parent. Nodes with the same parents are called twins or siblings.

The **network model** is similar to the hierarchical model, but records are organized differently. Unlike the hierarchical model, each record in the network model can have multiple parent and child records.

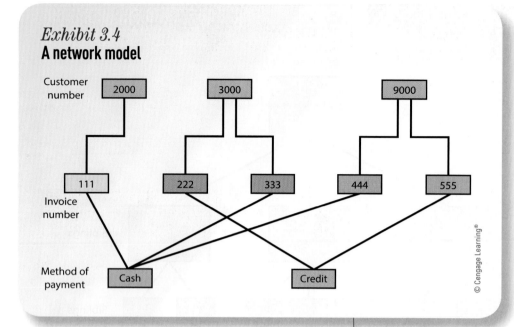

Exhibit 3.4
A network model

Customer number: 2000, 3000, 9000

Invoice number: 111, 222, 333, 444, 555

Method of payment: Cash, Credit

© Cengage Learning®

3-2a The Relational Model

A **relational model** uses a two-dimensional table of rows and columns of data. Rows are records (also called *tuples*), and columns are fields (also referred to as *attributes*). To begin designing a relational database, you must define the logical structure by defining each table and the fields in it. For example, the Students table has fields for StudentID, StudentFirstName, StudentLastName, and so forth. The collection of these definitions is stored in the data dictionary. The **data dictionary** can also store other definitions, such as data types for fields, default values for fields, and validation rules for data in each field, as described in the following list:

- *Field name*—Student name, admission date, age, and major
- *Field data type*—Character (text), date, and number

A **relational model** uses a two-dimensional table of rows and columns of data. Rows are records (also called *tuples*), and columns are fields (also referred to as *attributes*).

The **data dictionary** stores definitions, such as data types for fields, default values, and validation rules for data in each field.

A **primary key** uniquely identifies every record in a relational database. Examples include student ID numbers, account numbers, Social Security numbers, and invoice numbers.

A **foreign key** is a field in a relational table that matches the primary key column of another table. It can be used to cross-reference tables.

Normalization improves database efficiency by eliminating redundant data and ensuring that only related data is stored in a table.

- *Default value*—The value entered if none is available; for example, if no major is declared, the value is "undecided."
- *Validation rule*—A rule determining whether a value is valid; for example, a student's age can not be a negative number.

In a relational database, every record must be uniquely identified by a **primary key**. Student ID numbers, Social Security numbers, account numbers, and invoice numbers are examples of primary keys. To establish relationships among tables so data can be linked and retrieved more efficiently, a primary key for one table can appear in other tables. In this case, it is called a **foreign key**. For example, a student ID number is the primary key for the Students table, and the code for a student's major (such as MKT, MIS, or FIN) is the primary key for the Majors table. Each student can have one major, and a number of students can be enrolled in each major. The primary key of the Majors table is a foreign key in the Students table.

To improve database efficiency, a process called **normalization** is used, which eliminates redundant data (e.g., ensuring customer names are stored in only one table) and ensures that only related data is stored in a table. Normalization can go through several stages, from first normal form (1NF) to fifth normal form (5NF). Typically, however, only stages 1NF through 3NF are used. For example, the following tasks are performed in the 1NF stage:

- Eliminate duplicated fields from the same table.
- Create separate tables for each group of related data.
- Identify each record with a unique field (the primary key).

Data stored in a relational model is retrieved from tables by using operations that pick and combine data from one or more tables. There are several operations, such as select, project, join, intersect, union, and

TABLE 3.1 DATA IN THE STUDENTS TABLE

Student ID	Name	Major	Age	GPA
111	Mary	MIS	18	4.00
222	Sue	CS	21	3.60
333	Debra	MGT	20	3.50
444	Bob	MKT	22	3.40
555	George	MIS	21	3.70

© Cengage Learning®

TABLE 3.2 RESULTS OF THE SELECT OPERATION

Student ID	Name	Major	Age	GPA
111	Mary	MIS	18	4.00
555	George	MIS	21	3.70

© Cengage Learning®

TABLE 3.3 RESULTS OF THE PROJECT OPERATION

Student ID	Name	Major	GPA
111	Mary	MIS	4.00
222	Sue	CS	3.60
333	Debra	MGT	3.50
444	Bob	MKT	3.40
555	George	MIS	3.70

© Cengage Learning®

TABLE 3.4 CUSTOMERS TABLE

Customer	Name	Address
2000	ABC	Broadway
3000	XYZ	Jefferson
9000	TRY	Madison

© Cengage Learning®

TABLE 3.5 INVOICES TABLE

Invoice	Customer	Amount	Payment
1110	2000	$2000.00	Cash
2220	3000	$4000.00	Credit
3330	3000	$1500.00	Cash
4440	9000	$6400.00	Cash
5550	9000	$7000.00	Credit

© Cengage Learning®

TABLE 3.6 JOINING THE INVOICES AND CUSTOMERS TABLES

Invoice	Customer	Amount	Payment	Name	Address
1110	2000	$2000.00	Cash	ABC	Broadway
2220	3000	$4000.00	Credit	XYZ	Jefferson
3330	3000	$1500.00	Cash	XYZ	Jefferson
4440	9000	$6400.00	Cash	TRY	Madison
5550	9000	$7000.00	Credit	TRY	Madison

© Cengage Learning®

difference. The first three are the most commonly used and are explained in the following paragraphs.

A select operation searches data in a table and retrieves records based on certain criteria (also called conditions). Table 3.1 shows data stored in the Students table. Using the select operation "Major=MIS," you can generate a list of only the students majoring in MIS, as Table 3.2 shows.

A project operation pares down a table by eliminating columns (fields) according to certain criteria. For example, say you need a list of students but do not want to include their ages. Using the project operation "PROJECT Student ID#, Name, Major, GPA (Table 3.1)," you can retrieve the data shown in Table 3.3. The "(Table 3.1)" in this statement means to use the data in Table 3.1.

A join operation combines two tables based on a common field (e.g., the primary key in the first table and the foreign key in the second table). Table 3.4 shows data in the Customers table, and Table 3.5 shows data in the Invoices table. The Customer# is the primary key for the Customers table and is a foreign key in the Invoices table; the Invoice# is the primary key for the Invoices table. Table 3.6 shows the table resulting from joining these two tables.

Now that you have learned about the components of a database and a common data model, you can examine the software used to manage databases.

3-3 COMPONENTS OF A DBMS

DBMS software includes these components, discussed in the following sections:

- Database engine
- Data definition
- Data manipulation
- Application generation
- Data administration

3-3a Database Engine

A database engine, the heart of DBMS software, is responsible for data storage, manipulation, and retrieval. It converts logical requests from users into their physical equivalents (e.g., reports) by interacting with other components of the DBMS (usually the data manipulation component). For example, say a marketing manager wants to see a list of the top three salespeople in the Southeast region (a logical request). The database engine interacts with the data manipulation component to find where these three names are stored and displays them on screen or in a printout (the physical equivalent). Because more than one logical view of data is possible, the database engine can retrieve and return data to users in many different ways.

3-3b Data Definition

The data definition component is used to create and maintain the data dictionary and define the structure of files in a database. Any changes to a database's structure, such as adding a field, deleting a field, changing a field's size, or changing the data type stored in a field, are made with this component.

3-3c Data Manipulation

The data manipulation component is used to add, delete, modify, and retrieve records from a database. Typically, a query language is used for this component. Many query languages are available, but Structured Query Language (SQL) and query by example (QBE) are two of the most widely used.

Structured Query Language (SQL) is a standard fourth-generation query language used by many

> A database engine, the heart of DBMS software, is responsible for data storage, manipulation, and retrieval.

DBMS packages, such as Oracle Database 11g and Microsoft SQL Server. SQL consists of several keywords specifying actions to take. The basic format of an SQL query is as follows:

SELECT *field* FROM *table or file* WHERE *conditions*

After the SELECT keyword, you list the fields you want to retrieve. After FROM, you list the tables or files from where the data is retrieved, and after WHERE, you list conditions (criteria) for retrieving the data. The following example retrieves the name, Social Security number, title, gender, and salary from the Employee and Payroll tables for all employees with the job title of engineer:

```
SELECT NAME, SSN, TITLE, GENDER,
    SALARY
FROM EMPLOYEE, PAYROLL
WHERE EMPLOYEE.SSN=PAYROLL.SSN AND
    TITLE="ENGINEER"
```

This query means that data in the NAME, SSN, TITLE, GENDER, and SALARY fields from the two tables EMPLOYEE and PAYROLL should be retrieved. Line 3 indicates on which field the EMPLOYEE and PAYROLL tables are linked (the SSN field) and specifies a condition for displaying data: only employees who are engineers. You can add many other conditions to SQL statements by using AND, OR, and NOT operators (discussed next).

With **query by example (QBE)**, you request data from a database by constructing a statement made up of query forms. With current graphical databases, you simply click to select query forms instead of having to remember keywords, as you do with SQL. You can add AND, OR, and NOT operators to the QBE form to fine-tune the query:

- *AND*—Means that all conditions must be met. For example, "Major=MIS AND GPA > 3.8" means a student must be majoring in MIS and have a GPA higher than 3.8 to be retrieved.

- *OR*—Means only one of the conditions must be met. For example, "Major=MIS OR GPA > 3.8" means that if a student has a GPA higher than 3.8, he can be majoring in another field, such as accounting. Alternatively, if a student is majoring in MIS, she can have a GPA of 3.8 or lower.

- *NOT*—Searches for records that do not meet the condition. For example, "Major NOT ACC" retrieves all students except accounting majors.

Structured Query Language (SQL) is a standard fourth-generation query language used by many DBMS packages, such as Oracle 11g and Microsoft SQL Server. SQL consists of several keywords specifying actions to take.

With **query by example (QBE)**, you request data from a database by constructing a statement made up of query forms. With current graphical databases, you simply click to select query forms instead of having to remember keywords, as you do with SQL. You can add AND, OR, and NOT operators to the QBE form to fine-tune the query.

3-3d Application Generation

The application generation component is used to design elements of an application using a database, such as data entry screens, interactive menus, and interfaces with other programming languages. These applications, for example, might be used to create a form or generate a report. If you are designing an order entry application for users, you could use the application generation component to create a menu system that makes the application easier to use. Typically, IT professionals and database administrators use this component.

3-3e Data Administration

The data administration component, also used by IT professionals and database administrators, is used for tasks such as backup and recovery, security, and change management. In addition, this component is used to determine who has permission to perform certain functions, often summarized as **create, read, update, and delete (CRUD)**.

In large organizations, database design and management is handled by the **database administrator (DBA)**, although with complex databases this task is sometimes handled by an entire department. The DBA's responsibilities include:

- Designing and setting up a database
- Establishing security measures to determine users' access rights
- Developing recovery procedures in case data is lost or corrupted
- Evaluating database performance
- Adding and fine-tuning database functions

3-4 RECENT TRENDS IN DATABASE DESIGN AND USE

Recent trends in database design and use include data-driven Web sites, natural language processing, distributed databases, and object-oriented databases. In addition to these trends, advances in artificial intelligence and natural language processing will have an impact on database design and use, such as improving user interfaces.[3] Chapter 13 covers natural language processing, and the other trends are discussed in the following sections.

3-4a Data-Driven Web Sites

With the popularity of e-commerce applications, data-driven Web sites are used more widely to provide dynamic content. A **data-driven Web site** acts as an interface to a database, retrieving data for users and allowing users to enter data in the database. Without this feature, Web site designers must edit the HTML code every time a Web site's data contents change. This type of site is called a "static" Web site. A data-driven Web site, on the other hand, changes automatically because it retrieves content from external dynamic data sources, such as MySQL, Microsoft SQL Server, Microsoft Access, Oracle, IBM DB2, and other databases.

A data-driven Web site improves access to information so users' experiences are more interactive, and it reduces the support and overhead needed to maintain static Web sites. A well-designed data-driven Web site is easier to maintain because most content changes require no change to the HTML code. Instead, changes are

© Dotshock/Shutterstock.com

Create, read, update, and delete (CRUD) refers to the range of functions that data administrators determine who has permission to perform certain functions.

Database administrators (DBA), found in large organizations, design and set up databases, establish security measures, develop recovery procedures, evaluate database performance, and add and fine-tune database functions.

A **data-driven Web site** acts as an interface to a database, retrieving data for users and allowing users to enter data in the database.

made to the data source, and the Web site adjusts automatically to reflect these changes. With a data-driven Web site, users can get more current information from a variety of data sources. Data-driven Web sites are useful for the following applications, among others:

- E-commerce sites that need frequent updates
- News sites that need regular updating of content
- Forums and discussion groups
- Subscription services, such as newsletters

3-4b Distributed Databases

The database types discussed so far use a central database for all users of an information system. However, in some situations, a **distributed database management system (DDBMS)**, in which data is stored on multiple servers placed throughout an organization, is preferable. Here are some of the reasons an organization would choose a distributed database[4]:

- The design better reflects the organization's structure. For example, an organization with several branch offices might find a distributed database more suitable because it allows faster local queries and can reduce network traffic.
- Local storage of data decreases response time but increases communication costs.
- Distributing data among multiple sites minimizes the effects of computer failures. If one database server goes down, it does not affect the entire organization.
- The number of users of an information system is not limited by one computer's capacity or processing power.
- Several small integrated systems might cost less than one large server.

A **distributed database management system (DDBMS)** stores data on multiple servers throughout an organization.

The **fragmentation** approach to a distributed DBMS addresses how tables are divided among multiple locations. There are three variations: horizontal, vertical, and mixed.

The **replication** approach to a distributed DBMS has each site store a copy of the data in the organization's database.

The **allocation** approach to a distributed DBMS combines fragmentation and replication, with each site storing the data it uses most often.

In **object-oriented databases**, both data and their relationships are contained in a single object. An object consists of attributes and methods that can be performed on the object's data.

- Accessing one central database server could increase communication costs for remote users. Storing some data at remote sites can help reduce these costs.
- Distributed processing, which includes database design, is used more widely now and is often more responsive to users' needs than centralized processing.
- Most importantly, a distributed database is not limited by the data's physical location.

There are three approaches to setting up a DDBMS, although these approaches can be combined[5]:

- *Fragmentation*—The **fragmentation** approach addresses how tables are divided among multiple locations. Horizontal fragmentation breaks a table into rows, storing all fields (columns) in different locations. Vertical fragmentation stores a subset of columns in different locations. Mixed fragmentation, which combines vertical and horizontal fragmentation, stores only site-specific data in each location. If data from other sites is needed, the DDBMS retrieves it.
- *Replication*—With the **replication** approach, each site stores a copy of data in the organization's database. Although this method can increase costs, it also increases availability of data, and each site's copy can be used as a backup for other sites.
- *Allocation*—The **allocation** approach combines fragmentation and replication. Generally, each site stores the data it uses most often. This method improves response time for local users (those in the same location as the database storage facilities).

Security issues are more challenging in a distributed database because of multiple access points from both inside and outside the organization. Security policies, scope of user access, and user privileges must be clearly defined, and authorized users must be identified. Distributed database designers should also keep in mind that distributed processing is not suitable for every situation, such as a company that has all its departments in one location.

3-4c Object-Oriented Databases

The relational model discussed previously is designed to handle homogenous data organized in a field-and-record format. Including different types of data, such as multimedia files, can be difficult, however. In addition, a relational database has a simple structure: Relationships between tables are based on a common value (the key). Representing more complex data relationships sometimes is not possible with a relational database.

To address these problems, **object-oriented databases** were developed. Like object-oriented

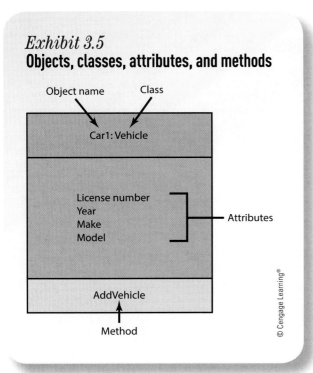

Exhibit 3.5
Objects, classes, attributes, and methods

Object name — Car1: Vehicle — Class

License number
Year
Make
Model — Attributes

AddVehicle

Method

© Cengage Learning®

programming, this data model represents real-world entities with database objects. An object consists of attributes (characteristics describing an entity) and methods (operations or calculations) that can be performed on the object's data. For example, as shown in Exhibit 3.5, a real-world car can be represented by an object in the Vehicle class. Objects in this class have attributes of year, make, model, and license number, for example. You can then use methods to work with data in a Vehicle object, such as the AddVehicle method to add a car to the database. Thinking of classes as categories or types of objects can be helpful.

Grouping objects along with their attributes and methods into a class is called **encapsulation**, which essentially means grouping related items into a single unit. Encapsulation helps handle more complex types of data, such as images and graphs. Object-oriented databases can also use **inheritance**, which means new objects can be created faster and more easily by entering new data in attributes. For example, you can add a BMW as a Vehicle object by having it inherit attributes and methods of the Vehicle class. You do not have to redefine an object—in other words, specifying all its attributes and methods—every time you want to add a new one.

This data model expands on the relational model by supporting more complex data management, so modeling real-world problems is easier. In addition, object-oriented databases can handle storing and manipulating all types of multimedia as well as numbers

and characters. Being able to handle many file types is useful in many fields. In the medical field, for example, doctors need to access X-ray images and graphs of vital signs in addition to patient histories consisting of text and numbers.

In contrast to the query languages used to interact with a relational database, interaction with an object-oriented database takes places via methods, which are called by sending a message to an object. Messages are usually generated by an event of some kind, such as pressing Enter or clicking the mouse button. Typically, a high-level language, such as C++, is used to create methods. Some examples of object-oriented DBMSs are Progress ObjectStore and Objectivity/DB.

3-5 DATA WAREHOUSES

A **data warehouse** is a collection of data from a variety of sources used to support decision-making applications and generate business intelligence.[6] Data warehouses store multidimensional data, so they are sometimes called "hypercubes." Typically, data in a data warehouse is described as having the following characteristics in contrast to data in a database:

- *Subject oriented*—Focused on a specific area, such as the home-improvement business or a university, whereas data in a database is transaction/function oriented

- *Integrated*—Comes from a variety of sources, whereas data in a database usually does not

- *Time variant*—Categorized based on time, such as historical information, whereas data in a database only keeps recent activity in memory

- *Type of data*—Captures aggregated data, whereas data in a database captures raw transaction data

- *Purpose*—Used for analytical purposes, whereas data in a database is used for capturing and managing transactions

encapsulation refers to the grouping into a class of various objects along with their attributes and methods—meaning, grouping related items into a single unit. This helps handle more complex types of data, such as images and graphs.

inheritance refers to new objects being created faster and more easily by entering new data in attributes.

A **data warehouse** is a collection of data from a variety of sources used to support decision-making applications and generate business intelligence.

Data Warehouse Applications at Marriott International

Marriott International, one of the world's largest hospitality companies, operates over 3,700 properties in 74 countries and territories.[7] The company, which encompasses 19 different brands, needed a campaign management tool to help it gather information from its various regions, brands, and franchises in order to effectively communicate and engage with its customers.

To help achieve this goal, Marriott built a data warehouse that provides sales and marketing managers with critical information about its customers in a timely manner. The system collects all sorts of relevant data about the customers. Using this data, statistical models, and various algorithms, the system is able to present offers to individual customers based on their past preferences and their transaction histories. Marriott now knows exactly which service has been used by each customer. The system provides the opportunity to tailor each offer to an individual customer. The new system includes Web-based tools for regional marketing managers and has enabled these managers to reduce campaign development from 6 weeks to 2 days. By using the new system, Marriott has exceeded its revenue goals, and it has been able to communicate more effectively with its customers by conducting more targeted communication.[8]

Designing and implementing a data warehouse is a complex task, but specific software is available to help. Oracle, IBM, Microsoft, Teradata, SAS, and Hewlett-Packard are market leaders in data-warehousing platforms. The information box above discusses how a data warehouse was used at Marriott International.

Exhibit 3.6 shows a data warehouse configuration with four major components: input; extraction, transformation,

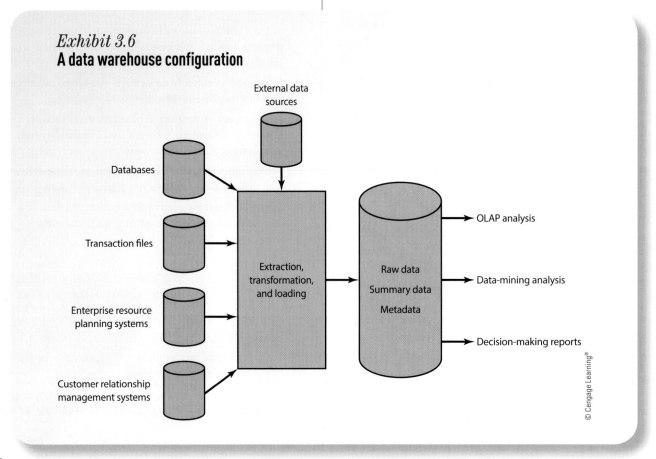

Exhibit 3.6
A data warehouse configuration

External data sources

Databases

Transaction files

Enterprise resource planning systems

Customer relationship management systems

Extraction, transformation, and loading

Raw data
Summary data
Metadata

OLAP analysis

Data-mining analysis

Decision-making reports

© Cengage Learning®

and loading (ETL); storage; and output. These components are explained in the following sections.

3-5a Input

Data can come from a variety of sources, including external data sources, databases, transaction files, enterprise resource planning (ERP) systems, and customer relationship management (CRM) systems. ERP systems collect, integrate, and process data that can be used by all functional areas in an organization. CRM systems collect and process customer data to provide information for improving customer service. (ERP and CRM systems are discussed in Chapter 11.) Together, these data sources provide the input a data warehouse needs to perform analyses and generate reports.

3-5b ETL

Extraction, transformation, and loading (ETL) refers to the processes used in a data warehouse. Extraction means collecting data from a variety of sources and converting it into a format that can be used in transformation processing. The extraction process can also parse (divide into pieces) data to make sure it meets the data warehouse's structural needs. For example, parsing can be used to separate the street number, street name, city, and state in an address if you want to find out how many customers live in a particular region of a city.

Transformation processing is done to make sure data meets the data warehouse's needs. Its tasks include the following:

- Selecting only certain columns or rows to load
- Translating coded values, such as replacing *Yes* with *1* and *No* with *2*
- Performing select, project, and join operations on data
- Sorting and filtering data
- Aggregating and summarizing data before loading it in the data warehouse

Loading is the process of transferring data to the data warehouse. Depending on the organization's needs and the data warehouse's storage capacity, loading might overwrite existing data or add collected data to existing data.

3-5c Storage

Collected information is organized in a data warehouse as raw data, summary data, or metadata. Raw data is information in its original form. Summary data gives users subtotals of various categories, which can be useful. For example, sales data for a company's southern regions can be added and represented by one summary number. However, maintaining both raw data (disaggregated data) and summary data (aggregated data) is a good idea for decision-making purposes, as you learned in Chapter 1. Metadata is information about data—its content, quality, condition, origin, and other characteristics. Metadata tells users how, when, and by whom data was collected and how data has been formatted and converted into its present form. For example, metadata in a financial database could be used to generate a report for shareholders explaining how revenue, expenses, and profits from sales transactions are calculated and stored in the data warehouse.

3-5d Output

As Exhibit 3.6 shows, a data warehouse supports different types of analysis and generates reports for decision making. The databases discussed so far support **online transaction processing (OLTP)** to generate reports such as the following[9]:

- Which product generated the highest sales last month?
- Which region generated the lowest sales last month?
- Which salespersons increased sales by more than 30 percent last quarter?

Data warehouses, however, use online analytical processing and data-mining analysis to generate reports. These are discussed in the following sections.

Online Analytical Processing

Online analytical processing (OLAP), unlike OLTP, is used to quickly answer multidimensional analytical queries, thereby generating business intelligence.

Extraction, transformation, and loading (ETL) refers to the processes used in a data warehouse. It includes extracting data from outside sources, transforming it to fit operational needs, and loading it into the end target (database or data warehouse).

Online transaction processing (OLTP) is used to facilitate and manage transaction-oriented applications, such as point-of-sale, data entry, and retrieval transaction processing. It generally uses internal data and responds in real time.

Online analytical processing (OLAP) generates business intelligence. It uses multiple sources of information and provides multidimensional analysis, such as viewing data based on time, product, and location.

Exhibit 3.7
Slicing and dicing data

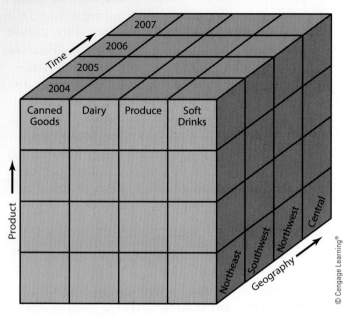

© Cengage Learning®

It uses multiple sources of information and provides multidimensional analysis, such as viewing data based on time, product, and location. For example, if you wanted to find out how Product X performed in the Northwest region during the previous quarter, you could use OLAP. Sometimes, this analysis is called slicing and dicing data. The hypercube in Exhibit 3.7 can be sliced in different directions. You can think of this hypercube as a multidimensional spreadsheet, with each side representing a dimension, such as region ("Geography" in the exhibit). The advantage of a hypercube is that it enables fast manipulations and calculations. In the hypercube in Exhibit 3.7, each smaller cube in a dimension represents a subdivision of data. Therefore, data in one of these smaller cubes could pertain to the sale of canned goods in the Northeast region in 2004. Each smaller cube can be subdivided further; for example, 2004 could be divided into financial quarters: Q1, Q2, Q3, and Q4. The number of cubes is determined by the "granularity" (specificity) of each dimension.

OLAP allows you to analyze information that has been summarized in multidimensional views. OLAP tools are used to perform trend analysis and sift through massive amounts of statistics to find specific information. These tools usually have a "drill-down" feature for accessing multilayer information. For example, an OLAP tool might access the first layer of information to generate a report on sales performance in a company's eight regions. If a marketing executive is interested in more information on the Northwest region, the OLAP tool can access the next layer of information for a more detailed analysis. OLAP tools are also capable of "drilling up," proceeding from the smallest unit of data to higher levels. For example, an OLAP tool might examine sales data for each region and then drill up to generate sales performance reports for the entire company.

Data-Mining Analysis

Data-mining analysis is used to discover patterns and relationships. For example, data-mining tools can be used to examine point-of-sale data to generate reports on customers' purchase histories. Based on this information, a company could better target marketing promotions to certain customers. Similarly, a company could mine demographic data from comment or warranty cards and use it to develop products that appeal to a certain customer group, such as teenagers or women over 30. When Netflix.com recommends movies to you based on your rental history, it is using information generated by data-mining tools. Netflix awarded a $1 million prize in September 2009 to the team that devised the best algorithm for substantially improving the accuracy of movie recommendations (*www.netflixprize.com*). American Express conducts the same type of analysis to suggest products and services to cardholders based on their monthly expenditures—patterns discovered by using data-mining tools. The following are typical questions you can answer by using data-mining tools:

- Which customers are likely to respond to a new product?
- Which customers are likely to respond to a new ad campaign?
- What product should be recommended to this customer based on his or her past buying patterns?

Data-mining analysis is used to discover patterns and relationships.

When Netflix.com recommends movies to you based on your rental history, it is using information generated by data-mining tools.

Vendors of data-mining software include SAP Business Objects (*www.sap.com*), SAS (*www.sas.com*), Cognos (*http://cognos.com*), and Informatica (*www.informatica.com*).

The information box below highlights the applications of data mining at Airline Industries.

DATA MINING AND AIRLINE INDUSTRIES

In the airline industry, data mining could improve customer service and, at the same time, increase revenue. Soon, an airline crew will know if, say, the customer in seat 19A is vegetarian or is allergic to a certain food. It will also know which passenger lost her bags on a previous flight. Airline marketing is becoming more sophisticated, based on passengers' online-browsing histories and the wealth of information available through social media. To sell products and services to their passengers, some airlines now combine the passenger's flight information (such as name, frequent flyer number, age, income, and address) with social media information (such as the number of "Likes" on the passenger's Facebook page). All this information could be accessible to flight attendants equipped with tablets and smartphones in the cabin. Data mining could improve the targeted in-flight sales by knowing the passenger's destination, income, and social media interests. For example, show tickets, helicopter tours, and tickets for boxing matches could be sold to passengers en route to Las Vegas. However, airlines must make every effort to protect the privacy of their passengers. There must be a balance between providing excellent customer service and protecting the passengers' personal information.[10]

Decision-Making Reports

A data warehouse can generate all types of information as well as reports used for decision making. The following are examples of what a data warehouse can allow you to do:

- Cross-reference segments of an organization's operations for comparison purposes—for example, compare personnel data with data from the finance department, even if they have been stored in different databases with different formats.
- Generate complex queries and reports faster and easier than when using databases.
- Generate reports efficiently using data from a variety of sources in different formats and stored in different locations throughout an organization.
- Find patterns and trends that cannot be found with databases.
- Analyze large amounts of historical data quickly.
- Assist management in making well-informed business decisions.
- Manage a high demand for information from many users with different needs and decision-making styles.

3-6 DATA MARTS

A **data mart** is usually a smaller version of a data warehouse, used by a single department or function. Data marts focus on business functions for a specific user group in an organization, such as a data mart for the Marketing Department. Despite being smaller,

A **data mart** is usually a smaller version of a data warehouse, used by a single department or function.

data marts can usually perform the same types of analysis as a data warehouse. Data marts have the following advantages over data warehouses:

- Access to data is often faster because of their smaller size.

- Response time for users is improved.

- They are easier to create because they are smaller and often less complex.

- They are less expensive.

- Users are targeted better, because a data mart is designed for a specific department or division; identifying their requirements and the functions they need is easier. A data warehouse is designed for an entire organization's use.

Data marts, however, usually have more limited scope than data warehouses, and consolidating information from different departments or functional areas (such as sales and production) is more difficult.

3-7 BUSINESS ANALYTICS

Business analytics (BA) uses data and statistical methods to gain insight into the data and provide decision makers with information they can act on. BA is increasingly used for data-driven decision making that leverages and explores the data in a database, data warehouse, or data mart system. Compared to business intelligence (BI), which was discussed earlier in this chapter, BA is more forward looking; it tells the user what is going to happen in the future rather than what has happened in the past.

BI can help determine what happened, what the problem is, and what decisions need to be made based on the available data. BA can help determine why this is happening, what it will mean if the sales trend continues, and what actions should be taken. BI uses dashboards, scorecards, OLAP, and query reports to support decision-making activities, whereas BA uses statistical analysis, data mining tools, and predictive modeling.

There are several types of BA methods. Two popular ones are descriptive and predictive analytics. Descriptive analytics reviews past events, analyzes the data, and provides a report indicating what happened in a given period and how to prepare for the future. Thus, it is a reactive strategy. Predictive analytics, as the name indicates, is a proactive strategy; it prepares a decision maker for future events. As an example, it might indicate that if a current sales trend continues, a company will need a certain number of salespeople in 2016.

The value that BA offers organizations is huge. Organizations that want to take full advantage of it, however, will need a chief analytics officer (CAO). This individual should sift through the enterprise-wide data in order to generate BI.[11]

Major providers of predictive analytics software include SAS, IBM, SAP, Microsoft, and Oracle. The information box below describes two real-life examples of its use.

> **Business analytics (BA)** uses data and statistical methods to gain insight into the data and provide decision makers with information they can act on.

PREDICTIVE ANALYTICS IN ACTION

Blue Cross and Blue Shield System (BCBS), which provides healthcare insurance to nearly 33 percent of Americans, is a successful user of predictive analytics. It has established a database that contains more than 100 million medical claims. To control increasing medical costs, BCBS uses predictive analytics to identify the risk factors that lead to several chronic diseases. It also uses predictive analytics to identify individuals who are at heightened risk of getting such diseases. The goal is to be able to use the data to get doctors to provide better, more targeted care for high-risk patients, reducing their need for expensive, long-term treatment.

Match.com, an online dating service, is also a successful user of predictive analytics. The company collects data from subscribers during the registration process and also by monitoring subscribers' interactions on the Web site. The goal is to find the best possible matches based on each subscriber's preferences. By using predictive analytics, Match.com matches its members based on not just their expressed preferences but on their behavior and interactions on the Web site.[12]

3-8 THE BIG DATA ERA

Companies such as MasterCard and Visa have been involved in big data for years, processing billions of transactions on a typical holiday weekend. But there has been an explosion in data growth throughout the business world in recent years. In fact, it has been estimated that the volume of business data worldwide doubles every 1.2 years.[13]

So what is big data? The result of the exponential growth of data in both public and private organizations, **big data** is data so voluminous that conventional computing methods are not able to efficiently process and manage it. There are three dimensions to big data, known as the 3 Vs: volume, variety, and velocity.[14]

Volume—This refers to the sheer quantity of transactions, measured in petabytes (1,024 terabytes) or exabytes (1,024 petabytes). Here are four examples of big data volume:

- All the packages shipped through the U.S. Postal Service the week before Christmas
- The sales of all Walmart stores on Black Friday
- All the items purchased from Amazon on Cyber Monday
- The number of tweets sent or received around the world per day

Variety—This refers to the combination of structured data (e.g., customers' product ratings between 1 and 5) and unstructured data (e.g., call center conversations or customer complaints about a service or product). Most data available on social networks is unstructured. Businesses combine data collected from the Internet and various handheld and mobile devices with location-related data and multimedia data. Machine-to-machine communication data, which is collected and transmitted automatically, also plays a major role in big data operations.

Velocity—This refers to the speed with which the data has to be gathered and processed. As an example, imagine a billboard that could display a particular ad as soon as a particular customer drives by it. The billboard would recognize the driver's face by comparing it to a huge database, integrate that data with the driver's social media data, finding out what the driver's favorites are based on the number of likes and dislikes on his/her Facebook page, then display the right ad. All of this would need to happen in a nanosecond; otherwise, the window of opportunity will be lost!

3-8a Who Benefits from Big Data?

Many industries could benefit from big data analytics and gain a competitive advantage in the following areas[15]:

- *Retail*—Customer relationship management, tailoring retail offerings to customer needs, offering personalized service, finding optimum store location and layout[16,17]
- *Financial services*—Risk analysis, fraud detection, attracting new customers
- *Advertising and public relations*—Targeted advertising, more effective messaging of the effectiveness of advertising campaigns, data driven PR[18]
- *Government*—National security, airport security, weapon systems, and counterterrorism
- *Manufacturing*—Product research, process and quality analysis, route and distribution optimization
- *Media and telecommunications*—Customer scoring, network optimization, effective media programming
- *Energy*—Smart grid, customer segmentation, energy saving[19]
- *Healthcare*—Pharmaceutical research, developing effective medical treatments, genome research[20]

3-8b Tools and Technologies of Big Data

Many technologies and applications have contributed to the growth and popularity of big data. Mobile and wireless technology, the popularity of social networks, and the enhanced power and sophistication of smartphones and handheld devices are among the key factors. Significant improvements in storage technology and substantial cost reduction and improved capabilities and affordability of analytics tools have made big data analytics accessible to nearly all types of organizations.

The most commonly used platform for big data analytics is the open-source Apache Hadoop, which uses the Hadoop Distributed File System (HDFS) to manage storage. Distributed databases, including NoSQL and Cassandra, are also commonly associated with big data projects. Examples of big data commercial platforms include: SAP Big Data Analytics (*www.sap.com/BigData*), Tableau (*www.tableausoftware.com*), SAS Big Data Analytics (*www.sas.com/big-data*), and QlikView (*www.qlikview.com*).

The information box on the next page highlights the application of big data at Express Scripts.

The Industry Connection box summarizes Oracle Corporation's database products and services.

> **Big data** is data so voluminous that conventional computing methods are not able to efficiently process and manage it.

BIG DATA IN ACTION

Based in St. Louis, Missouri, Express Scripts provides pharmacy benefits management. It processes nearly 1.5 billion prescriptions for approximately 100 million consumers per year. The company constantly analyzes its massive amount of data on prescriptions and insurance claims in order to speed up delivery, reduce errors, and increase profitability. Big data analytics has enabled Express Scripts to quickly find out if customers are filling their prescriptions through mail order, which is cheaper, or going to a retail pharmacy. For customers not using mail order, the company can intervene and provide them with cost options.[21] By using analytics, the company is also able to be proactive and present customers who are taking drugs on a long-term basis with cost alternatives.

Express Scripts has recently expanded into predictive analytics with a system called Screen Rx. The goal of the system is to screen and identify those patients with chronic diseases (such as high blood pressure, diabetes, or cholesterol) who are not sticking with their prescriptions. This system uses 400 factors (such as family history, past diseases, gender, and where the customer lives) to identify these patients and then offer proactive recommendations. Non-adherence is the most expensive healthcare-related problem in the United States, with an annual cost of over $317 billion. The Screen Rx system is able to bring this cost down by intervening proactively.[22]

Industry Connection: Oracle Corporation[23]

Oracle offers database software and services. It is a major vendor of software for enterprise-level information management, and it was the first software company to offer Web-based products. In addition, Oracle offers "single-user versions" of some of its database products. The following list describes some of Oracle's database software and services:

Oracle Database 11g: A relational DBMS that runs on Windows, Linux, and UNIX platforms and includes a variety of features for managing transaction processing, business intelligence, and content management applications.

Oracle OLAP: An option in Oracle Database 11g Enterprise Edition that is a calculation engine for performing analyses such as planning, budgeting, forecasting, sales, and marketing reports. It improves the performance of complex queries with multidimensional data.

Oracle PeopleSoft: An enterprise-level product for customer/supplier relationship management and human capital management (HCM).

Oracle Fusion Middleware: A suite of products for service-oriented architecture (SOA), business process management, business intelligence, content management, identity management, and Web 2.0.

Other Oracle products and services include Oracle E-Business Suite as well as Siebel customer-relationship management applications.

STUDY TOOLS 3

LOCATED AT BACK OF THE TEXTBOOK

☐ Rip out Chapter Review Card

LOCATED AT WWW.CENGAGE.COM/LOGIN

☐ Review Key Terms Flash Cards (Print or Online)

☐ Download audio and visual summaries to review

☐ Complete both Practice Quizzes to prepare for tests

☐ Play "Beat the Clock" and "Quizbowl" to master concepts

☐ Complete "Crossword Puzzle" to review key terms

☐ Watch video on "Method" for a real company example

KEY TERMS

allocation 54
big data 61
business analytics (BA) 60
create, read, update, and delete (CRUD) 53
data dictionary 50
data hierarchy 45
data mart 59
data model 49
data warehouse 55
database 45
database administrator (DBA) 53
database management system (DBMS) 46
data-driven Web site 53
data-mining analysis 58
distributed database management system
  (DDBMS) 54
encapsulation 55
extraction, transformation, and loading
  (ETL) 57

foreign key 50
fragmentation 54
hierarchical model 49
indexed sequential access method (ISAM) 48
inheritance 55
logical view 48
network model 49
normalization 50
object-oriented databases 54
online analytical processing (OLAP) 57
online transaction processing (OLTP) 57
physical view 48
primary key 50
query by example (QBE) 52
random access file structure 48
relational model 50
replication 54
sequential access file structure 48
Structured Query Language (SQL) 52

REVIEWS AND DISCUSSIONS

1. Define "database"?

2. What is a DBMS? What are its major components?

3. What are three methods for accessing files?

4. What are the differences between a logical view and a physical view of data?

5. Which companies benefit the most from data-driven Web sites?

6. What are the differences between OLAP and OLTP?

7. What are two major types of business analytics?

8. What are the 3 Vs in big data?

1. After reading the information presented in this chapter and other sources, write a two-page paper that explains BI. Identify three companies (in addition to those mentioned in this chapter) that have been using BI, and explain the applications of BI in these companies.

2. After reading the information presented in this chapter and other sources, write a one-page paper that identifies three companies (in addition to those mentioned in this chapter) that are using data mining tools. Explain how data mining has helped these companies with their bottom lines.

3. After reading the information presented in this chapter and other sources, write a one-page paper that identifies a service company that is using a data warehouse. How has the data warehouse helped this company achieve its goals? (Hint: Many universities are using this technology.)

4. After reading the information presented in this chapter and other sources, write a two-page paper that explains BA. Identify three companies (in addition to those mentioned in this chapter) that have been using BA, and explain the applications of BA in these companies.

5. After reading the information presented in this chapter and other sources, write a two-page paper that identifies three companies in the retail business that have been using big data. Explain how big data is helping these companies improve the efficacy of their operations.

6. The sample table below shows 11 of the students enrolled in an MIS course. Organize the data in a relational format, and use Microsoft Access to list all ACC majors, all ACC majors with a GPA higher than 3.7, all students who are MIS or ACC majors, and all students who aren't ACC majors. Repeat this assignment, this time using Excel, and generate the same results.

SAMPLE TABLE

Fname	Lname	ID	Major	GPA	Age	Status
Mary	Jones	111-1	MIS	2.79	19	Freshman
Tom	Smith	222-7	ACC	3.60	22	Senior
Alan	Bidgoli	333-9	ACC	3.86	21	Junior
Brian	Stark	444-6	MKT	3.45	20	Sophomore
Steve	Kline	555-6	MGT	3.75	24	Junior
Susan	Brown	666-1	MIS	3.90	21	Junior
Moury	Bidgoli	777-2	MIS	3.90	22	Junior
Janet	Jeffrey	888-0	MGT	4.00	20	Junior
Andy	Thomas	999-2	MKT	3.65	29	Senior
Jack	Tucker	234-1	ACC	3.92	23	Senior

ARE YOU READY TO MOVE ON?

1. In a database, a file is a group of related records, and a record is a group of related fields. True or False?

2. A physical view of data involves how the data is stored on and retrieved from storage media, such as hard disks, magnetic tapes, or CDs. True or False?

3. The 3 Vs of big data are variety, volume, and vicinity. True or False?

4. All of the following are outputs of a data warehouse except:
   a. big data
   b. OLAP analysis
   c. data-mining analysis
   d. decision-making reports

5. Which of the following is not among the components of a DBMS?
   a. database engine
   b. magnetic tape
   c. application generation
   d. data administration

6. All of the following are a type of a data model used for designing a database except:
   a. object-oriented
   b. hierarchical
   c. indexed
   d. relational

CASE STUDY 3-1

Data Mining Helps Students Enroll in Courses with Higher Chances of Success

Austin Peay State, a university near Nashville, Tennessee, is applying a data-mining approach to higher education. Before students register for classes, a robot looks at their profiles and transcripts and recommends courses in which they are likely to be successful or have higher chances of success. The software was developed by Tristan Denley, a former math professor who is now the university's provost.

The software takes an approach similar to the ones Netflix, eHarmony, and Amazon use to make their recommendations. It compares a student's transcripts with those of past students who had similar grades and SAT scores. When a student logs in, the program offers 10 "Course Suggestions for You." This recommendation is based on the student's major and other information related to that student. The goal is to steer students toward courses in which they will make better grades. According to Denley, students who

follow the recommendations do substantially better. In the fall of 2011, 45 percent of the classes that students were taking had been on their top 10 recommendations list. This data-mining concept is catching on. Three other Tennessee colleges now use Denley's software. Institutions outside the state are developing their own versions of the idea.[24]

Answer the following questions:

1. Which other companies are using approaches similar to the one used by Austin Peay State?
2. Based on which data does the system makes a course recommendation to a student?
3. How many courses are recommended to a student for possible selection?
4. According to the case study, are any other Tennessee schools using this approach?

CASE STUDY 3-2

Data Mining Tools at Pandora Radio

With more than 200 million registered users, Pandora Radio is a personalized Internet radio service that helps you find new music based on your past and current favorites.[25] (The service is also available to mobile devices—as an app.) The success of Pandora Radio's business model derives from applying data-mining tools to the Music Genome Project, which is a vast database of songs that a team

of experts has broken down into their various components: melody, rhythm, vocals, lyrics, and so on. Listeners begin by entering their favorite songs, artists, or genres, creating customized "stations." Then, Pandora Radio mines its database to find songs that are similar. Another data-mining tool that

Pandora uses is the like/dislike (thumbs up/thumbs down) option that accompanies each song the site suggests. These responses are also factored into which songs the Web site decides to play for the user.[26,27,28]

Answer the following questions:

1. How does Pandora Radio recommend music to its listeners?
2. How are listeners able to create their own customized stations?
3. What are some variables that Pandora Radio uses to recommend a song?

4 Personal, Legal, Ethical, and Organizational Issues of Information Systems

LEARNING OUTCOMES

After studying this chapter, you should be able to:

4-1 Discuss information privacy and methods for improving the privacy of information.

4-2 Explain the privacy of e-mail, data collection issues, and censorship.

4-3 Discuss the ethical issues related to information technology.

4-4 Discuss the principles of intellectual property and issues related to the infringement of intellectual property.

4-5 Discuss information system issues that affect organizations, including the digital divide, electronic publishing, and the connection between the workplace and employees' health.

4-6 Describe green computing and the ways it can improve the quality of the environment.

After you finish this chapter, go to **PAGE 80** for the **STUDY TOOLS**

Tetra Images/Alamy

This chapter examines privacy issues (including the privacy of e-mail), censorship, and data collection on the Web. It then discusses ethical issues of information technology, as well as intellectual property and copyright laws. Finally, it reviews some broader issues of information technologies, including the digital divide, electronic publishing, impacts on the workplace, health effects, and green computing.

4-1 PRIVACY ISSUES

Information technologies have brought many benefits, but they have also created concerns about privacy in the workplace. For example, employers now search social networking sites, such as Facebook or MySpace, to find background information on applicants, and this information can influence their hiring decisions. In fact, some recruiters and hiring managers use the extracted information to assign the potential employee a numeric rank between 0 and 100.[1]

Is this use of social networking sites legal or ethical? What about users' privacy? Because information posted on these sites is often considered public domain, you should be careful about what you post in case it comes back to haunt you. See the adjoining information box for an example of what could happen.

With employee-monitoring systems, managers can also supervise employees' performance—the number of errors they make, their work speeds, and their time away from the desk. Naturally, this monitoring has made some workers concerned about their privacy. See the information box on employee monitoring on the next page.

Healthcare organizations, financial institutions, legal firms, and even online-ordering firms gather a great deal of personal data and enter it in databases. Misuse and abuse of this information can have

SOCIAL NETWORKING SITES AND PRIVACY ISSUES

Stacy Snyder, a former student at Millersville University of Pennsylvania, posted on MySpace a photo of herself wearing a pirate's hat while drinking. The photo was captioned "Drunken Pirate." Although Snyder was of legal drinking age at the time, Millersville administrators considered the image unprofessional and refused to grant her a degree in education and a teaching certificate. Instead, she was given a degree in English. Did the university violate Stacy's privacy?[2]

> With employee-monitoring systems, managers can supervise employees' performance—the number of errors they make, their work speeds, and their time away from the desk.

EMPLOYEE MONITORING: IMPROVING PRODUCTIVITY OR INVASION OF PRIVACY

Workers in logistics and delivery trucks have long been monitored by their supervisors through GPS-type devices. Today, Telematics, a more advanced version of GPS, can both perform the monitoring task and reduce a truck speed if the truck gets too close to the next vehicle. In general, employers now have a myriad of software tools with which to monitor their employees. For example, they can monitor every keystroke that their employees enter into their computers or handheld devices. In addition, managers can view their employees' photos, text messages, e-mails (sent or received), call logs, and Web site visits. Using these monitoring tools, companies are able to track their employees' locations, check their phone calls, find out if a particular driver is wearing his seat belt, or even find out if the driver is tailgating other cars.

Without telling his drivers, a supervisor for Accurid Pest Solutions in southern Virginia installed GPS tracking software on the company-owned smartphones of 5 of its 18 drivers. The software allowed the supervisor, through his computer, to see a map of the drivers' movements, the number of stops, and the duration of the stops. The software disclosed that one driver was frequently visiting the same address. It was later revealed that the driver had been seeing a woman during work hours; he was fired.

Employees see these monitoring practices as an invasion of their privacy, but employers believe that they improve productivity, customer service, and safety as well as reduce theft and the loss of corporate secrets. Some employers tell their employees they are being monitored; some do not. However, employees should assume that they are always being monitored and that such practices are legal.[3]

serious consequences. For this reason, organizations should establish comprehensive security systems (discussed in Chapter 5) to protect their employees' or clients' privacy.

Some "information paranoia" is valid, because information about almost every aspect of people's lives is now stored on various databases, and misuse of extremely sensitive information (such as medical records) could prevent someone from getting employment, health insurance, or housing. Laws are in place to prevent these problems, but taking legal action is often costly, and by that point, the damage has often already been done.

You can probably give examples of things you expect to be private, such as your personal mail, your bank account balances, and your phone conversations. Defining privacy is difficult, however. In terms of electronic information, most people believe they should be able to keep their personal affairs to themselves and should be told how information about them is being used. Based on this definition, many practices of government agencies, credit agencies, and marketing companies using databases would represent an invasion of privacy. Unfortunately, information technologies have increased the ease of access to information for hackers as well as for legitimate organizations.

The recent revelation of widespread surveillance and spying actions by the National Security Agency (NSA) as well as the news stories related to the activities of this U.S. governmental agency have brought the privacy discussion to the forefront.[4,5]

The number of databases is increasing rapidly. In the United States, for example, the top three credit-rating companies—Experian, Equifax, and TransUnion—have records on nearly every person in the United States. Although these organizations and agencies are reputable and supply information only to people using it for its intended purpose, many small companies buy information from credit-rating companies and use it in ways that were never intended.

This action is clearly illegal, but enforcement of federal laws has been lax. You may have noticed the effects of this problem if you recently joined an organization and then began receiving mail from other organizations to whom you have not given your address.

Advances in computer technology have made it easy to do what was once difficult or impossible. Information in databases can now be cross-matched to create profiles of people and predict their behavior, based on their transactions with educational, financial, and government institutions. This information is often used for direct marketing and for credit checks on potential borrowers or renters.

The most common way to index and link databases is by using Social Security numbers (typically obtained from credit bureaus), although names are sometimes used to track transactions that do not require Social Security numbers, such as credit card purchases, charitable contributions, and movie rentals. Direct marketing companies are major users of this information. You may think that the worst result of this information sharing is an increase in junk mail (postal mail or e-mail), but there are more serious privacy issues. Should information you give to a bank to help establish a credit record be repackaged (i.e., linked with other databases) and used for other purposes?

In 1977, the U.S. government began linking large databases to find information. The Department of Health, Education, and Welfare decided to look for people collecting welfare who were also working for the government. (Collecting welfare while being employed is illegal.) By comparing records of welfare payments with the government payroll, the department was able to identify these workers. In this case, people abusing the system were discovered, so this use of databases was useful.

The Department of Housing and Urban Development, which keeps records on whether mortgage borrowers are in default on federal loans, previously made this information available to large banking institutions, such as Citibank, which added it to their credit files. This action led Congress to pass the first of several laws intended to protect people's rights of privacy with regard to their credit records.

Several federal laws now regulate the collecting and using of information on people and corporations, but the laws are narrow in scope and contain loopholes. For example, the 1970 Fair Credit Reporting Act prohibits credit agencies from sharing information with anyone but "authorized customers." An authorized customer, however, is defined as anyone with a "legitimate need," and the act does not specify what a legitimate need is.

There are three important concepts regarding the Web and network privacy: **acceptable use policies**, **accountability**, and **nonrepudiation**. To guard against possible legal ramifications and the consequences of using the Web and networks, an organization usually establishes an acceptable use policy, which is a set of rules specifying the legal and ethical use of a system and the consequences of noncompliance. Having a clear, specific policy can help prevent users from taking legal action against an organization, as in cases of termination. Most organizations have new employees sign an acceptable use policy before they can access the network. The second concept, accountability, refers to issues involving both the user's and the organization's responsibilities and liabilities. As for nonrepudiation, this is basically a method for binding all the parties to a contract; it is covered in more detail in Chapter 5.

Because of concerns about privacy, hardware or software controls should be used to determine what personal information is provided on the Web. Chapter 5 explains these controls in more detail, but to minimize the invasion of privacy, users and organizations should adhere to the following guidelines:

- Conduct business only with Web sites with privacy policies that are easy to find, read, and understand.

- Limit access to your personal information to those who have authorization.

- Any organization creating, maintaining, using, or disseminating records of personal data must ensure the data's reliability and take precautions to prevent misuse of the data.

- Any data collection must have a stated purpose. Organizations should keep collected information only as long as it is needed for the stated purpose.

- There must be a way for people to prevent personal information that was gathered about them for one purpose from being used for other purposes or being disclosed to others without their consent.

- Organizations should monitor data collection and entry and should use verification procedures to ensure data accuracy; they should also collect only the data that is necessary.

An **acceptable use policy** is a set of rules specifying the legal and ethical use of a system and the consequences of noncompliance.

accountability refers to issues involving both the user's and the organization's responsibilities and liabilities.

nonrepudiation is a method for binding all the parties to a contract.

- Records kept on an individual should be accurate and up to date. Organizations must correct or delete incorrect data and delete data when it is no longer needed for the stated purpose.
- Users should be able to review their records and correct any inaccuracies.
- The existence of record-keeping systems storing personal data should not be kept secret. In addition, there must be a way for people to find out what information about them has been stored and how it is used.
- Organizations must take all necessary measures to prevent unauthorized access to data and misuse of data.

© Paul Paladin/Shutterstock.com

Privacy-protection software can take many forms. For example, to guard against cookies, which record your navigations around the Web, you can use the cookie control features contained in your browser. There are also commercial vendors that address this problem—for example, Anonymizer, Inc., which is discussed in the Industry Connection box at the end of this chapter. Using privacy-protection software has some drawbacks, however. For example, eBay often has to contend with sellers who, by using different user accounts, are able to bid on their own items, thereby inflating the prices. Currently, eBay can trace these sellers' user accounts, but privacy-protection software would make this tracking impossible.

4-1a E-mail

Although e-mail is widely used, it presents some serious privacy issues. One issue is junk e-mail, also known as **spam**—unsolicited e-mail sent for advertising purposes. Because sending these e-mails is so inexpensive, even a small response—a fraction of a percent, for example—is a worthwhile return on the investment. Usually, spam is sent in bulk using automated mailing software, and many spammers sell their address lists. For these reasons, the volume of spam can rise to an unmanageable level quickly, clogging users' inboxes and preventing access to legitimate e-mails.

> Whether an e-mail is distributed through the Web or through a company network, people should assume that others have access to their messages.

Another privacy concern is ease of access. Whether an e-mail is distributed through the Web or through a company network, people should assume that others have access to their messages. In addition, many organizations have policies stating that any e-mails sent on company-owned computers are the organization's property and that the organization has the right to access them. In other words, employees often have no right to privacy, although there is a lot of controversy over this point and several lawsuits have resulted.

Spamming has also created decency concerns, because these e-mails often contain explicit language or nudity and can be opened by children. Here are some 2012 statistics for e-mail and spam[6]:

- 4.3 billion—Number of e-mail clients worldwide
- 144 billion—Number of e-mails per day worldwide
- 69—Percentage of all e-mails that were spam
- 61—Percentage of all e-mails that were considered unnecessary
- 22—Percentage of e-mail that was malicious

The information box on the next page highlights other privacy issues related to e-mail.

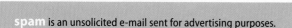
spam is an unsolicited e-mail sent for advertising purposes.

E-MAIL AND CORPORATE DATA LEAKAGE

Employees' lack of knowledge about e-mail or their reluctance to follow company policy can have significant consequences for an organization. For example, pressing "Reply All" instead of "Reply" can result in private information being sent to recipients who were not supposed to have access to that information. Also, printing out confidential e-mails, which may end up in wastebaskets, hotel rooms, or airport terminals (or linger in the printer's memory), may introduce risks. Sending and receiving e-mails through smartphones, tablets, and other handheld devices further adds to the vulnerability of corporate private information.

A survey of professional e-mail users conducted by VaporStream revealed some interesting statistics. Of those surveyed, 10 percent had unintentionally leaked private information, 73.7 percent had e-mailed information in violation of organizational polices, 28 percent had done so intentionally, and 45.3 percent had sent e-mails that were forwarded to someone they did not intend to see it. Additionally, 20 percent indicated that an e-mail they had sent had come back to haunt them. User education and enforcing the organization's e-mail policy can significantly reduce the risk of privacy invasion.[7]

4-1b Data Collection on the Web

The number of people shopping online is increasing rapidly because of convenience, the array of choices, and lower prices. Many customers, however, are reluctant to make online purchases because of concerns about hackers getting access to their credit card numbers and charging merchandise to their accounts. To lessen consumers' concerns, many credit card companies reimburse fraudulent charges. In addition, other electronic payment systems are being developed, such as e-wallets and smart cards, that reduce the risks of exposing consumers' information on the Web (discussed in Chapter 8).

Some Web sites require you to enter your name, address, and employment information before you are allowed to use the site. Privacy issues include the concern that this personal information will be sold to telemarketing firms, and consumers do not want to be bombarded with spam. Also, some consumers are concerned about their computers' contents being searched while they are connected to the Web, and personal information could be used without their consent for solicitation and other purposes.

Information that users provide on the Web can be combined with other information and technologies to produce new information. For example, a financial profile can be created by collecting a person's employment information. Two commonly used technologies for data collection are cookies and log files.

Cookies are small text files with unique ID tags that are embedded in a Web browser and saved on the user's hard drive. Whenever a user accesses the same domain, the browser sends the saved information to the Web server. Sometimes, cookies are useful or innocuous, such as those used by a Web page that welcomes you or those used by a Web site that remembers your personal information for online ordering. Typically, users rely on Web sites to keep this information from being compromised. Cookies also make it possible for Web sites to customize pages for users, such as Amazon.com recommending books based on your past purchases.

Other times, cookies can be considered an invasion of privacy, and some people believe their information should be collected only with their consent. Cookies provide information about the user's location and computer equipment, and this information can be used for unauthorized purposes, such as corporate espionage.

For these reasons, many users disable cookies by installing a cookie manager, which can eliminate existing cookies and prevent additional cookies from being saved to a user's hard drive. Popular Web browsers such as Internet Explorer, Chrome, and Firefox provide a range of options for accepting and restricting cookies.

Log files, which are generated by Web server software, record a user's actions on a Web site. Sometimes, users give incorrect information on purpose—on chatting or dating sites, for example, or when opening e-mail accounts. If the information collected is not accurate, the result could be identity misrepresentation. For example, if someone claims to be younger on an online

Cookies are small text files with unique ID tags that are embedded in a Web browser and saved on the user's hard drive.

Log files, which are generated by Web server software, record a user's actions on a Web site.

dating site, any demographic data collected would be flawed. Similarly, if a TV network collects data on viewing trends through online surveys and people supply answers that are not truthful, any analyses the network attempts to conduct would not be accurate. Therefore, data collected on the Web must be used and interpreted with caution.

4-2 ETHICAL ISSUES OF INFORMATION TECHNOLOGIES

Companies such as Enron, Arthur Andersen, WorldCom, and Tyco have highlighted the ethics issues that corporations face in the 21st century. In essence, ethics means doing the right thing, but what is "right" can vary from one culture to another and even from one person to another.[8]

The distinction between what is legal and what is illegal is usually clear, but drawing a line between what is ethical and what is unethical is more difficult, as discussed in the adjoining information box.

Exhibit 4.1 shows a grid that can be used for assessing whether an action is legal and/or ethical.

UNETHICAL ISSUES AT THE NEWS OF THE WORLD

News of the World and another British tabloid newspaper's employees were accused of phone hacking, police bribery, and exercising improper influence in the pursuit of publishing stories. When the authorities first conducted investigations, from 2005 to 2007, they discovered that the phone hacking activities were done mostly on celebrities, politicians, and members of the British royal family. More recent investigations, conducted in 2011, revealed that the phones of a murdered school girl, the relatives of deceased British soldiers, and some victims in the 7/7 London bombings had been accessed. Once all this information was released to the public, people began to question the ethics of the company owned by Rupert Murdoch. According to an article posted on the *Daily Mail's* Web site, Mail Online, hundreds of thousands of what the newspaper considered "unhelpful" e-mails were deleted in order to hide the evidence; computers were also destroyed. As a result of this scandal, James Murdock, the chairman of *News of the World*, announced on July 7, 2011, that the newspaper would publish its last edition and close. *News of the World* had been publishing for 168 years. Closing this company created a loss of nearly 200 jobs. Several former employees have been arrested, jailed, or made appearances in court in connection with the phone hacking scandal.[9]

Review the following situations and try to determine where they might fall in the Exhibit 4.1 grid:

1. You make two copies of a software package you just bought and sell one to a friend.

2. You make two copies of a software package you just bought for personal use, in case the original software fails and you need a backup.

The distinction between what is legal and what is illegal is usually clear, but drawing a line between what is ethical and what is unethical is more difficult.

Exhibit 4.1
Ethical versus legal grid

	Legal	Illegal
Ethical	I	II
Unethical	III	IV

© Cengage Learning®

3. A banker uses the information a client enters in a loan application to sell other financial products to this client.

4. A credit card company sells its customers' mailing addresses to its competitors.

5. A supervisor fires a programmer who has intentionally spread viruses to the organization's network.

Statement 1 above is clearly illegal and unethical (quadrant IV). Statement 2 is ethical because you made the copy for your own use, but some software vendors who prohibit making copies might consider it illegal (quadrant II). Statements 3 and 4 are legal but not ethical (quadrant III). In statement 5, the supervisor's behavior is both legal and ethical. The supervisor has a clear legal reason for firing the programmer, and allowing the programmer to continue working there would not be ethical. As a future knowledge worker, watch your own actions, and make sure you behave both legally and ethically. Be careful about decisions you make that might affect coworkers so you can help maintain an ethical working environment.

Some information systems professionals believe that information technology offers many opportunities for unethical behavior, particularly because of the ease of collecting and disseminating information. Cybercrime, cyberfraud, identity theft, and intellectual property theft (discussed later in this chapter) are on the rise. Approximately 12.6 million people were victims of identity theft in 2012, and two identities are stolen every five seconds.[10]

Many experts believe management can reduce employees' unethical behavior by developing and enforcing codes of ethics. Many associations promote the ethically responsible use of information systems and technologies and have developed codes of ethics for their members. The Association for Computing Machinery (ACM), for example, has a code of ethics and professional conduct that includes the following general moral imperatives[11]:

- 1.1: Contribute to society and human well-being.
- 1.2: Avoid harm to others.
- 1.3: Be honest and trustworthy.
- 1.4: Be fair, and take action not to discriminate.
- 1.5: Honor property rights, including copyrights and patents.
- 1.6: Give proper credit for intellectual property.
- 1.7: Respect the privacy of others.
- 1.8: Honor confidentiality.

As a knowledge worker, you should consider the following questions and statements before making a work-related decision:

- Does this decision comply with my organization's values?
- How will I feel about myself after making this decision?
- If I know this decision is wrong, I must not make it.
- If I am not sure about this decision, I must ask my supervisor before making it.
- Is the decision right?
- Is the decision fair? How will I feel if somebody else makes this decision on my behalf?

TEN COMMANDMENTS OF COMPUTER ETHICS[12]

Thou shalt not use a computer to harm other people.

Thou shalt not interfere with other people's computer work.

Thou shalt not snoop around in other people's files.

Thou shalt not use a computer to steal.

Thou shalt not use a computer to bear false witness.

Thou shalt not use or copy software for which you have not paid.

Thou shalt not use other people's computer resources without authorization.

Thou shalt not appropriate other people's intellectual output.

Thou shalt think about the social consequences of the program you write.

Thou shalt use a computer in ways that show consideration and respect.

- Is the decision legal?
- Would I want everyone to know about this decision after I make it?

The information box on the previous page presents the "Ten Commandments of Computer Ethics" created by the Computer Ethics Institute. You should follow these guidelines whenever you use any type of information technology for personal and/or professional reasons.

4-2a Censorship

No organization controls the whole Web, so who decides what content should be on it? Two types of information are available on the Web: public and private. Public information, posted by an organization or public agency, can be censored for public policy reasons—such as not allowing military secrets to be published, lest the information fall into enemy hands. Public information can also be censored if the content is deemed offensive to a political, religious, or cultural group. However, private information—what is posted by a person—is not subject to censorship because of your constitutional freedom of expression. Of course, whether or not something can be censored depends in part on who is doing the censoring. For example, if you agree to abide by an organization's (e.g., a company's or an Internet service provider's) terms of service or policies and then post something that violates that, you might be censored or denied access.

Another type of censorship is restricting access to the Web. Some countries, such as Burma, China, and Singapore, restrict or forbid their citizens' access to the Web, or try to censor the information posted on the Web. These governments believe that the racist, pornographic, and politically extreme content of some Web sites could affect national security. In other countries, only employees of multinational corporations have direct access to the Web.

Most experts believe that Internet neutrality (also known as "net neutrality") must be practiced in all cases. According to this principle, Internet service providers (ISPs) and government agencies should treat all data on the Internet equally—that is, they should not block traffic, charge different rates, or discriminate in any way based on user, content, Web site, types of equipment in use, telecommunication provider, platform, or application.[13]

Although U.S. citizens do not want the government controlling Web access, many parents are concerned about what their children are exposed to while using the Web, such as pornography, violence, and adult language.

Another concern is children searching for information on the Web. If a search includes keywords such as *toys, pets, boys,* or *girls,* for example, the results could list pornography sites. Guidelines for Web use have been published to inform parents of the benefits and hazards of the Web, and parents can use these to teach their children to use good judgment while on the Web. For example, Microsoft posts a guideline called "Four Things You Can Do to Help Protect Kids Online" (*www .microsoft.com/security/family-safety/ childsafety-steps.aspx*).

In addition, many parents use programs such as CyberPatrol, CyberSitter, Net Nanny, and SafeSurf to prevent their children's access to certain Web sites. Web browser software has also been developed to improve children's security. For example, a Web browser may accept e-mail only from an address that uses the same Web browser software. This helps ensure that children receive e-mail only from

© Zurijeta/Shutterstock.com

> Public information, posted by an organization or public agency, can be censored for public policy reasons.

other children. Another possibility is creating different levels of user access, similar to movie ratings, to prevent children from accessing controversial or pornographic information. This system could use techniques such as requiring passwords or using biometrics, including fingerprints or retinal scans (discussed in Chapter 5).

4-2b Intellectual Property

Intellectual property is a legal umbrella covering protections that involve copyrights, trademarks, trade secrets, and patents for "creations of the mind" developed by people or businesses.[14] Intellectual property can be divided into two categories: industrial property (inventions, trademarks, logos, industrial designs, and so on) and copyrighted material, which covers literary and artistic works.

Generally, copyright laws protect tangible material, such as books, drawings, and so forth. However, they also cover online materials, including Web pages, HTML code, and computer graphics, as long as the content can be printed or saved on a storage device. Copyright laws give only the creator exclusive rights, meaning no one else can reproduce, distribute, or perform the work without permission.[15]

Copyright laws do have some exceptions, however, usually under the Fair Use Doctrine. This exception means you can use copyrighted material for certain purposes, such as quoting passages of a book in literary reviews. There are limits on the length of material

you can use. In addition, some copyrighted material can be used to create new work, particularly for educational purposes. Checking copyright laws carefully before using this material is strongly recommended. The United States Copyright Office (*www.copyright.gov*) offers detailed information on copyright issues. Exhibit 4.2 shows its home page.

Other intellectual property protections include trademarks and patents. A trademark protects product names and identifying marks (logos, for example). A patent protects new processes. (Note that laws governing trademarks, patents, and copyrights in the United States might not apply in other countries.) The length of a copyright varies based on the type of work. In general, copyrights last the author's lifetime plus 70 years and do not need to be renewed; patents last 20 years (14 years for design patents).

An organization can benefit from a patent in at least three ways[16]:

- It can generate revenue by licensing its patent to others.
- It can use the patent to attract funding for further research and development.
- It can use the patent to keep competitors from entering certain market segments.

Another copyright concern is software piracy, but the laws covering it are very straightforward. The 1980 revisions to the Copyright Act of 1976 include computer programs, so both people and organizations can be held liable for unauthorized duplication and use of copyrighted programs. Sometimes, contracts are used to supplement copyrights and give the software originator additional protection. For example, a software vendor might have a university sign a contract specifying how many people can use the software. Companies also make use of laws on trade

Exhibit 4.2
United States Copyright Office home page

Courtesy of the US Copyright Office

> **intellectual property** is a legal umbrella covering protections that involve copyrights, trademarks, trade secrets, and patents for "creations of the mind" developed by people or businesses.

secrets, which cover ideas, information, and innovations, as extra protection.

Most legal issues related to information technologies in the United States are covered by the Telecommunications Act of 1996, the Communications Decency Act (CDA), and laws against spamming. The CDA was partially overturned in the 1997 *Reno v. ACLU* case, in which the U.S. Supreme Court unanimously voted to strike down the CDA's anti-indecency provisions, finding they violated the freedom of speech provisions of the First Amendment. To avoid the following legal risks, organizations should have an Internet use policy[17]:

- *Risk 1*—If employees download pornographic materials to their office computers over the corporate network, the organization could be liable for harassment charges as well as infringement of privacy and even copyright laws.

- *Risk 2*—Indecent e-mail exchanges among employees can leave the corporation open to discrimination and sexual harassment charges.

- *Risk 3*—Employees using the corporate network to download and distribute unlicensed software can leave the corporation open to serious charges of copyright infringement and other legal issues.

One aspect of intellectual property that has attracted attention recently is **cybersquatting**, which is registering, selling, or using a domain name to profit from someone else's trademark. Often, it involves buying domains containing the names of existing businesses

VERIZON'S CYBERSQUATTING SUIT

In June 2008, Verizon sued OnlineNic, accusing it of trademark infringement and illegal cybersquatting. According to Verizon, OnlineNic registered domain names containing Verizon trademarks. The registered names included myverizonwireless.com, iphoneverizonplans.com, and verizon-cellular.com, among others, and Verizon was concerned about the names misleading consumers. Verizon won this suit and was awarded a $33 million judgment.[18]

and then selling the names for a profit. A variation of cybersquatting is typosquatting, also called URL hijacking. This technique relies on typographical errors made by Web users when typing a Web site address into a Web browser. If the user accidentally enters an incorrect Web site address, he or she may be directed to an alternative Web site owned by a typosquatter. In such a case, the legitimate site does not get the intended traffic. Examples include *whitehouse.gov* instead of *whitehouse.org* and *goggle.com* instead of *google.com*. To guard against this, major Web sites register variations of their names so the traffic gets redirected to the main site if typographical errors are made. The information box called "Verizon's Cybersquatting Suit" describes such a case.

One aspect of intellectual property that has attracted attention recently is cybersquatting, which is registering, selling, or using a domain name to profit from someone else's trademark.

4-2c Social Divisions and the Digital Divide

Some believe that information technology and the Internet have created a **digital divide** between the information rich and the information poor. Although prices have been decreasing steadily, computers still are not affordable for many people. In addition, a type

cybersquatting is registering, selling, or using a domain name to profit from someone else's trademark.

Information technology and the Internet have created a **digital divide**. Computers still are not affordable for many people. The digital divide has implications for education.

of economic "red-lining" can occur when companies installing coaxial and fiber-optic cables for Internet connections focus on higher-income communities, where more residents are expected to use the Internet.[19]

Children, in particular, are often victims of the digital divide. Those without computers or Web access at home, as well as students who cannot afford computer equipment, are at a disadvantage and can often fall behind in their education. Students without access to the wide array of resources on the Web have more difficulty writing papers and learning about topics that interest them. Interactive and virtual reality educational games available on the Web can widen the gap more. Increasing funding for computer equipment at schools and adding more computers in public places, such as libraries, can help offset this divide. Some schools have even started loaner programs so students can have a portable computer to use after school hours.

© Monkey Business Images/Shutterstock.com

4-3 THE IMPACT OF INFORMATION TECHNOLOGY IN THE WORKPLACE

Although information technology has eliminated some clerical jobs, it has created many new jobs (described in Chapter 1) for programmers, systems analysts, database and network administrators, network engineers, Webmasters, Web page developers, e-commerce specialists, chief information officers (CIOs), and technicians. In e-commerce, jobs for Web designers, Java programmers, and Web troubleshooters have been created, too. Some argue that the eliminated jobs have been clerical and the created jobs have been mostly technical, requiring extensive training. Others believe that information technologies have reduced production costs and, therefore, improved and increased consumers' purchasing power, resulting in a stronger economy.

Information technologies have a direct effect on the nature of jobs. Telecommuting, also known as virtual work, has enabled some people to perform their jobs from home. With telecommunications technology, a worker can send and receive data to and from the main office, and organizations can use the best and most cost-effective human resources in a large geographical region. Table 4.1 lists some benefits and drawbacks of telecommuting.

TABLE 4.1 POTENTIAL BENEFITS AND DRAWBACKS OF TELECOMMUTING

Potential Benefits	Potential Drawbacks
Can care for small children or elderly parents and spend more time with family	Can become a workaholic (no hard boundaries between "at work" and "at home")
Have fewer restrictions on clothing for work, thereby saving the expense of work wear	No regulated work routine
No commute, so distance and time factors are reduced as well as the effects of car emissions on air quality	Less interaction with coworkers
Able to work in more pleasant surroundings	No separation between work and home life
Increased productivity	Potential legal issues about workers' injuries
Decreased neighborhood crime because of more people being home during the day	Family interruptions and household distractions
Easier work environment for employees with disabilities	Lack of necessary supplies or equipment
Reduced costs for office space and utilities	Could create a two-tiered workforce—telecommuters and on-site workers—that affects promotions and raises
Reduced employee turnover and absenteeism	
Able to find and hire people with special skills, regardless of where they are located	
Fewer interruptions from coworkers	

© Cengage Learning®

By handling repetitive and boring tasks, information technologies have made many jobs more interesting, resulting in more worker satisfaction. Information technologies have also led to "job deskilling." This occurs when skilled labor is eliminated by high technology or when a job is downgraded from a skilled to a semi-skilled or unskilled position. It usually takes place when a job is automated or when a complex job is fragmented into a sequence of easily performed tasks. An example is when a computer-aided design (CAD) program performs the technical tasks that used to be performed by a designer. On the other hand, information technologies have created "job upgrading," as when clerical workers use computers for word-processing tasks. This upgrading makes it possible to add new tasks to employees' responsibilities, too; for example, clerical workers could be responsible for updating the company's Web site. Job upgrading has some limitations, however. Even with information technologies, training clerical workers to write programs for the company Web site would be difficult, for instance.

With information technologies, one skilled worker might be capable of doing the job of several workers. For example, with mail-merge programs, an office worker can generate thousands of letters, eliminating the need for additional workers. Information technologies can also make workers more efficient—being able to send a message throughout an organization by using e-mail instead of interoffice memos, for example. Similarly, mass-marketing efforts for new product announcements have been streamlined, reducing the expense and personnel needed to reach millions of customers.

Another impact of information technology is the creation of **virtual organizations**, which are networks of independent companies, suppliers, customers, and manufacturers connected via information technologies so they can share skills and costs and have access to each other's markets.[20] A virtual organization does not need central offices or an organizational hierarchy for participants to contribute their expertise. Advantages of virtual organizations include the following[21]:

virtual organizations are networks of independent companies, suppliers, customers, and manufacturers connected via information technologies so they can share skills and costs and have access to each other's markets.

- Each participating company can focus on what it does best, thus improving the ability to meet customers' needs.
- Because skills are shared among participating companies, the cost of hiring additional employees is reduced.
- Companies can respond to customers faster and more efficiently.
- The time needed to develop new products is reduced.
- Products can be customized more to respond to customers' needs.

In 2001, Dell, Microsoft, and Unisys Corporation created a partnership to design a voting system for several U.S. states. Microsoft offered software, Dell offered hardware, and Unisys served as the systems integrator. This example illustrates the principle of virtual organizations—the idea that several organizations working together can do what one organization cannot.

4-3a Information Technology and Health Issues

Although there have been reports of health problems caused by video display terminals (VDTs), no conclusive study indicates that VDTs are the cause, despite all the complaints. Work habits can cause some physical problems, however, and so can the work environment in which computers are used—static electricity, inadequate ventilation, poor lighting, dry air, unsuitable furniture, and too few rest breaks.

The increasing popularity of touchscreens on smartphones, tablets, and some PCs may result in more stress-related injuries of the users' hands, arms, back, and eyes. Because these devices can be accessed almost anywhere and are used in any number of body positions (most of which involve poor posture), users should exercise caution and take frequent breaks.[22]

Other reports of health problems related to computer equipment include vision problems, such as fatigue, itching, and blurriness; musculoskeletal problems (back strain and wrist pain); skin problems, such as rashes; reproductive problems, such as miscarriages; and stress-related problems (headaches and depression). Ergonomics experts believe that using better-designed furniture as well as flexible or wireless keyboards, correct lighting, special monitors for workers with vision problems, and so forth can solve many of these problems.

HEALTH AND SOCIAL ISSUES OF ONLINE GAMING

The online games World of Warcraft and EverQuest, both of them massively multiplayer online role-playing games (MMORPG), have been blamed for a host of problems, including poor academic performance, divorce, suicide, and the death of a child because of parental neglect. Mental health professionals believe the fantasy worlds in online games can be addictive and affect marriages and careers. According to Dr. Timothy Miller, a clinical psychologist, it is a growing problem with teenage and young adult males.[23]

In 2011, according to Global Times, a 33-year-old Chinese man died after playing online games at an Internet café for 27 days in a row. That's nearly 650 hours of playing while barely eating or sleeping.[24] Also in 2011, an Xbox addict died from a blood clot. According to the report, he had been playing games on his Xbox for up to 12 hours a day.[25] In 2007, a 26-year-old man died after gaming for 15 days straight. In 2005, a 28-year-old man died after 50 hours of playing Starcraft.

Another recent health issue is the amount of time some people spend on the Web playing games, participating in chat rooms, and other activities. Although the Internet can provide valuable educational resources, too much time on the Web can create psychological, social, and health problems, especially for young people. The information box called "Health and Social Issues of Online Gaming" discusses some of these problems.

4-4 GREEN COMPUTING

Green computing is computing that promotes a sustainable environment and consumes the least amount of energy. Information and communications technology (ICT) generates approximately 2 percent of the world's carbon dioxide emissions, roughly the same amount as the aviation industry.[26] Although ICT is a part of the problem, it could also be part of the solution. Many IT applications and tools can help reduce carbon dioxide emissions. Green computing not only helps an organization save on energy costs, it improves the quality of the environment in which you live and work.

Green computing involves the design, manufacture, use, and disposal of computers, servers, and computing devices (such as monitors, printers, storage devices, and networking and communications equipment) in such a way that there is minimal impact on the environment.[27] It is one of the methods for combating global warming. In some states, certain computer

manufacturers collect a fee from their customers, called an advance recovery fee, in order to dispose of the computer after its useful life. A successful green computing strategy cannot be fully implemented without the cooperation of both the private and the public sector. Furthermore, both employees and top management must be involved.

There are a number of ways to pursue a green computing strategy. Some can be easily done with no cost to the organization. Others are more challenging and require an initial investment. Here are some of the ways that green computing can be achieved:

- Design products that last longer and are modular in design so certain parts can be upgraded without replacing the entire system.

- Design search engines and other computing routines that are faster and consume less energy.

- Replace several underutilized smaller servers with one large server using a virtualization technique. In this case, multiple operating systems are hosted on a single hardware platform and can share this hardware platform. IBM's Project Big Green is an example of virtualization, with energy savings of approximately 42 percent for an average data center.[28]

Green computing involves the design, manufacture, use, and disposal of computers, servers, and computing devices (such as monitors, printers, storage devices, and networking and communications equipment) in such a way that there is minimal impact on the environment.

- Use computing devices that consume less energy and are biodegradable.

- Allow certain employees to work from their homes, resulting in fewer cars on the roads (discussed earlier in this chapter).

- Replace actual face-to-face meetings with meetings over computer networks (discussed in Chapter 6).

- Use video conferencing, electronic meeting systems, and groupware (discussed in Chapter 12). These technologies can also reduce business travel.

- Use a virtual world (discussed in Chapter 14). This technology can also reduce face-to-face meetings, resulting in less travel.

- Use cloud computing as promoted by companies such as Amazon.com (discussed in Chapter 14). This platform can also reduce energy consumption.

- Turn off idle PCs, recycle computer-related materials, and encourage carpool and nonmotorized transportation for employees.

The Industry Connection highlights the online privacy services offered by Anonymizer, Inc.

Industry Connection: Anonymizer, Inc.[29]

Anonymizer, Inc., provides online privacy services so users can browse the Web anonymously and securely. Its features include hiding users' IP addresses and removing cookies, spyware, and adware. Although using Anonymizer can slow down connection and surfing speeds, many consumers, businesses, and government agencies take advantage of its products and services, which include the following:

Digital Shredder Lite: Erases traces of your Web activity and Windows use, including Internet history, cache, and cookies as well as recently opened files, the Temp folder, and the Recycle Bin.

Nyms: Enable you to create and destroy alias e-mail addresses to protect your real e-mail address from spamming and phishing attempts.

Antispyware: Removes adware and spyware that have accumulated on your computer and prevents new spyware and adware from being installed on your computer.

Enterprise Web-harvesting tools: Web harvesting is collecting data from Web sites, usually for competitive intelligence. Anonymizer helps businesses protect their corporate information from Web-harvesting tools.

STUDY TOOLS 4

LOCATED AT BACK OF THE TEXTBOOK
- [ ] Rip out Chapter Review Card

LOCATED AT WWW.CENGAGE.COM/LOGIN
- [ ] Review Key Terms Flash Cards (Print or Online)
- [ ] Download audio and visual summaries to review

- [ ] Complete both Practice Quizzes to prepare for tests
- [ ] Play "Beat the Clock" and "Quizbowl" to master concepts
- [ ] Complete "Crossword Puzzle" to review key terms
- [ ] Watch video on "Method" for a real company example

KEY TERMS

acceptable use policy 69
accountability 69
cookies 71
cybersquatting 76
digital divide 76
green computing 79

intellectual property 75
log files 71
nonrepudiation 69
spam 70
virtual organizations 78

REVIEWS AND DISCUSSIONS

1. Define "privacy."
2. How has information technology invaded our privacy?
3. What are three risks associated with e-mail?
4. How can information technology lead to unethical decision-making?
5. What are three examples of intellectual properties?

6. What is green computing? What are three examples of information technologies that could promote green computing?
7. What are two advantages and two disadvantages of telecommuting?
8. What are two physiological issues related to using information technology?

PROJECTS

1. After reading the information presented in this chapter and other sources, write a two-page paper that compares and contrasts the privacy policies of Google and Facebook. In your opinion, which company offers a more comprehensive privacy policy? Which one is easier to find and implement?

2. After reading the information presented in this chapter and other sources, write a one-page paper that describes net neutrality. Identify two situations in which the principle of net neutrality has been violated.

3. The admission office of Southern State University needs a simple and easy-to-follow Web policy document to send to its incoming freshmen. The document should tell students how to behave while on the Web. After reading the information presented in this chapter and other sources, create a document that includes a 10-item bullet list highlighting the key issues that incoming students should be aware of.

4. An oil company in central Texas with over 2,000 employees needs a document regarding green computing to send to all its key employees. After reading the information presented in this chapter and other sources, create a one-page document that includes an eight-item bullet list highlighting the key issues that all division managers should be aware of.

5. Southern Tech, a major distributor of electronic devices, needs your advice. With over 1,500 employees who use tablets and smartphones to perform their jobs, the company needs a document highlighting key heath-related issues, which it will distribute to its employees to improve efficiency and reduce work-related injuries. After reading the information presented in this chapter and other sources, create a one-page document that includes a five-item bulleted list highlighting key health-related issues that employees need to aware of.

6. An Internet company in Northern California is receiving frequent requests from employees who want to telecommute. The company's CTO wants to be flexible and accommodate as many employees as possible. At the same time, the CTO wants to achieve productivity goals and keep to a minimum any legal issues that may arise from this new work alternative. After reading the information presented in this chapter and other sources, create a one-page document that includes three guidelines that telecommuters need to follow and three guidelines that the company needs to follow to achieve both personal and organizational objectives.

1. Cybersquatting is registering, selling, or using a domain name to profit from someone else's trademark. True or False?

2. To guard against cookies, which record your navigations around the Web, you can use the cookie control features contained in your browser. True or False?

3. Using computing devices that consume less energy and are biodegradable is not a part of a green computing strategy. True or False?

4. Which of the following is not an advantage of telecommuting?

   a. The ability to work in more pleasant surroundings

   b. Increased productivity

   c. Decreased neighborhood crime

   d. Less interaction with coworkers

5. Which of the following is not part of a green computing strategy?

   a. Eliminating all cloud computing projects

   b. Using computing devices that consume less energy and are biodegradable

   c. Allowing certain employees to work from their homes, resulting in fewer cars on the roads

   d. Replacing actual face-to-face meetings with computer and video conferencing

6. Which of the following is not part of the ACM code of ethics?

   a. Contribute to society and human well-being.

   b. Avoid harm to others.

   c. Do not give proper credit for intellectual property.

   d. Be honest and trustworthy.

CASE STUDY 4-1

Telecommuting with a New Twist

According to the *Wall Street Journal,* the number of corporate employees who work from home at least 1 day a month rose 23 percent a year between 2007 and 2011, when it reached 22.8 million. Those who work from home only 1 or 2 days a month are leading the trend, rising by an average 69.5 percent every year in the same period, reaching 3.3 million in 2011.[30] Working from home offers some clear advantages for both employees and employers. It also offers some disadvantages, such as lack of control and monitoring; however, this is changing. Using computer-monitoring

© Steve Heap/Shutterstock.com

software, employers are now able to check if employees are working or slacking off. One such software application is InterGuard (Awareness Technologies).[31] To make telecommuting more productive, some employers set targets for each employee to accomplish in a given period. Other employers review summaries of different Web sites that employees have visited and the amount of time that employees have spent on various Web sites.

Some other employers track projects and schedule meetings on shared calendars in order to monitor their employees. Still other employers require "virtual face time" via e-mail, instant messaging, video chat, or calls. As reported in the *Wall Street Journal* article, Gartner, Inc., a technology-research firm, predicts that the use of computer-security monitoring programs will rise to 60 percent of employers by 2015, up from fewer than 10 percent in 2012. Because using this type of technology may create privacy issues, employers should tell their employees that they are being monitored and that this is legal.[32]

Answer the following questions:

1. According to the *Wall Street Journal*, how many people worked from home in 2011?
2. What are some methods that employers use to monitor their employees?
3. What is an example of a software tool that is used for employee monitoring?
4. Are there any ethical or privacy issues involved when employers monitor their employees? Discuss.

CASE STUDY 4-2
Privacy and Other Legal Issues at Google

Two Google services, Google Maps and Google Books, have created legal and privacy issues for the company. Google Maps' Street View option shows 360-degree street-level imagery, which enables users to see homes, businesses, streets, and other scenery. After inquiries by German privacy authorities, Google announced that it had been unknowingly collecting private data from unencrypted Wi-Fi networks. The collected material included a list of people with certain medical conditions, e-mail addresses, video files, and lists of visited Web pages. According to Google, the data collection was an accident due to a programming error.[33,34]

© iStockphoto.com/Robert Corse

In another case, a woman claimed that by following Google Maps she got run over by a car.[35] And in still another case, a couple was upset because they had been trying for 3 years to get Google to alter the directions the service supplies for Round Valley State Park. The couple owns one of the three log cabins positioned at the back of the park, and because of incorrect directions supplied by Google Maps, they had annually been subjected to dozens of people coming into their yard or knocking on their door looking for an entrance to the park.[36]

Google Books has created some legal issues as well. By searching the full text of books, the service attempts to provide an online catalog of all the books contained in its vast database, but some people have claimed that the application has violated copyright protections.[37]

Although most of these lawsuits and technical issues have been resolved, it remains to be seen whether Google can steer clear of privacy and legal issues.

Answer the following questions:

1. What are some business applications of Google Maps' Street View?
2. How has Google Maps' Street View created privacy and other legal issues?
3. What is Google Books, and how has it created copyright protection and other legal issues?

5 | Protecting Information Resources

LEARNING OUTCOMES

After studying this chapter, you should be able to:

5-1 Describe information technologies that could be used in computer crimes.

5-2 Describe basic safeguards in computer and network security.

5-3 Explain the major security threats.

5-4 Describe security and enforcement measures.

5-5 Summarize the guidelines for a comprehensive security system, including business continuity planning.

After you finish this chapter, go to **PAGE 105** for the **STUDY TOOLS**

Information technologies can be misused to invade users' privacy and to commit computer crimes.

This chapter begins by discussing the impact of information technology tools on privacy and how these tools can be used to commit computer crimes. The chapter discusses security issues and measures related to computer and network environments. A comprehensive security system can protect an organization's information resources, which are its most important asset after human resources. The chapter then discusses major types of security threats and a variety of measures designed to protect against those threats. Some organizations even use the resources of a

Computer Emergency Response Team (CERT) to handle threats and their effects. Finally, the chapter covers the principles behind devising a comprehensive security system and the use of business continuity planning to help an organization recover from a disaster.

5-1 RISKS ASSOCIATED WITH INFORMATION TECHNOLOGIES

Information technologies can be misused to invade users' privacy and commit computer crimes. The following sections describe some of these misuses and discuss related privacy issues. Keep in mind, however, that you can minimize or prevent many of these risks by installing operating system updates regularly, using antivirus and antispyware software, and using e-mail security features.

5-1a The Costs of Cyber Crime to the U.S. Economy

Hackers, computer criminals, and cyber criminals, both domestic and international, could cost the U.S. economy over $100 billion and 500,000 jobs per year, according to a 2013 report by the Center for Strategic and International Studies (CSIS), a Washington D.C. think tank. The costs will include stolen identities, intellectual property, and trade secrets as well as the damage done to companies' and individuals' reputations. The total cost will also include the expense of enhancing and upgrading a company's network security after an attack. The CSIS report went further and included the opportunity costs associated with downtime and lost trust as well as the loss of sensitive business information. Job losses would include manufacturing jobs as well as jobs where stolen trade secrets and other intellectual properties resulted in jobs being moved overseas. Actually, the total cost may even be higher than the CSIS report projects, given that businesses often do not reveal or admit certain cybercrimes or do not even realize the amount of damage that has been caused by computer criminals and cyber criminals.[1]

5-1b Spyware and Adware

Spyware is software that secretly gathers information about users while they browse the Web. This information could be used for malicious purposes. Spyware can also interfere with users' control of their computers, through such methods as installing additional software and redirecting Web browsers. Some spyware changes computer settings, resulting in slow Internet connections, changes to users' default home pages, and loss of functions in other programs. To protect against spyware, you should install antivirus software that also checks for spyware or you should install antispyware

software, such as Spy Sweeper, CounterSpy, STOPzilla, and Spyware Doctor.

Adware is a form of spyware that collects information about the user (without the user's consent) to determine which advertisements to display in the user's Web browser. In addition to antivirus software, an ad-blocking feature should be installed in your Web browser to protect against adware.

5-1c Phishing and Pharming

Phishing is sending fraudulent e-mails that seem to come from legitimate sources, such as a bank or university. The e-mails usually direct recipients to false Web sites that look like the real thing for the purpose of capturing personal information, such as Social Security numbers, passwords, bank account numbers, and credit card numbers.

Pharming is similar to phishing in that Internet users are directed to fraudulent Web sites with the intention of stealing their personal information, such as Social Security numbers, passwords, bank account numbers, and credit card numbers. The difference is that pharmers usually hijack an official Web site address by hacking a Domain Name System server (discussed in Chapter 7), then alter the legitimate Web site IP address so that users who enter the correct Web address are directed to the pharmers's fraudulent Web site.

5-1d Keystroke Loggers

Keystroke loggers monitor and record keystrokes and can be software or hardware devices. Sometimes, companies use these devices to track employees' use of e-mail and the Internet, and this use is legal. However, keystroke loggers can be used for malicious purposes, too, such as collecting the credit card numbers that users enter while shopping online. Some antivirus and antispyware programs guard against software keystroke loggers, and utilities are available to install as additional protection.

5-1e Sniffing and Spoofing

Sniffing is capturing and recording network traffic. Although it can be done for legitimate reasons, such as monitoring network performance, hackers often use it to intercept information. **Spoofing** is an attempt to gain access to a network by posing as an authorized user in order to find sensitive information, such as passwords and credit card information. Spoofing also happens when an illegitimate program poses as a legitimate one.

Spyware is software that secretly gathers information about users while they browse the Web.

Adware is a form of spyware that collects information about the user (without the user's consent) to determine which advertisements to display in the user's Web browser.

Phishing is sending fraudulent e-mails that seem to come from legitimate sources, such as a bank or university.

Similar to phishing, **pharming** is directing Internet users to fraudulent Web sites with the intention of stealing their personal information, such as Social Security numbers, passwords, bank account numbers, and credit card numbers. The difference is that pharmers usually hijack an official Web site address, then alter its IP address so that users who enter the correct Web address are directed to the pharmers's fraudulent Web site.

Keystroke loggers monitor and record keystrokes and can be software or hardware devices.

Sniffing is capturing and recording network traffic.

Spoofing is an attempt to gain access to a network by posing as an authorized user in order to find sensitive information, such as passwords and credit card information.

5-1f Computer Crime and Fraud

Computer fraud is the unauthorized use of computer data for personal gain, such as transferring money from another's account or charging purchases to someone else's account. Many of the technologies discussed previously can be used for committing computer crimes. In addition, social networking sites, such as Facebook and MySpace, have been used for committing computer crimes.

In addition to phishing, pharming, and spoofing, computer crimes include the following:

- Denial-of-service attacks, which inundate a Web site or network with e-mails and other network traffic so that it becomes overloaded and cannot handle legitimate traffic
- Identity theft, such as stealing Social Security numbers for unauthorized use (The information box below discusses identity theft at the Internal Revenue Service.)
- Software piracy and other infringements of intellectual property
- Distributing child pornography
- E-mail spamming
- Writing or spreading viruses, worms, Trojan programs, and other malicious code
- Stealing files for industrial espionage
- Changing computer records illegally

© Carlos A. Oliveras/Shutterstock.com

- Virus hoaxes, in which individuals intentionally spread false statements or information through the Internet in such a way that readers believe they are true

Another computer crime is sabotage, which involves destroying or disrupting computer services. Computer criminals change, delete, hide, or use computer files for personal gain. Usually called hackers, many of them break into computer systems for personal satisfaction, but others seek financial gain. Surprisingly, most computer crimes are committed by company insiders, which makes protecting information resources even more difficult.

> **Computer fraud** is the unauthorized use of computer data for personal gain.

IDENTITY THEFT AT THE INTERNAL REVENUE SERVICE

In 2011 alone, the Internal Revenue Service (IRS) sent more than $5 billion in refund checks to identity thieves who had filed fraudulent claims. It was estimated at the time that another $21 billion would be lost to identity theft in the succeeding 5 years. Tampa and Miami were the two top cities from where fraudulent tax returns originated, the perpetrators usually stealing the identities of dead people, children, or someone else who normally does not file a tax return. In 2011, the IRS detected about 940,000 fraudulent returns; however, it was estimated that another 1.5 million cases went undetected. In one case, a single address in Lansing, Michigan, was used to file 2,137 separate tax returns totaling $3.3 million. To combat this problem, the IRS needs access to third-party information in order to verify returns. Also, the timing of when employees can file their returns and when employers submit their withholding and income information needs to be synchronized. The IRS currently uses new ID theft-screening filters that will not issue refunds until the IRS can verify a taxpayer's identity. There is another system in place that flags returns filed with Social Security numbers of individuals who have died. As of April 2012, the new ID theft-screening filters system had stopped approximately $1.3 billion in potentially fraudulent refunds.[2]

TYPES OF HACKERS

Script kiddies—Inexperienced, usually young hackers who use programs that others have developed to attack computer and network systems and deface Web sites.

Black hats—Hackers who specialize in unauthorized penetration of information systems. They attack systems for profit, fun, or political motivation or as part of a social cause. These penetration attacks often involve modifying and destroying data.

White hats—Also known as ethical hackers, these are computer security experts who specialize in penetration testing and other testing methods to ensure that a company's information systems are secure.

5-2 COMPUTER AND NETWORK SECURITY: BASIC SAFEGUARDS

Computer and network security has become critical for most organizations, especially in recent years, with hackers becoming more numerous and more adept at stealing and altering private information. The various types of hackers are described in the information box above. To break into computers and networks, hackers use a variety of tools, such as sniffers, password crackers, rootkits, and many others; all can be found free on the Web. Also, journals such as *Phrack* and *2600: The Hacker Quarterly* offer hackers informative tips. A rootkit is a software application that hides its presence on the computer, which makes it nearly undetectable by common anti-malware software. The information box below highlights the widespread problem of hacking.

A comprehensive security system protects an organization's resources, including information, computer, and network equipment. The information an organization needs to protect can take many forms: e-mails, invoices transferred via electronic data interchange (EDI), new product designs, marketing campaigns, and financial statements. Security threats involve more than stealing data; they include such actions as sharing passwords with coworkers, leaving a computer unattended while logged on to the network, or even spilling coffee on a keyboard. A comprehensive security system includes hardware, software, procedures, and personnel that collectively protect information resources and keep intruders and hackers at bay. There are three

© Jirsak/Shutterstock.com

NEARLY ALL ORGANIZATIONS GET HACKED

Recently, several major private and public-sector organizations have been hacked, including Target, Neiman Marcus, Adobe, RSA, Lockheed Martin, Oak Ridge National Laboratories, and the International Monetary Fund. Ponemon Research conducted a survey of 583 U.S. companies, ranging from small organizations with less than 500 employees to enterprises with workforces of more than 75,000. Ninety percent of the respondents indicated their organizations' computers and network systems had been compromised by hackers at least once in the previous 12 months; nearly 60 percent reported two or more breaches in the past year. Over half the respondents indicated they had little confidence in their organization's ability to avoid further attacks. Roughly half blamed a lack of resources for their security problems, and about the same number said network complexity was their main challenge to implementing security protections.[3]

important aspects of computer and network security: confidentiality, integrity, and availability, collectively referred to as the *CIA triangle*.[4]

Confidentiality means that a system must not allow the disclosing of information by anyone who is not authorized to access it. In highly secure government agencies, such as the Department of Defense, the CIA, and the IRS, confidentiality ensures that the public cannot access private information. In businesses, confidentiality ensures that private information, such as payroll and personnel data, is protected from competitors and other organizations. In the e-commerce world, confidentiality ensures that customers' data cannot be used for malicious or illegal purposes.

Integrity refers to the accuracy of information resources within an organization. In other words, the security system must not allow data to be corrupted or allow unauthorized changes to a corporate database. In financial transactions, integrity is probably the most important aspect of a security system, because incorrect or corrupted data can have a huge impact. For example, imagine a hacker breaking into a financial network and changing a customer's balance from $10,000 to $1,000—a small change, but one with a serious consequence. Database administrators and Webmasters are essential in this aspect of security. In addition, part of ensuring integrity is identifying authorized users and granting them access privileges.

Availability means that computers and networks are operating and authorized users can access the information they need. It also means a quick recovery in the event of a system failure or disaster. In many cases, availability is the most important aspect for authorized users. If a system is not accessible to users, the confidentiality and integrity aspects cannot be assessed.

The Committee on National Security Systems (CNSS) has proposed another model, called the *McCumber cube*. John McCumber created this framework for evaluating information security. Represented as a three-dimensional cube (see Exhibit 5.1), it defines nine characteristics of information security.[5] The McCumber cube is more specific than the CIA triangle and helps designers of security systems consider many crucial issues for improving the effectiveness of security measures. Note that this model includes the different states in which information can exist in a system: transaction, storage, and processing.

In addition, a comprehensive security system must provide three levels of security:

- *Level 1*—Front-end servers, those available to both internal and external users, must be protected against unauthorized access. Typically, these systems are e-mail and Web servers.

- *Level 2*—Back-end systems (such as users' workstations and internal database servers) must be protected to ensure confidentiality, accuracy, and integrity of data.

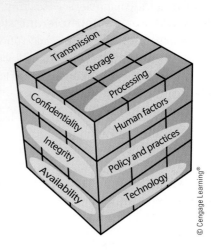

Exhibit 5.1
McCumber cube

Confidentiality means that a system must prevent disclosing information to anyone who is not authorized to access it.

Integrity refers to the accuracy of information resources within an organization.

Availability means that computers and networks are operating, and authorized users can access the information they need. It also means a quick recovery in the event of a system failure or disaster.

- *Level 3*—The corporate network must be protected against intrusion, denial-of-service attacks, and unauthorized access.

When planning a comprehensive security system, the first step is designing **fault-tolerant systems**, which use a combination of hardware and software for improving reliability—a way of ensuring availability in case of a system failure. Commonly used methods include the following:

- *Uninterruptible power supply (UPS)*—This backup power unit continues to provide electrical power in the event of blackouts and other power interruptions and is most often used to protect servers. It performs two crucial tasks: It serves as a power source to continue running the server (usually for a short period), and it safely shuts down the server. More sophisticated UPS units can prevent users from accessing the server and send an alert to the network administrator.

- *Redundant array of independent disks (RAID)*—As you learned in Chapter 2, a RAID system is a collection of disk drives used to store data in multiple places. RAID systems also store a value called a *checksum*, used to verify that data has been stored or transmitted without error. If a drive in the RAID system fails, data stored on it can be reconstructed from data stored on the remaining drives. RAID systems vary in cost, performance, and reliability.

- *Mirror disks*—This method uses two disks containing the same data; if one fails, the other is available, allowing operations to continue. Mirror disks are usually a less expensive, level-1 RAID system and can be a suitable solution for small organizations.

5-3 SECURITY THREATS: AN OVERVIEW

Computer and network security are important to prevent loss of, or unauthorized access to, important information resources. Some threats can be controlled completely or partially, but some cannot be controlled. For example, you can control power fluctuations and blackouts to some degree by using surge suppressors and UPSs, but you cannot control whether natural disasters strike. You can, however, minimize the effects of a natural disaster by making sure fire suppression systems are up to code or by making structural changes to your organization's facility for earthquake protection—such as bolting the foundation.

Threats can also be categorized by whether they are unintentional (such as natural disasters, a user's accidental deletion of data, and structural failures) or intentional. Intentional threats include hacker attacks and attacks by disgruntled employees—such as spreading a virus on the company network. The following sections describe the most common intentional threats.

5-3a Intentional Threats

Intentional computer and network threats include:

- Viruses
- Worms
- Trojan programs
- Logic bombs
- Backdoors
- Blended threats (e.g., a worm launched by Trojan)
- Rootkits
- Denial-of-service attacks
- Social engineering

Viruses

Viruses are the most well-known computer and network threats. (You have probably heard of the I Love You and Michelangelo viruses, for example.) They are a type of malware (short for *malicious software*), which is any program or file that is harmful to computers or networks.

A recent study estimates the annual cost of cybercrime to the global economy is $1 trillion and that malware is being introduced at a rate of 55,000 pieces per day.[6] However, estimating the dollar cost of viruses can be difficult. Many organizations are reluctant to report their losses because they do not want to publicize how vulnerable they are.

A **virus** consists of self-propagating program code that is triggered by a specified time or event. When the program or operating system containing the virus is used, the virus attaches itself to other files, and the cycle continues. The seriousness of viruses varies, ranging from

Fault-tolerant systems ensure availability in the event of a system failure by using a combination of hardware and software.

A **virus** consists of self-propagating program code that is triggered by a specified time or event. When the program or operating system containing the virus is used, the virus attaches itself to other files, and the cycle continues.

A recent study estimates the cost of cybercrime to the global economy is $1 trillion and that malware is being introduced at a rate of 55,000 pieces per day.

playing a prank, such as displaying a funny (but usually annoying) image on the user's screen, to destroying programs and data.

Viruses can be transmitted through a network or through e-mail attachments. Some of the most dangerous ones come through bulletin boards or message boards because they can infect any system using the board. Experts believe that viruses infecting large servers, such as those used by air traffic control systems, pose the most risk to national security.

There are times that virus hoaxes are spread as well. These reports about viruses that turn out not to exist can cause panic and even prompt organizations to shut down their networks. In some ways, virus hoaxes can cause as much damage as real viruses.

The following list describes some of the indications that a computer might be infected by a virus:

- Some programs have suddenly increased in size.
- Files have been corrupted, or the user is unable to open some files.
- Hard disk free space is reduced drastically.
- The keyboard locks up, or the screen freezes.
- Available memory dips down more than usual.
- Disk access is slow.
- The computer takes longer than normal to start.
- There is unexpected disk activity, such as the disk drive light flashing even though the user is not trying to save or open a file.
- There are unfamiliar messages on the screen.

Installing and updating an antivirus program is the best measure against viruses. Widely used antivirus programs include McAfee Virus Scan (*www.mcafee.com/us*), Norton Antivirus (*www.norton.com*), and Trend Micro (*www.trendmicro.com*). You can even download free or low-cost programs on the Internet. Most computers now have antivirus software already installed, but you should check for the most current version of the

© JMiks/Shutterstock.com

antivirus software. New viruses are released constantly, so use automatic updating to make sure your computer's protection is current.

Worms

A **worm** travels from computer to computer in a network, but it does not usually erase data. Unlike a virus, it is an independent program that can spread itself without having to be attached to a host program. It might corrupt data, but it usually replicates itself into a full-blown version that eats up computing resources, eventually bringing a computer or network to a halt. Well-known worms include Code Red, Melissa, and Sasser. Conficker, a recent worm, has infected millions of Windows computers.

A **worm** travels from computer to computer in a network, but it does not usually erase data. Unlike viruses, worms are independent programs that can spread themselves without having to be attached to a host program.

Trojan Programs

A **Trojan program** (named after the Trojan horse that the Greeks used to enter Troy during the Trojan War) contains code intended to disrupt a computer, network, or Web site, and it is usually hidden inside a popular program. Users run the popular program, unaware that the malicious program is also running in the background. Disgruntled programmers possibly seeking revenge with an organization have created many Trojan programs. These programs can erase data and wreak havoc on computers and networks, but they do not replicate themselves, as viruses and worms do.

Logic Bombs

A **logic bomb** is a type of Trojan program used to release a virus, worm, or other destructive code. Logic bombs are triggered at a certain time (sometimes the birthday of a famous person) or by a specific event, such as a user pressing the Enter key or running a certain program.

Backdoors

A **backdoor** (also called a *trapdoor*) is a programming routine built into a system by its designer or programmer. This routine enables the designer or programmer to bypass system security and sneak back into the system later to access programs or files. Usually, system users are not aware that a backdoor has been activated; a user logon or combination of keystrokes can be used to activate backdoors.

Blended Threats

A **blended threat** is a security threat that combines the characteristics of computer viruses, worms, and other malicious codes with vulnerabilities found on public and private networks. Blended threats search for vulnerabilities in computer networks and then take advantage of these vulnerabilities by embedding malicious codes in the server's HTML files or by sending unauthorized e-mails from compromised servers with a worm attachment. They may launch a worm through a Trojan horse or launch a denial-of-service (DoS) attack at a targeted IP address. Their goal is not just to start and transmit an attack but to spread it. A multilayer security system, as discussed in this chapter, can guard against blended threats.

Denial-of-Service Attacks

A **denial-of-service (DoS)** attack floods a network or server with service requests to prevent legitimate users' access to the system. Think of it as 5,000 people surrounding a store and blocking customers who want to enter; the store is open, but it cannot provide service to legitimate customers. Typically, DoS attackers target Internet servers (usually Web, FTP, or mail servers), although any system connected to the Internet running TCP services is subject to attack.

In 2013, a DoS attack hit the online banking sites of dozens of U.S. and Canadian banks, including Bank of America, Citigroup, Wells Fargo, U.S. Bancorp, PNC, Capital One, Fifth Third Bank, BB&T, and HSBC.[7,8] This particular assault was a distributed denial-of-service (DDoS) attack, which is when hundreds or thousands of computers work together to bombard a Web site with thousands of requests for information in a short period, causing it to grind to a halt. Because DDoS attacks come from multiple computers, they are difficult to trace.

Recently, emergency-service providers and many other organizations have been targeted by a new type of DoS attack, called a TDoS (telephony denial of service) attack. These attacks use high volumes of automated calls to tie up a target phone system, halting incoming and outgoing calls. In some cases, the attacker demands a ransom. If the ransom is not paid, the TDoS begins, perhaps lasting for several hours.[9]

A **Trojan program** contains code intended to disrupt a computer, network, or Web site, and it is usually hidden inside a popular program. Users run the popular program, unaware that the malicious program is also running in the background.

A **logic bomb** is a type of Trojan program used to release a virus, worm, or other destructive code. Logic bombs are triggered at a certain time (sometimes the birthday of a famous person) or by a specific event, such as a user pressing the Enter key or running a certain program.

A **backdoor** (also called a *trapdoor*) is a programming routine built into a system by its designer or programmer. It enables the designer or programmer to bypass system security and sneak back into the system later to access programs or files.

A **blended threat** is a security threat that combines the characteristics of computer viruses, worms, and other malicious codes with vulnerabilities found on public and private networks.

A **denial-of-service (DoS)** attack floods a network or server with service requests to prevent legitimate users' access to the system.

Social Engineering

In the context of security, **social engineering** means using "people skills"—such as being a good listener and assuming a friendly, unthreatening air—to trick others into revealing private information. This is an attack that takes advantage of the human element of security systems. Social engineers use the private information they

look up the phone number of a receptionist he or she can call and pretend to be someone else in the organization. Shoulder surfing—that is, looking over someone's shoulder—is the easiest form of collecting information. Social engineers use this technique to observe an employee entering a password or a person entering a PIN at an ATM, for example.

> Social engineers often search through dumpsters or trash cans looking for discarded material—such as phone lists and bank statements—that they can use to help break into a network.

have gathered to break into servers and networks and steal data, thus compromising the integrity of information resources. Social engineers use a variety of tools and techniques to gather private information, including publicly available sources of information—Google Maps, company Web sites, newsgroups, and blogs, for example.

In addition, two commonly used social-engineering techniques are called *dumpster diving* and *shoulder surfing*. Social engineers often search through dumpsters or trash cans looking for discarded material (such as phone lists and bank statements) that they can use to help break into a network. For example, a social engineer might

In addition to these intentional threats, loss or theft of equipment and computer media is a serious problem, particularly when a computer or flash drive contains confidential data. The information box below discusses this problem and offers some protective measures.

> In the context of security, **social engineering** means using "people skills"—such as being a good listener and assuming a friendly, unthreatening air—to trick others into revealing private information. This is an attack that takes advantage of the human element of security systems.

PROTECTING AGAINST DATA THEFT AND DATA LOSS

Memory sticks, PDAs, CDs, USB flash drives, smartphones, and other portable storage media pose a serious security threat to organizations' data resources. Theft or loss of these devices is a risk, of course, but disgruntled employees can also use these devices to steal company data. The following guidelines are recommended to protect against these potential risks[10]:

- Do a risk analysis to determine the effects of confidential data being lost or stolen.
- Ban portable media devices and remove or block USB ports, floppy drives, and CD/DVD-ROM drives, particularly in organizations that require tight security. This measure might not be practical in some companies, however.
- Make sure employees have access only to data they need for performing their jobs, and set up rigorous access controls.
- Store data in databases instead of in spreadsheet files, for better access control.
- Have clear, detailed policies about what employees can do with confidential data, including whether data can be removed from the organization.
- Encrypt data downloaded from the corporate network.

5-4 SECURITY MEASURES AND ENFORCEMENT: AN OVERVIEW

In addition to backing up data and storing it securely, organizations can take many other steps to guard against threats. A comprehensive security system should include the following:

- Biometric security measures
- Nonbiometric security measures
- Physical security measures
- Access controls
- Virtual private networks
- Data encryption
- E-commerce transaction security measures
- Computer Emergency Response Team

5-4a Biometric Security Measures

Biometric security measures use a physiological element that is unique to a person and cannot be stolen, lost, copied, or passed on to others. The following list describes some biometric devices and measures, some of which can be seen in Exhibit 5.2:

- *Facial recognition*—Identify users by analyzing the unique shape, pattern, and positioning of facial features.
- *Fingerprints*—Scan users' fingerprints and verify them against prints stored in a database.
- *Hand geometry*—Compare the length of each finger, the translucence of fingertips, and the webbing between fingers against stored data to verify users' identities.
- *Iris analysis*—Use a video camera to capture an image of the user's iris, then use software to compare the data against stored templates.
- *Palm prints*—Use the palm's unique characteristics to identify users. A palm reader uses near-infrared light to capture a user's vein pattern, which is unique to each individual. This is compared to a database that contains existing patterns. This method is often used by law enforcement agencies.

- *Retinal scanning*—Scan the retina using a binocular eye camera, then check against data stored in a database.
- *Signature analysis*—Check the user's signature as well as deviations in pen pressure, speed, and length of time used to sign the name.
- *Vein analysis*—Analyze the pattern of veins in the wrist and back of the hand without making any direct contact with the veins.
- *Voice recognition*—Translate words into digital patterns, which are recorded and examined for tone and pitch. Using voice to verify user identity has one advantage over most other biometric measures: It can work over long distances via ordinary telephones. A well-designed voice-recognition security system can improve the security of financial transactions conducted over the phone.

Although biometric techniques are effective security measures, they might not be right for all organizations. Some drawbacks of biometrics are high cost, users' reluctance, and complex installation. However, with improvements being made to address these drawbacks, biometrics can be a viable alternative to traditional security measures. The information box below presents a real-life application of biometrics.

5-4b Nonbiometric Security Measures

The three main nonbiometric security measures are callback modems, firewalls, and intrusion detection systems.

BIOMETRICS AT PHOEBE PUTNEY MEMORIAL HOSPITAL

Phoebe Putney Memorial Hospital, a 443-bed community hospital in Albany, Georgia, needed to improve its electronic health record (EHR) system. Doctors and nurses were complaining about the number of passwords required to access clinical records, so the hospital switched to fingerprint scanners, which, along with a single sign-on application, made the EHR system both easier to use and more secure. With the scanners, it is possible to audit usage, thereby ensuring that only authorized users have access to sensitive information. Another advantage of fingerprint scanners: Fingerprints do not get lost, like smart cards.[11]

Biometric security measures use a physiological element that is unique to a person and cannot be stolen, lost, copied, or passed on to others.

Exhibit 5.2
Examples of biometric devices

© Pmphoto/Shutterstock.com

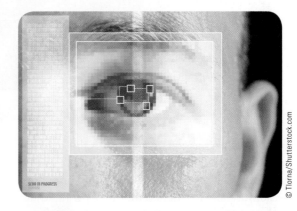
© Tlorna/Shutterstock.com

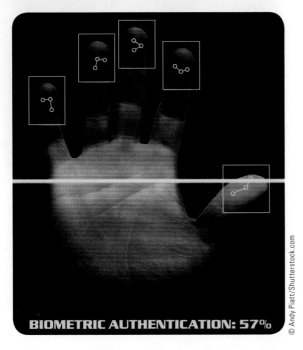
BIOMETRIC AUTHENTICATION: 57%
© Andy Piatt/Shutterstock.com

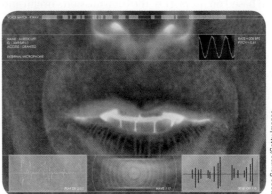
Image Source/Getty Images

Callback Modems

A **callback modem** verifies whether a user's access is valid by logging the user off (after he or she attempts to connect to the network) and then calling the user back at a predetermined number. This method is useful in organizations with many employees who work off-site and who need to connect to the network from remote locations.

Firewalls

A **firewall** is a combination of hardware and software that acts as a filter or barrier between a private network and external computers or networks, including the Internet. A network administrator defines rules for access, and all other data transmissions are blocked. An effective firewall should protect data going from

A **callback modem** verifies whether a user's access is valid by logging the user off (after he or she attempts to connect to the network) and then calling the user back at a predetermined number.

A **firewall** is a combination of hardware and software that acts as a filter or barrier between a private network and external computers or networks, including the Internet. A network administrator defines rules for access, and all other data transmissions are blocked.

Exhibit 5.3
Basic firewall configuration

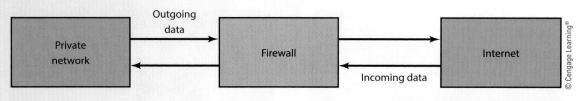

© Cengage Learning®

the network as well as data coming into the network. Exhibit 5.3 shows a basic firewall configuration.

A firewall can examine data passing into or out of a private network and decide whether to allow the transmission based on users' IDs, the transmission's origin and destination, and the transmission's contents. Information being transmitted is stored in what's called a *packet*, and after examining a packet, a firewall can take one of the following actions:

- Reject the incoming packet.
- Send a warning to the network administrator.
- Send a message to the packet's sender that the attempt failed.
- Allow the packet to enter (or leave) the private network.

The main types of firewalls are packet-filtering firewalls, application-filtering firewalls, and proxy servers. Packet-filtering firewalls control data traffic by configuring a router to examine packets passing into and out of the network. The router examines the following information in a packet: source IP address and port, destination IP address and port, and protocol used. Based on this information, rules called *packet filters* determine whether a packet is accepted, rejected, or dropped. For example, a packet filter can be set up to deny packets coming from specific IP addresses. A packet-filtering firewall informs senders if packets are rejected but does nothing if packets are dropped; senders have to wait until their requests time out to learn that the packets they sent were not received.

In addition, these firewalls record all incoming connections, and packets that are rejected might be a warning sign of an unauthorized attempt. Packet-filtering firewalls are somewhat inefficient, however, because they have to examine packets one by one, and they might be difficult to install. In addition,

they cannot usually record every action taking place at the firewall, so network administrators could have trouble finding out whether and how intruders are trying to break into the network.

Application-filtering firewalls are generally more secure and flexible than packet-filtering firewalls, but they are also more expensive. Typically, they are software that is installed on a host computer (a dedicated workstation or server) to control use of network applications, such as e-mail, Telnet, and FTP. In addition to checking which applications are requested, these firewalls monitor the time when application requests take place. This information can be useful, because many unauthorized attempts take place after normal work hours. Application-filtering firewalls also filter viruses and log actions more effectively than packet-filtering firewalls, which helps network administrators spot potential security breaches. Because of all the application-filtering that these firewalls do, however, they are often slower than other types of firewalls, which can affect network performance.

A proxy server, shown in Exhibit 5.4, is software that acts as an intermediary between two systems—between

Exhibit 5.4
Proxy server

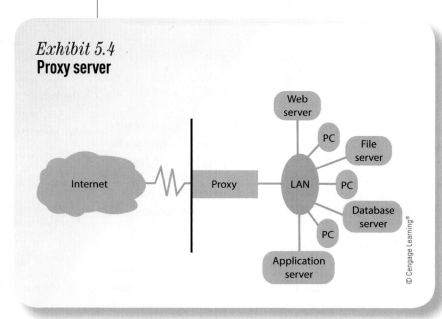

© Cengage Learning®

network users and the Internet, for example. It is often used to help protect the network against unauthorized access from outside the network by hiding the network addresses of internal systems. A proxy server can also be used as a firewall that scans for malware and viruses, speeds up network traffic, or takes some load off internal servers (which firewalls cannot do). It can also block requests from certain servers.

Although firewalls can do a lot to protect networks and computers, they do not offer complete security. Sophisticated hackers and computer criminals can circumvent almost any security measure. For example, some hackers use a technique called IP spoofing to trick firewalls into treating packets as coming from legitimate IP addresses. This technique is the equivalent of forgery. To provide comprehensive security for data resources, firewalls should be used along with other security measures. Other guidelines for improving a firewall's capabilities include the following:

- Identify what data must be secured, and conduct a risk analysis to assess the costs and benefits of a firewall.

- Compare a firewall's features with the organization's security needs. For example, if your organization uses e-mail and FTP frequently, make sure the application-filtering firewall you are considering can handle these network applications.

- Compare features of packet-filtering firewalls, application-filtering firewalls, and proxy servers to determine which of these types addresses your network's security needs the best.

- Examine the costs of firewalls, and remember that the most expensive firewall is not necessarily the best. Some inexpensive firewalls might be capable of handling everything your organization needs.

- Compare the firewall's security with its ease of use. Some firewalls emphasize accuracy and security rather than ease of use and functionality. Determine what is most important to your organization when considering the trade-offs.

- Check the vendor's reputation, technical support, and update policies before making a final decision. As the demand for firewalls has increased, so has the number of vendors, and not all vendors are equal. Keep in mind that you might have to pay more for a product from a vendor with a good reputation that offers comprehensive technical support.

Another alternative is to build a firewall instead of purchase one. This option might be more expensive (and requires having an employee with the necessary skills), but the customized features and flexibility offered by a firewall developed in-house could outweigh the cost.

Intrusion Detection Systems

Firewalls protect against external access, but they leave networks unprotected from internal intrusions. An **intrusion detection system (IDS)** can protect against both external and internal access. It is usually placed in front of a firewall and can identify attack signatures, trace patterns, generate alarms for the network administrator, and cause routers to terminate connections with suspicious sources. These systems can also prevent DoS attacks. An IDS monitors network traffic and uses the "prevent, detect, and react" approach to security. Although it improves security, it requires a great deal of processing power and can affect network performance. It might also need additional configuration to prevent it from generating false positive alarms.

A number of third-party tools are available for intrusion detection. The vendors listed in Table 5.1 offer comprehensive IDS products and services.

5-4c Physical Security Measures

Physical security measures primarily control access to computers and networks, and they include devices for securing computers and peripherals from theft.

An **intrusion detection system (IDS)** can protect against both external and internal access. It is usually placed in front of a firewall and can identify attack signatures, trace patterns, generate alarms for the network administrator, and cause routers to terminate connections with suspicious sources.

Physical security measures primarily control access to computers and networks, and they include devices for securing computers and peripherals from theft.

TABLE 5.1 IDS VENDORS

Vendor	URL
Enterasys Networks	www.enterasys.com
Cisco Systems	www.cisco.com
IBM Internet Security Systems	www.iss.net
Juniper Networks	www.juniper.net/us/en
Check Point Software Technologies	www.checkpoint.com

© Cengage Learning®

Exhibit 5.5
Common physical security measures

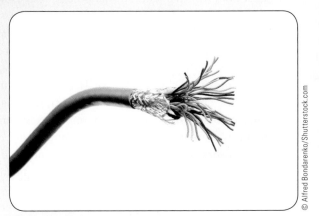

Cable shielding

Lock for securing a computer

As shown in Exhibit 5.5, common physical security measures can include the following:

- *Cable shielding*—Braided layers around the conductor cable protect it from electromagnetic interference (EMI), which could corrupt data or data transmissions.

- *Corner bolts*—An inexpensive way to secure a computer to a desktop or counter, these often have locks as an additional protection against theft.

- *Electronic trackers*—These devices are secured to a computer at the power outlet. If the power cord is disconnected, a transmitter sends a message to an alarm that goes off or to a camera that records what happens.

- *Identification (ID) badges*—These are checked against a list of authorized personnel, which must be updated regularly to reflect changes in personnel.

- *Proximity-release door openers*—These are an effective way to control access to the computer room.

A small radio transmitter is placed in authorized employees' ID badges, and when they come within a predetermined distance of the computer room's door, a radio signal sends a key number to the receiver, which unlocks the door.

- *Room shielding*—A nonconductive material is sprayed in the computer room, which reduces the number of signals transmitted or confines the signals to the computer room.

- *Steel encasements*—These fit over the entire computer and can be locked.

With the increasing popularity of laptops, theft has become a major security risk. Laptops can store confidential data, so a variety of security measures should be used. For example, a cable lock on the laptop could be combined with a fingerprint scan to make sure only the laptop's owner can access files. The information box on the next page discusses this security threat in more detail.

LOST AND STOLEN LAPTOPS

With wireless connections now available in many public places, laptops are more popular than ever. However, they can easily be lost or stolen. And replacing the laptop is not the only problem. You also have to replace the data stored on it, which can be a quite serious loss. In 2006, an employee of the U.S. Department of Veteran Affairs lost a laptop that contained personal information regarding 26 million veterans. The same year, an employee of the American Institute of Certified Public Accountants (AICPA) lost a laptop that stored the Social Security numbers of AICPA's members. If unauthorized users gain access to this kind of confidential information, identity theft and other crimes can result. To make laptops more secure, consider the following recommendations[12]:

- Install cable locks on laptops, and use biometric security measures.
- Make sure confidential data is stored on laptops only when absolutely necessary.
- Use logon passwords, screensaver passwords, and passwords for confidential files.
- Encrypt data stored on the laptop.
- Install security chips that disable a laptop if unauthorized users try to access it. Some chips send out an audio distress signal and a GPS alert showing the laptop's location.

5-4d Access Controls

Access controls are designed to protect systems from unauthorized access in order to preserve data integrity. The following sections describe two widely used access controls: terminal resource security and passwords.

Terminal Resource Security

Terminal resource security is a software feature that erases the screen and signs the user off automatically after a specified length of inactivity. This method of access control prevents unauthorized users from using an unattended computer to access the network and data. Some programs also allow users to access data only during certain times, which reduces break-in attempts during off hours.

Passwords

A **password** is a combination of numbers, characters, and symbols that is entered to allow access to a system. A password's length and complexity determines its vulnerability to discovery by unauthorized users. For example, *p@s$w0rD* is much harder to guess than *password*. The human element is one of the most notable weaknesses of password security, because users can forget passwords or give them to an unauthorized user (intentionally or unintentionally). To increase the effectiveness of passwords, follow these guidelines:

- Change passwords frequently.
- Passwords should be eight characters or longer.

- Passwords should be a combination of uppercase and lowercase letters, numbers, and special symbols, such as @ or $.
- Passwords should not be written down.
- Passwords should not be common names, such as the user's first or last name, obvious dates (such as birthdays or anniversaries), or words that can be found in a dictionary.
- Passwords should not be increased or decreased sequentially or follow a pattern (for example, 222ABC, 224ABC, 226ABC).
- Before employees are terminated, make sure their passwords have been deleted.

> **Access controls** are designed to protect systems from unauthorized access in order to preserve data integrity.
>
> A **password** is a combination of numbers, characters, and symbols that is entered to allow access to a system.

5-4e Virtual Private Networks

A **virtual private network (VPN)** provides a secure "tunnel" through the Internet for transmitting messages and data via a private network (see Exhibit 5.6). It is often used so remote users have a secure connection to the organization's network. VPNs can also be used to provide security for extranets, which are networks set up between an organization and an external entity, such as a supplier (discussed in more detail in Chapter 7). Data is encrypted before it is sent through the tunnel with a protocol, such as Layer Two Tunneling Protocol (L2TP) or Internet Protocol Security (IPSec). The cost of setting up a VPN is usually low, but transmission speeds can be slow, and lack of standardization can be a problem.

Typically, an organization leases the media used for a VPN on an as-needed basis, and network traffic can be sent over the combination of a public network (usually the Internet) and a private network. VPNs are an alternative to private leased lines or dedicated Integrated Services Digital Network (ISDN) lines and T1 lines.

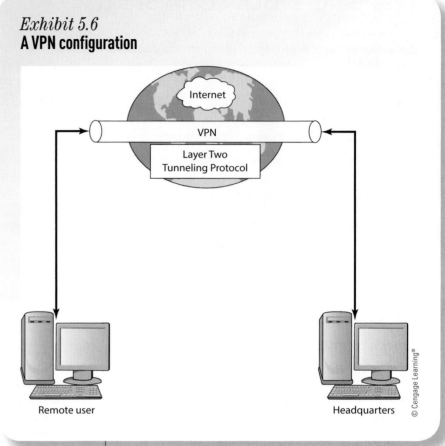

Exhibit 5.6
A VPN configuration

Internet

VPN

Layer Two Tunneling Protocol

Remote user

Headquarters

© Cengage Learning®

A **virtual private network (VPN)** provides a secure "tunnel" through the Internet for transmitting messages and data via a private network.

Data encryption transforms data, called plaintext or cleartext, into a scrambled form called ciphertext that cannot be read by others.

5-4f Data Encryption

Data encryption transforms data, called *plaintext* or *cleartext*, into a scrambled form called *ciphertext* that cannot be read by others. The rules for encryption, known as the encryption algorithm, determine how simple or complex the transformation process should be. The receiver then unscrambles the data by using a decryption key.

There are many different encryption algorithms used. One of the oldest encryption algorithms, used by Julius Caesar, is a simple substitution algorithm in which each letter in the original message is replaced by the letter three positions farther in the alphabet. For example, the

One of the oldest encryption algorithms, used by Julius Caesar, is a simple substitution algorithm in which each letter in the original message is replaced by the letter three positions farther in the alphabet.

word *top* is transmitted as *wrs*. Exhibit 5.7 shows a simple example of encryption with a substitution algorithm.

A commonly used encryption protocol is **Secure Sockets Layer (SSL)**, which manages transmission security on the Internet. Next time you purchase an item online, notice that the *http* in the browser address bar changes to *https*. The *https* indicates a Secure HTTP connection over SSL. You might also see a padlock icon in the status bar at the bottom to indicate that your information has been encrypted and hackers cannot intercept it. A more recent cryptographic protocol is **Transport Layer Security (TLS)**, which ensures data security and integrity over public networks, such as the Internet. Similar to SSL, TLS encrypts the network segment used for performing transactions. In addition to being encryption protocols, SSL and TLS have authentication functions.

As mentioned, encryption algorithms use a key to encrypt and decrypt data. The key's size varies from 32 bits to 168 bits; the longer the key, the harder the encryption is to break. There are two main types of encryption: asymmetric (also called *public key encryption*) and symmetric, which will be explained next, but first you need to understand PKI. A **PKI (public key infrastructure)** enables users of a public network such as the Internet to securely and privately exchange data through the use of a pair of keys—a public one and a private one—that is obtained from a trusted authority and shared through that authority.

Asymmetric encryption uses two keys: a public key known to everyone and a private or secret key known only to the recipient. A message encrypted with a public key can be decrypted only with the same algorithm used by the public key and requires the recipient's private key, too. Anyone intercepting the message cannot decrypt it, because he or she does not have the private key.

This encryption usually works better for public networks, such as the Internet. Each company conducting transactions or sending messages gets a private key and a public key; a company keeps its private key and publishes its public key for others to use. One of the first public key algorithms, RSA (named after its creators—Rivest, Shamir, and Adleman), is still widely used today. The main drawback of asymmetric encryption is that it is slower and requires a large amount of processing power.

In **symmetric encryption** (also called *secret key encryption*), the same key is used to encrypt and decrypt

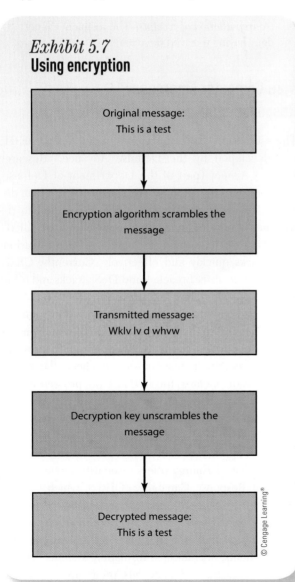

Exhibit 5.7
Using encryption

Original message:
This is a test

↓

Encryption algorithm scrambles the message

↓

Transmitted message:
Wklv lv d whvw

↓

Decryption key unscrambles the message

↓

Decrypted message:
This is a test

© Cengage Learning®

Secure Sockets Layer (SSL) is a commonly used encryption protocol that manages transmission security on the Internet.

Transport Layer Security (TLS) is a cryptographic protocol that ensures data security and integrity over public networks, such as the Internet.

A **PKI (public key infrastructure)** enables users of a public network such as the Internet to securely and privately exchange data through the use of a pair of keys—a public one and a private one—that is obtained from a trusted authority and shared through that authority.

Asymmetric encryption uses two keys: a public key known to everyone and a private or secret key known only to the recipient. A message encrypted with a public key can be decrypted only with the same algorithm used by the public key and requires the recipient's private key, too. Anyone intercepting the message cannot decrypt it because he or she does not have the private key.

In **symmetric encryption** (also called secret key encryption), the same key is used to encrypt and decrypt the message. The sender and receiver must agree on the key and keep it secret.

the message. The sender and receiver must agree on the key and keep it secret. Advanced Encryption Standard (AES), a symmetric encryption algorithm with a 56-bit key, is the one used by the U.S. government. The problem with symmetric encryption is that sharing the key over the Internet is difficult.

Encryption can also be used to create digital signatures that authenticate senders' identities and verify that the message or data has not been altered. Digital signatures are particularly important in online financial transactions. They also provide nonrepudiation, discussed in the next section. Here is how they work: You encrypt a message with your private key and use an algorithm that hashes the message and creates a message digest. The message digest cannot be converted back to the original message, so anyone intercepting the message cannot read it. Then you use your private key to encrypt the message digest, and this encrypted piece is called the *digital signature*.

You then send the encrypted message and digital signature. The recipient has your public key and uses it to decrypt the message, and then uses the same algorithm that you did to hash the message and create another version of the message digest. Next, the recipient uses your public key to decrypt your digital signature and get the message digest you sent. The recipient then compares the two message digests. If they match, the message was not tampered with and is the same as the one you sent.

5-4g E-Commerce Transaction Security Measures

In e-commerce transactions, three factors are critical for security: authentication, confirmation, and nonrepudiation. Authentication is important because the person using a credit card number in an online transaction is not necessarily the card's legitimate owner, for example. Two factors are important: what the receiver knows to be accurate and what the sender is providing. Passwords and personal information, such as your mother's maiden name, your Social Security number, or your date of birth, can be used for authentication. Physical proof, such as a fingerprint or retinal scan, works even better.

Confirmation must also be incorporated into e-commerce transactions—to verify orders and receipt of shipments, for example. When an electronic document, such as a payment, is sent from a customer to a vendor, a digitally signed confirmation with the vendor's private key is returned to verify the transaction was carried out.

Nonrepudiation is needed in case a dispute over a transaction is raised. Digital signatures are used for this and serve to bind partners in a transaction. With this process, the sender receives proof of delivery and the receiver is assured of the sender's identity. Neither party can deny sending or receiving the information.

E-commerce transaction security is concerned with the following issues:

- *Confidentiality*—How can you ensure that only the sender and intended recipient can read the message?
- *Authentication*—How can the recipient know that data is actually from the sender?
- *Integrity*—How can the recipient know that the data's contents have not been changed during transmission?
- *Nonrepudiation of origin*—The sender cannot deny having sent the data.
- *Nonrepudiation of receipt*—The recipient cannot deny having received the data.

5-4h Computer Emergency Response Team

The Computer Emergency Response Team (CERT) was developed by the Defense Advanced Research Projects Agency (part of the Department of Defense) in response to the 1988 Morris worm attack, which disabled 10 percent of the computers connected to the Internet. Many organizations now follow the CERT model to form teams that can handle network intrusions and attacks quickly and effectively. Currently, CERT focuses on security breaches and DoS attacks and offers guidelines on handling and preventing these incidents. CERT also conducts a public awareness campaign and researches Internet security vulnerabilities and ways to improve security systems. Network administrators and e-commerce site managers should check the CERT Coordination Center for updates on protecting network and information resources. Exhibit 5.8 shows the CERT Coordination Center home page (*www.cert.org/certcc.html*).

In addition, the Office of Cyber Security at the Department of Energy offers a security service: Cyber Incident Response Capability (CIRC), which you can learn more about at *http://energy.gov/cio/office-chief-information-officer/services/incident-management*. CIRC's main function is to provide information on security incidents, including information systems' vulnerabilities, viruses, and malicious programs. CIRC also

Exhibit 5.8
CERT Coordination Center home page

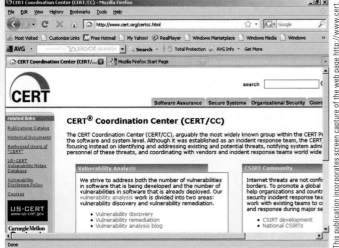

organization's intranet. Tests and certificates should be given to participants at the end of training sessions. In addition, making sure management supports security training is important to help promote security awareness throughout the organization.

Organizations should understand the principles of the Sarbanes-Oxley Act of 2002 (described in the information box below) and conduct a basic risk analysis before establishing a security program. This analysis often makes use of financial and budgeting techniques, such as return on investment (ROI), to determine which resources are most important and should have the strongest protection. This information can also help organizations weigh the cost of a security system.

provides awareness training, analysis of threats and vulnerabilities, and other services.

5-5 GUIDELINES FOR A COMPREHENSIVE SECURITY SYSTEM

An organization's employees are an essential part of the success of any security system, so training employees about security awareness and security measures is important. Some organizations use a classroom setting for training, and others conduct it over the

SARBANES-OXLEY AND INFORMATION SECURITY

Section 404 of the Sarbanes-Oxley Act of 2002 requires IT professionals to document and test the effectiveness of security measures protecting information technology and systems, including general computer controls, application controls, and system software controls. IT professionals must be familiar with this act and incorporate it into their organizations' security policies. In addition, companies must set up a security system that protects vital records and data and prevents them from being destroyed, lost, corrupted, or altered by unauthorized users. The purpose of this act is to maintain data integrity and availability of business operations, which are particularly critical in financial organizations.[13]

> An organization's employees are an essential part of the success of any security system, so training on security awareness and security measures is important.

The following steps should be considered when developing a comprehensive security plan[14]:

1. Set up a security committee with representatives from all departments as well as upper management. The committee's responsibilities include the following:
   - Developing clear, detailed security policy and procedures
   - Providing security training and security awareness for key decision makers and computer users
   - Periodically assessing the security policy's effectiveness
   - Developing an audit procedure for system access and use
   - Overseeing enforcement of the security policy
   - Designing an audit trail procedure for incoming and outgoing data

2. Post the security policy in a visible place, or post copies next to all workstations.

3. Raise employees' awareness of security problems.

4. Revoke terminated employees' passwords and ID badges immediately to prevent attempts at retaliation.

5. Keep sensitive data, software, and printouts locked in secure locations.

6. Exit programs and systems promptly, and never leave logged-on workstations unattended.

7. Limit computer access to authorized personnel only.

8. Compare communication logs with communication billing periodically. The log should list all outgoing calls with users' name, call destination, and time of call. Investigate any billing discrepancies.

9. Install antivirus programs, and make sure they are updated automatically.

10. Install only licensed software purchased from reputable vendors.

11. Make sure fire protection systems and alarms are up to date, and test them regularly.

12. Check environmental factors, such as temperature and humidity levels.

13. Use physical security measures, such as corner bolts on workstations, ID badges, and door locks.

14. Install firewalls and intrusion detection systems. If necessary, consider biometric security measures.

These steps should be used as a guideline. Not every organization needs to follow every step; however, some might need to include even more to fit their needs.

5-5a Business Continuity Planning

To lessen the effects of a natural disaster or a network attack or intrusion, planning the recovery is important. This should include **business continuity planning**, which outlines procedures for keeping an organization operational. A disaster recovery plan lists the tasks that must be performed to restore damaged data and equipment as well as steps to prepare for disaster, such as the following:

- Back up all files.
- Periodically review security and fire standards for computer facilities.
- Periodically review information from CERT and other security agencies.
- Make sure staff members have been trained and are aware of the consequences of possible disasters and steps to reduce the effects of disasters.
- Test the disaster recovery plan with trial data.
- Identify vendors of all software and hardware used in the organization, and make sure their mailing addresses, phone numbers, and Web site addresses are up to date.
- Document all changes made to hardware and software.
- Get a comprehensive insurance policy for computers and network facilities. Periodically review the policy to make sure coverage is adequate and up to date.
- Set up alternative sites to use in case of a disaster. Cold sites have the right environment for computer equipment (such as air conditioning and humidity controls), but no equipment is stored in them. Hot sites, on the other hand, have all the needed equipment and are ready to go.
- Investigate using a colocation facility, which is rented from a third party and usually contains telecommunication equipment.
- Check sprinkler systems, fire extinguishers, and halon gas systems.
- Keep backups in off-site storage, periodically test data recovery procedures, and keep a detailed record of machine-specific information, such as model and serial number. Backup facilities can be shared to reduce costs.
- Keep a copy of the disaster recovery plan off site.
- Go through a mock disaster to assess response time and recovery procedures.

business continuity planning outlines procedures for keeping an organization operational in the event of a natural disaster or network attack.

If disaster strikes, organizations should follow these steps to resume normal operations as soon as possible:

1. Put together a management crisis team to oversee the recovery plan.
2. Contact the insurance company.
3. Restore phone lines and other communication systems.
4. Notify all affected people, including customers, suppliers, and employees.
5. Set up a help desk to assist affected people.
6. Notify the affected people that recovery is underway.
7. Document all actions taken to regain normality so you know what worked and what did not work; revise the disaster recovery plan, if needed.

The Industry Connection highlights the McAfee Corporation, which offers several security products and services.

Industry Connection: McAfee Corporation[15]

McAfee (now a part of Intel Corporation) is a leading vendor of antivirus software and uses the Internet as a distribution medium for its products and services, although products can also be purchased through other outlets, such as retailers. In addition to antivirus software, McAfee offers network management software that includes virus scanning, firewalls, authentication, and encryption capabilities. McAfee also has an online bug-tracking system. The following list describes some popular McAfee products:

Internet Security: Includes antivirus, antispyware, antispam, antiphishing, identity protection, parental controls, data backup, and other features.

VirusScan Plus: Offers antivirus, antispyware, and firewall features, and Web site safety ratings.

Total Protection: Includes features for antivirus, antispyware, antispam, antiphishing, two-way firewall, advanced Web site safety ratings, identity protection, parental controls, and data backup.

McAfee also offers several products and services for free. Selected free services include:

FreeScan: Searches for the most recent viruses and displays a detailed list of any infected files.

World Virus Map: Shows where recent viruses are infecting computers worldwide.

Virus Removal Tools: Used to remove viruses and repair damage.

Security Advice Center: Offers tips and advice on keeping your computer and network safe and preventing attacker intrusions.

Free PC and Internet Security Newsletter: Includes virus alerts, special offers, and breaking news.

Internet Connection Speedometer: Tests your Internet connection to see how fast or slow it is.

STUDY TOOLS 5

LOCATED AT BACK OF THE TEXTBOOK
- ☐ Rip out Chapter Review Card

LOCATED AT WWW.CENGAGE.COM/LOGIN
- ☐ Review Key Terms Flash Cards (Print or Online)
- ☐ Download audio and visual summaries to review

- ☐ Complete both Practice Quizzes to prepare for tests
- ☐ Play "Beat the Clock" and "Quizbowl" to master concepts
- ☐ Complete "Crossword Puzzle" to review key terms
- ☐ Watch video on "Method" for a real company example

KEY TERMS

access controls 99
adware 86
asymmetric encryption 101
availability 89
backdoor 92
biometric security measures 94
blended threat 92
business continuity planning 104
callback modem 95
computer fraud 87
confidentiality 89
data encryption 100
denial-of-service (DoS) attack 92
fault-tolerant systems 90
firewall 95
integrity 89
intrusion detection system (IDS) 97
keystroke loggers 86

logic bomb 92
password 99
pharming 86
phishing 86
physical security measures 97
PKI (public key infrastructure) 101
Secure Sockets Layer (SSL) 101
sniffing 86
social engineering 93
spoofing 86
spyware 86
symmetric encryption 101
Transport Layer Security (TLS) 101
Trojan program 92
virtual private network (VPN) 100
virus 90
worm 91

REVIEWS AND DISCUSSIONS

1. What are three examples of technology tools that could be used for committing crimes?

2. What is the CIA triangle?

3. What are three types of hackers?

4. What are three examples of security threats?

5. What are three advantages of biometric security measures compared to non-biometric measures?

6. What are three types of firewalls? Generally speaking, which type offers higher security?

7. What are two types of encryption?

8. What are the differences between cold and hot sites used for disaster recovery?

PROJECTS

1. The computer lab of a local college needs a one-page document that it will distribute to its incoming students to increase their security awareness. After reading the information presented in this chapter and other sources, prepare the document, which should include a 10-item bullet list of the things that students must remember to reduce the risks of using information technology.

2. After reading the information presented in this chapter and other sources, write a one-page paper about three high-profile companies that faced security breaches in 2013. Identify a couple of the vulnerabilities that enabled hackers to break into these companies' systems. [Hint: Target Corporation is an example.]

3. Denial of Service (DoS) is among the security threats that have been on the rise in recent years. After reading the information presented in this chapter and other sources, write a one-page paper that outlines four recommendations for dealing with this security threat.

4. After reading the information presented in this chapter and other sources, write a one-page paper that lists three anti-virus software applications; include one advantage and one disadvantage of each. Which one do you recommend for the student lab mentioned in Project 1? What are the bases for your recommendation?

5. After reading the information presented in this chapter and other sources, write a one-page paper that identifies three companies (in addition to those mentioned in this book) that are using biometric security measures for authentication. Why has each company chosen this alternative over other security measures?

6. After reading the information presented in this chapter and other sources, write a two-page paper that offers five recommendations for improving the security of online transations. List two companies that provide preventive measures for e-commerce sites. What are included in their offerings? What are the costs?

ARE YOU READY TO MOVE ON?

1. Mirror disks are usually more expensive than RAID systems. True or False?

2. An intrusion detection system (IDS) can protect against both external and internal access. True or False?

3. Cable shielding is an example of a biometric measure. True or False?

4. Which of the following is not part of the CIA triangle?
   a. Confidentiality
   b. Integrity
   c. Correctness
   d. Availability

5. Which of the following technology tools or techniques is not usually used for committing crimes?
   a. Spyware and adware
   b. Proxy servers
   c. Keystroke loggers
   d. Phishing and pharming

6. Which of the following strategies is not considered a physical security measure?
   a. Corner bolts
   b. Firewalls
   c. Electronic trackers
   d. Cable shielding

CASE STUDY 5-1

Computer Viruses Target Medical Devices

Medical devices that are controlled by computer software—from heart monitors and pacemakers to mammogram and X-ray machines—are new targets for computer viruses and malware.[16] This could put patients at risk, although no injuries or deaths have been reported so far. The Food and Drug Administration is warning the manufacturers of medical devices about the problem and is requesting them to review the parts of their security plans that are related to these devices when they seek approval from the government agency.

Ariel Skelley Blend Images/Newscom

A Department of Veterans Affairs report has shown that 327 devices at VA hospitals have been infected by malware since 2009. In January 2010, a VA catheterization laboratory was temporarily closed due to infected computer equipment that is used to open blocked arteries. And in a case at a private Boston hospital, computer viruses exposed sensitive patient data by sending it to outside servers. The increased applications of electronic record systems as a part of the 2009 stimulus package is adding to this risk.

Manufacturers must improve the security features of these devices, making them more difficult for hackers to break into. And there needs to be close coordination between the manufacturers and healthcare providers to further enhance security. Also, hospitals and medical facilities must make sure that all the software running these devices is up to date and any updates have been installed. Finally, these devices must be blocked from Internet access.[17]

Answer the following questions:

1. What are three examples of devices that could be attacked by computer viruses?
2. What are the risks related to using electronic health records in hospitals and medical facilities?
3. What are three pieces of advice for reducing the risk associated with using these devices?

CASE STUDY 5-2

Security Breach at Sony's Playstation Network

Some call it the largest breach of confidential user information in history. The attack occurred on April 20, 2011. It resulted in the loss of more than 75 million customers' crucial information from Sony's PlayStation Network (PSN) and Qriocity music and video service. The stolen information included name, address (city, state, and zip), country, e-mail address, birthday, PSN password, login name, and possibly credit card information. It is believed that hackers exploited a weakness in the PlayStation 3's encryption system and accessed the public key required to run any software on the machine.[18] This would cost Sony over $170 million, not including the loss of reputation and trust that the company had enjoyed for many years. Sony has been unable to determine who attacked its networks. The company did not even know that so much of

© iStockphoto.com/LPETTET

its customer information had been stolen until an external security consultant discovered the theft a week after the incident took place. According to Sony, credit card information was encrypted. To conduct a thorough investigation and improve and rebuild the security of the network services for the future, Sony disconnected PlayStation Network and Qriocity services for several days. This situation underscores that, given the growth of cybersecurity threats, console makers must beef up their security systems. Furthermore, console users must be aware of these threats when they are online.[19,20]

Answer the following questions:

1. Which networks at Sony were attacked?
2. Which information was stolen?
3. How do companies such as Sony, Microsoft, and Apple protect their customers' critical information?

PART 2

6 | Data Communication: Delivering Information Anywhere and Anytime

LEARNING OUTCOMES

After studying this chapter, you should be able to:

6-1 Describe major applications of a data communication system.

6-2 Explain the major components of a data communication system.

6-3 Describe the major types of processing configurations.

6-4 Explain the three types of networks.

6-5 Describe the main network topologies.

6-6 Explain important networking concepts, such as bandwidth, routing, routers, and the client/server model.

6-7 Describe wireless and mobile technologies and networks.

6-8 Discuss the importance of wireless security and the techniques used.

6-9 Summarize the convergence phenomenon and its applications for business and personal use.

After you finish this chapter, go to **PAGE 132** for the **STUDY TOOLS**

© Goodluz/Shutterstock.com

This chapter explains the role of data communication systems in delivering information for decision making, although you can see applications of data communication systems everywhere, from within your own home to multinational corporations. The chapter starts with the basics of data communication systems, including components, processing configurations, and types of networks and topologies. The chapter also covers important concepts in data communication, such as bandwidth, routing, routers, and the client/server model. Next, the chapter gives an overview of wireless and mobile networks and the technologies they use. Finally, the chapter takes a look at a growing phenomenon—the convergence of voice, video, and data—and its importance in the business world.

6-1 DEFINING DATA COMMUNICATION

Data communication is the electronic transfer of data from one location to another. An information system's effectiveness is measured in part by how efficiently it delivers information, and a data communication system is what enables an information system to carry out this function. In addition, because most organizations collect and transfer data across large geographic distances, an efficient data communication system is critical. A data communication system can also improve the flexibility of data collection and transmission. For example, many workers use portable devices, such as laptops, smartphones, and other handheld devices, to communicate with the office at any time and from any location.

Data communication is also the basis of virtual organizations, discussed in Chapter 4. By using the capabilities of a data communication system, organizations are not limited by physical boundaries. They can collaborate with other organizations, outsource certain functions to reduce costs, and provide customer services via data communication systems.

E-collaboration is another main application of data communication. Decision makers can be located

> **Data communication** is the electronic transfer of data from one location to another.

throughout the world but can still collaborate with their colleagues no matter where they are.

6-1a Why Managers Need to Know About Data Communication

Data communication has become so woven into the fabric of corporate activity that separating an organization's core functions from the data communication systems that enable and support them is difficult. When a new product is introduced, for example, the executives who are the key decision makers might be scattered throughout the world in a multinational corporation. However, they can use data communication systems to collaborate and coordinate their efforts to introduce the new product in a timely manner.

Data communication applications can enhance decision makers' efficiency and effectiveness in many ways. For example, data communication applications support just-in-time delivery of goods, which reduces inventory costs and improves the competitive edge. As you learned in previous chapters, many large corporations, such as Walmart, The Home Depot, and UPS, use data communication technologies to stay ahead of their competitors. As mentioned, data communication systems also make virtual organizations possible, and these can cross geographic boundaries to develop products more quickly and effectively.

Data communication systems also enable organizations to use e-mail and electronic file transfer to improve efficiency and productivity. A communication network, a crucial part of an organization's information system, shortens product and service development life cycles and delivers information to those who need it faster and more efficiently. Here are some of the ways data

> **Data communication systems make virtual organizations possible, and these can cross geographic boundaries to develop products more quickly and effectively.**

communication technologies affect the workplace:

- Online training for employees can be provided via virtual classrooms. In addition, employees get the latest technology and product information immediately.
- Internet searches for information on products, services, and innovation keep employees up to date.
- The Internet and data communication systems facilitate lifelong learning, which will be an asset for knowledge workers of the future.
- Boundaries between work and personal life are less clear-cut as data communication is more available in both homes and businesses. The increase in telecommuters is an example of this trend.
- Web and video conferencing are easier, which can reduce the costs of business travel.

Managers need a clear understanding of the following areas of data communication:

- The basics of data communication and networking
- The Internet, intranets, and extranets
- Wired and wireless networks
- Network security issues and measures
- Organizational and social effects of data communication
- Globalization issues
- Applications of data communication systems

E-collaborations and virtual meetings are other important applications of data communication systems for managers. These applications are cost effective and improve customer service. One example of an e-collaboration tool is GoToMeeting, described in the information box below.

GoToMeeting: Example of an E-collaboration Tool

GoToMeeting is a Web-conferencing service hosted by Citrix Online, a division of Citrix Systems. Capable of running on both PCs and Macs, it offers high-definition video conferencing that enables you to see your fellow meeting attendees. It also allows you to show them the applications and files that are currently running on your desktop. The other attendees can share their desktops as well. Users can either chat with all the attendees or speak privately with an individual participant. They also have the option of choosing free VoIP, phone conferencing, or both. GoToMeeting allows users to save, replay, post, or e-mail important interactions and presentations, with audio. And it offers comprehensive security and privacy features. Meetings cannot be viewed by anyone except meeting attendees

WebEx (Cisco) and My Web Conferences (MyTrueCloud) are two alternatives to GoToMeeting.[1]

6-2 # BASIC COMPONENTS OF A DATA COMMUNICATION SYSTEM

A typical data communication system includes the following components:

- Sender and receiver devices
- Modems or routers
- Communication medium (channel)

Before examining these components, you need to review some basic concepts in data communication. **Bandwidth** is the amount of data that can be transferred from one point to another in a certain time period, usually one second. It is often expressed as bits per second (bps). Other measurements include kilobits per second (Kbps), megabits per second (Mbps), and gigabits per second (Gbps). **Attenuation** is the loss of power in a signal as it travels from the sending device to the receiving device.

Data transmission channels are generally divided into two types: broadband and narrowband. In **broadband** data transmission, multiple pieces of data are sent simultaneously to increase the transmission rate. As of May, 2013, 70 percent of the U.S. population that was 18 and older had broadband access. However, smartphones are widening the broadband gap between seniors and young adults.[2]

Narrowband is a voice-grade transmission channel capable of transmitting a maximum of 56,000 bps, so only a limited amount of information can be transferred in a specific period of time.

Before a communication link can be established between two devices, they must be synchronized, meaning that both devices must start and stop communicating at the same point. Synchronization is handled with **protocols**, rules that govern data communication, including error detection, message length, and transmission speed. Protocols also help ensure compatibility between different manufacturers' devices.

6-2a Sender and Receiver Devices

A sender and receiver device can take various forms:

- *Input/output device, or "thin client"*—Used only for sending or receiving information; it has no processing power.
- *Smart terminal*—An input/output device that can perform certain processing tasks but is not a full-featured computer. This type of device is often used on factory floors and assembly lines for collecting data and transmitting it to the main computer system.
- *Intelligent terminal, workstation, or personal computer*—These serve as input/output devices or as stand-alone systems. Using this type of device, a remote computer can perform certain processing tasks without the main computer's support. Generally, an intelligent terminal is considered a step up from a smart terminal.
- *Netbook computer*—A low-cost, diskless computer used to connect to the Internet or a LAN. Netbook computers run software off servers and save data to servers. According to Forrester Research, however, the iPad and other tablet devices will significantly reduce the demand for netbooks.[3]
- *Minicomputers, mainframes, and supercomputers*—These process data and send it to other devices or receive data that has been processed elsewhere, process it, then transmit it to other devices.
- *Smartphones, mobile phones, MP3 players, PDAs, and game consoles*—Smartphones, briefly described in Chapter 1, are mobile phones with advanced capabilities, such as e-mail and Web-browsing, and most have a built-in keyboard or an external USB keyboard. A video game console is an electronic device for playing video games. It receives instructions from a game player and produces a video display signal on a monitor such as a television screen or a computer monitor.

6-2b Modems

A **modem** (short for "modulator-demodulator") is a device that connects a user to the Internet. Not all Internet connections require a modem; for example, wireless users connect via access points, and satellite users use a satellite dish. However, dial-up, digital subscriber line (DSL), and cable access require modems to connect.

> **Bandwidth** is the amount of data that can be transferred from one point to another in a certain time period, usually one second.
>
> **Attenuation** is the loss of power in a signal as it travels from the sending device to the receiving device.
>
> In **broadband** data transmission, multiple pieces of data are sent simultaneously to increase the transmission rate.
>
> **Narrowband** is a voice-grade transmission channel capable of transmitting a maximum of 56,000 bps, so only a limited amount of information can be transferred in a specific period of time.
>
> **Protocols** are rules that govern data communication, including error detection, message length, and transmission speed.
>
> A **modem** (short for "modulator-demodulator") is a device that connects a user to the Internet.

When phone lines are used for Internet connections, an analog modem is necessary to convert a computer's digital signals to analog signals that can be transferred over analog phone lines. In today's broadband world, however, analog modems are rarely used. Instead, DSL or cable modems are common. **Digital subscriber line (DSL)**, a common carrier service, is a high-speed service that uses ordinary phone lines. With DSL connections, users can receive data at up to 7.1 Mbps and send data at around 1 Mbps, although the actual speed is determined by proximity to the provider's location. Also, different providers might offer different speeds. Cable modems, on the other hand, use the same cable that connects to TVs for Internet connections; they can usually reach transmission speeds of about 16 Mbps.

6-2c Communication Media

Communication media, or channels, connect sender and receiver devices. They can be conducted (wired or guided) or radiated (wireless), as shown in Exhibit 6.1.

Conducted media provide a physical path along which signals are transmitted, including twisted pair copper cable, coaxial cable, and fiber optics. Twisted pair copper cable consists of two copper lines twisted around each other and either shielded or unshielded; it is used in the telephone network and for communication within buildings. Coaxial cables are thick cables that can be used for both data and voice transmissions. They are used mainly for long-distance telephone transmissions and local area networks. Fiber-optic cables are glass tubes (half the diameter of a human hair) surrounded by concentric layers of glass, called "cladding," to form a light path through wire cables. At the core is the central piece of glass that carries the light. Surrounding the core is a second layer of glass that keeps the light from escaping the core. And

> **Digital subscriber line (DSL)**, a common carrier service, is a high-speed service that uses ordinary phone lines.
>
> **Communication media**, or channels, connect sender and receiver devices. They can be conducted or radiated.
>
> **Conducted media** provide a physical path along which signals are transmitted, including twisted pair copper cable, coaxial cable, and fiber optics.

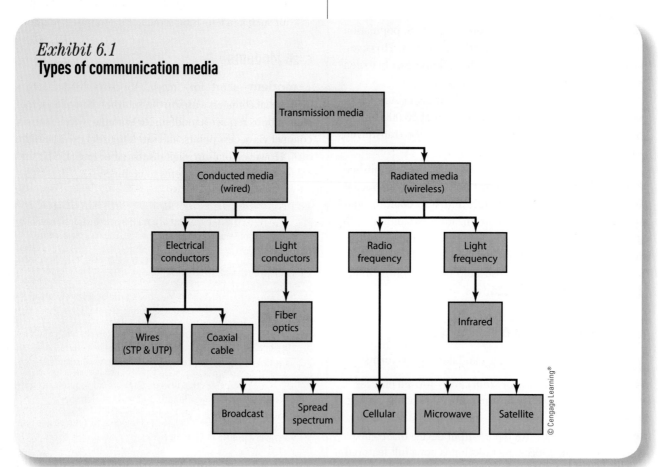

Exhibit 6.1
Types of communication media

© Cengage Learning®

6-3a Centralized Processing

In a **centralized processing** system, all processing is done at one central computer. In the early days of computer technology, this type of processing was justified because data-processing personnel were in short supply, hardware and software were expensive, and only large organizations could afford computers. The main advantage of this configuration is being able to exercise tight control over system operations and applications. The main disadvantage is lack of responsiveness to users' needs, because the system and its users could be located far apart from each other. This configuration is not used much now.

6-3b Decentralized Processing

In **decentralized processing**, each user, department, or division has its own computer (sometimes called an "organizational unit") for performing processing tasks. A decentralized processing system is certainly more responsive to users than a centralized processing system. Nevertheless, decentralized systems have some drawbacks, including lack of coordination among organizational units, the high cost of having many systems, and duplication of efforts.

6-3c Distributed Processing

Distributed processing solves two main problems—the lack of responsiveness in centralized processing and the lack of coordination in decentralized processing—by maintaining centralized control and decentralizing operations. Processing power is distributed among several locations. For example, in the retail industry, each store's network does its own processing but is under the centralized control of the store's headquarters. Databases and input/output devices can also be distributed.

around both of these lies the buffer, an outer layer of plastic, which provides protection and strength. Fiber-optic cables have a higher capacity, smaller size, lighter weight, lower attenuation, and higher security than other cable types; they also have the highest bandwidth of any communication medium.

Radiated media use an antenna for transmitting data through air or water. Some of these media are based on "line of sight" (an open path between sending and receiving devices or antennas), including broadcast radio, terrestrial microwave, and satellite. Satellites link ground-based microwave transmitters/receivers, known as Earth stations, and are commonly used in long-distance telephone transmissions and TV signals. Terrestrial microwave systems use Earth-based transmitters and receivers and are often used for point-to-point links between buildings.

A communication medium can be a point-to-point or a multipoint system. In a point-to-point system, only one device at a time uses the medium. In a multipoint system, several devices share the same medium, and a transmission from one device can be sent to all other devices sharing the link.

6-3 PROCESSING CONFIGURATIONS

Data communication systems can be used in several different configurations, depending on users' needs, types of applications, and responsiveness of the system. During the past 60 years, three types of processing configurations have emerged: centralized, decentralized, and distributed.

Radiated media use an antenna for transmitting data through air or water.

In a **centralized processing** system, all processing is done at one central computer.

In **decentralized processing**, each user, department, or division has its own computer (sometimes called an "organizational unit") for performing processing tasks.

Distributed processing maintains centralized control and decentralized operations. Processing power is distributed among several locations.

The advantages of distributed processing include:

- Accessing unused processing power is possible.
- Modular design means computer power can be added or removed, based on need.
- Distance and location are not limiting.
- It is more compatible with organizational growth because workstations can be added easily.
- Fault tolerance is improved because of the availability of redundant resources.
- Resources, such as high-quality laser printers, can be shared to reduce costs.
- Reliability is improved because system failures can be limited to only one site.
- The system is more responsive to user needs.

The **Open Systems Interconnection (OSI) model** is a seven-layer architecture for defining how data is transmitted from computer to computer in a network, from the physical connection to the network to the applications that users run. It also standardizes interactions between network computers exchanging information.

The disadvantages of distributed processing include:

- There may be more security and privacy challenges.
- There may be incompatibility between the various pieces of equipment.
- Managing the network can be challenging.

6-3d Open Systems Interconnection Model

The **Open Systems Interconnection (OSI) model** is a seven-layer architecture for defining how data is transmitted from computer to computer in a network. OSI also standardizes interactions between network computers exchanging information. Each layer in the architecture performs a specific task (see Exhibit 6.2):

- *Application layer*—Serves as the window through which applications access network services. It performs different tasks, depending on the application, and provides services that support users' tasks, such as file transfers, database access, and e-mail.
- *Presentation layer*—Formats message packets.

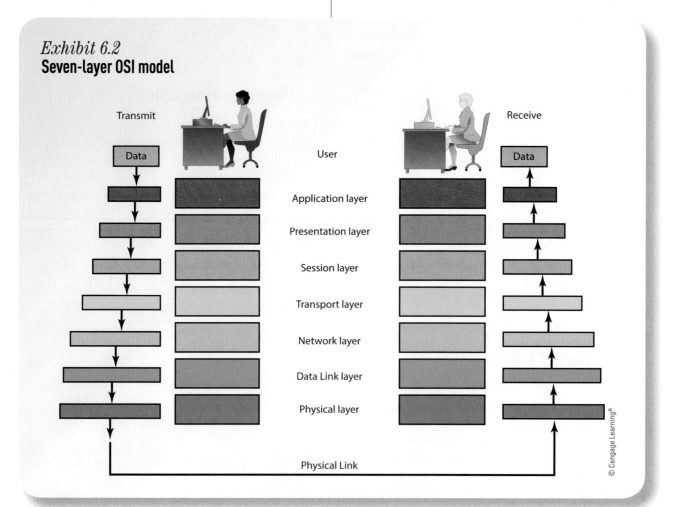

Exhibit 6.2
Seven-layer OSI model

Transmit

Data

Receive

Data

User

Application layer

Presentation layer

Session layer

Transport layer

Network layer

Data Link layer

Physical layer

Physical Link

© Cengage Learning®

- *Session layer*—Establishes a communication session between computers.
- *Transport layer*—Generates the receiver's address and ensures the integrity of messages by making sure packets are delivered without error, in sequence, and with no loss or duplication. This layer provides methods for controlling data flow, ordering received data, and acknowledging received data.
- *Network layer*—Routes messages.
- *Data Link layer*—Oversees the establishment and control of the communication link.
- *Physical layer*—Specifies the electrical connections between computers and the transmission medium; defines the physical medium used for communication. This layer is primarily concerned with transmitting binary data, or bits, over a communication network.

6-4 TYPES OF NETWORKS

There are three major types of networks: local area networks, wide area networks, and metropolitan area networks. In all these networks, computers are usually connected to the network via a **network interface card (NIC)**, a hardware component that enables computers to communicate over a network. A NIC, also called an "adapter card," is the physical link between a network and a workstation, so it operates at the OSI model's Physical and Data Link layers. NICs are available from many vendors; and the most common types of local area networks, Ethernet and token ring, can use NICs from almost any vendor. In addition, to operate a server in a network, a network operating system (NOS) must be installed, such as Windows Server 2012 or Novell Enterprise Server.

6-4a Local Area Networks

A **local area network (LAN)** connects workstations and peripheral devices that are in close proximity (see Exhibit 6.3). Usually, a LAN covers a limited geographical area, such as a building or campus, and one company owns it. Its data transfer speed varies from 100 Mbps to 10 Gbps.

LANs are used most often to share resources, such as peripherals, files, and software. They are also used to integrate services, such as e-mail and file sharing. In a LAN environment, there are two basic terms to remember: *Ethernet* and *Ethernet cable*. Ethernet

A **network interface card (NIC)** is a hardware component that enables computers to communicate over a network.

A **local area network (LAN)** connects workstations and peripheral devices that are in close proximity.

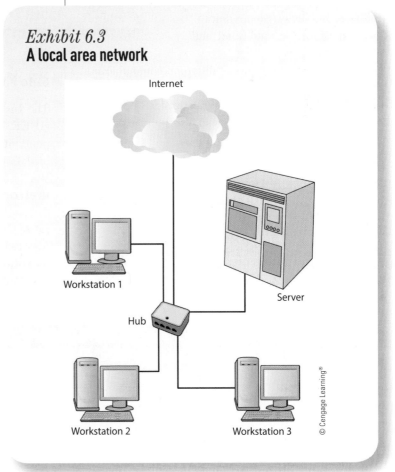

Exhibit 6.3
A local area network

Internet

Workstation 1

Server

Hub

Workstation 2

Workstation 3

© Cengage Learning®

is a standard communication protocol embedded in software and hardware devices used for building a LAN. An Ethernet cable is used to connect computers, hubs, switches, and routers to a network. Assessing information needs and careful planning are important in setting up a LAN.

6-4b Wide Area Networks

A **wide area network (WAN)** can span several cities, states, or even countries, and it is usually owned by several different parties (see Exhibit 6.4). The data transfer speed depends on the speed of its interconnections (called "links") and can vary from 28.8 Kbps to 155 Mbps. For example, a WAN can be useful for a company headquartered in Washington, D.C., with 30 branch offices in 30 states. The WAN makes it possible for all branch offices to communicate with headquarters and send and receive information.

A WAN can use many different communication media (coaxial cables, satellite, and fiber optics) and terminals of different sizes and sophistication (PCs, workstations, and mainframes); it can also be connected to other networks.

Exhibit 6.4
A wide area network

© Cengage Learning®

A **wide area network (WAN)** can span several cities, states, or even countries, and it is usually owned by several different parties.

A **metropolitan area network (MAN)** is designed to handle data communication for multiple organizations in a city and sometimes nearby cities as well.

6-4c Metropolitan Area Networks

The Institute of Electrical and Electronics Engineers (IEEE) developed specifications for a public, independent, high-speed network that connects a variety of data communication systems, including LANs and WANs, in metropolitan areas. This network, called a **metropolitan area network (MAN)**, is designed to handle data communication for multiple organizations in a city and sometimes nearby cities as well (see Exhibit 6.5). The data transfer speed varies from 34 Mbps to 155 Mbps.

Table 6.1 compares these three types of networks.

TABLE 6.1 COMPARISON OF LAN, WAN, AND MAN NETWORKS

Network type	Ownership	Data transfer speed	Scope
LAN	Usually one party	100 Mbps to 10 Gbps	A building or a campus
WAN	More than one party	28.8 Kbps to 155 Mbps	Intercity to international
MAN	One to several parties	34 Mbps to 155 Mbps	One city to several contiguous cities

© Cengage Learning®

Exhibit 6.5
A metropolitan area network

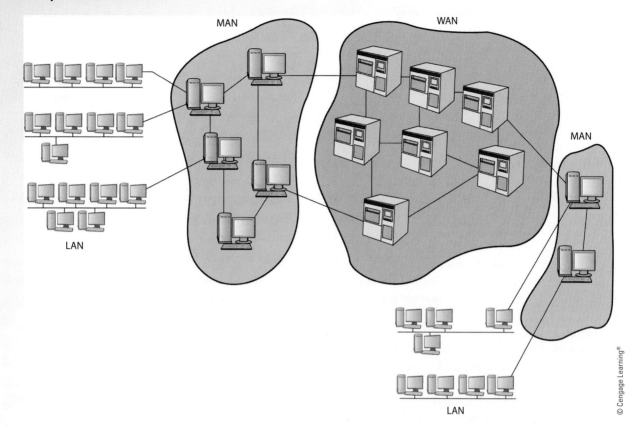

© Cengage Learning®

6-5 NETWORK TOPOLOGIES

A **network topology** represents a network's physical layout, including the arrangement of computers and cables. Five common topologies are discussed in the following sections: star, ring, bus, hierarchical, and mesh.

6-5a Star Topology

The **star topology** usually consists of a central computer (host computer, often a server) and a series of nodes (typically, workstations or peripheral devices). The host computer supplies the main processing power. If a node fails, it does not affect the network's operation, but if the host computer fails, the entire network goes down.

Advantages of the star topology include:

- Cable layouts are easy to modify.
- Centralized control makes detecting problems easier.
- Nodes can be added to the network easily.
- It is more effective at handling heavy but short bursts of traffic.

Disadvantages of the star topology include:

- If the central host fails, the entire network becomes inoperable.
- Many cables are required, which increases cost.

A **network topology** represents a network's physical layout, including the arrangement of computers and cables.

The **star topology** usually consists of a central computer (host computer, often a server) and a series of nodes (typically, workstations or peripheral devices).

In a ring topology, if any link between nodes is severed, the entire network is affected, and failure of a single node disrupts the entire network.

6-5b Ring Topology

In a **ring topology**, no host computer is required because each computer manages its own connectivity. Computers and devices are arranged in a circle so each node is connected to two other nodes: its upstream neighbor and its downstream neighbor. Transmission is in one direction, and nodes repeat a signal before passing it to the downstream neighbor. If any link between nodes is severed, the entire network is affected, and failure of a single node disrupts the entire network. A token ring is a common implementation of the ring topology. It is a LAN protocol specified in the IEEE 802.5 in which all stations are connected in a ring and each station can directly receive transmissions only from its immediate neighbor. Permission to transmit is granted by a message (token) that circulates around the ring. Modern ring topologies, such as Fiber Distributed Data Interface (FDDI), are capable of bidirectional transmission (clockwise and counterclockwise), which prevents the problems caused by a single node failure.

A ring topology needs less cable than a star topology, but it is similar to a star topology in that it is better for handling heavy but short bursts of traffic. Also, diagnosing problems and modifying the network are more difficult than with a star topology.

6-5c Bus Topology

The **bus topology** (also called "linear bus") connects nodes along a network segment, but the ends of the cable are not connected, as they are in a ring topology. A hardware device called a terminator is used at each end of the cable to absorb the signal. Without a terminator, the signal would bounce back and forth along the length of the cable and prevent network communication.

A node failure has no effect on any other node. Advantages of the bus topology include:

- It is easy to extend.
- It is very reliable.
- The wiring layout is simple and uses the least amount of cable of any topology, which keeps costs down.
- It handles steady (even) traffic well.

Disadvantages of the bus topology include:

- Fault diagnosis is difficult.
- The bus cable can be a bottleneck when network traffic is heavy.

6-5d Hierarchical Topology

A **hierarchical topology** (also called a "tree") combines computers with different processing strengths in different organizational levels. For example, the bottom level might consist of workstations, with minicomputers in the middle and a server at the top. Companies that are geographically dispersed and organized hierarchically are good candidates for this type of network. Failure of nodes at the bottom might not have a big impact on network performance, but the middle nodes and especially the top node (which controls the entire network) are crucial for network operation.

Traditional mainframe networks also use a hierarchical topology. The mainframe computer is at the top, front-end processors (FEPs) are at the next

---

In a **ring topology**, no host computer is required because each computer manages its own connectivity.

The **bus topology** (also called "linear bus") connects nodes along a network segment, but the ends of the cable are not connected, as they are in a ring topology.

A **hierarchical topology** (also called a "tree") combines computers with different processing strengths in different organizational levels.

level, controllers and multiplexers are at the next level, and terminals and workstations are at the bottom level. A **controller** is a hardware and software device that controls data transfer from a computer to a peripheral device (examples are a monitor, a printer, or a keyboard) and vice versa. A **multiplexer** is a hardware device that allows several nodes to share one communication channel. The intermediate-level devices (FEPs and controllers) reduce the host's processing load by collecting data from terminals and workstations.

The hierarchical topology offers a great deal of network control and lower cost, compared to a star topology. Its disadvantages include that network expansion may pose a problem, and there could be traffic congestion at the root and higher-level nodes.

6-5e Mesh Topology

In a **mesh topology** (also called "plex" or "interconnected"), every node (which can differ in size and configuration from the others) is connected to every other node. This topology is highly reliable. Failure of one or a few nodes does not usually cause a major problem in network operation, because many other nodes are available. However, this topology is costly and difficult to maintain and expand.

6-6 MAJOR NETWORKING CONCEPTS

The following sections explain important networking concepts, including protocols, TCP/IP, routing, routers, and the client/server model.

6-6a Protocols

As mentioned earlier, protocols are agreed-on methods and rules that electronic devices use to exchange information. People need a common language to communicate, and the same is true of computer and other electronic devices. Some protocols deal with hardware connections, and others control data transmission and file transfers. Protocols also specify the format of message packets sent between computers. In today's networks, multiple protocol support is becoming more important, as networks need to support protocols of computers running different operating systems, such as Mac OS, Linux/UNIX, and Windows. The following section describes the most widely used network protocol, TCP/IP.

6-6b Transmission Control Protocol/ Internet Protocol

Transmission Control Protocol/Internet Protocol (TCP/IP) is an industry-standard suite of communication protocols. TCP/IP's main advantage is that it enables interoperability—in other words, it allows the linking of devices running on many different platforms. TCP/IP was originally intended for Internet communication, but because it addressed issues such as portability, it also became the standard protocol for UNIX network communication.

Two of the major protocols in the TCP/IP suite are Transmission Control Protocol (TCP), which operates at the OSI model's Transport layer, and Internet Protocol (IP), which operates at the OSI model's Network layer. TCP's primary functions are establishing a link between hosts, ensuring message integrity, sequencing and acknowledging packet delivery, and regulating data flow between source and destination nodes.

IP is responsible for packet forwarding. To perform this task, it must be aware of the available data link protocols and the optimum size of each packet. After it recognizes the size of each packet, it must be able to divide data into packets of the correct size. An IP address consists of 4 bytes in IPv4 or 16 bytes in IPv6 (32 bits or 128 bits) and is divided into two parts: a network address and a node address. Computers on the same network must use the same network address, but each computer must have a unique node address. IP networks combine network and node addresses into one IP address; for example, 131.255.0.0 is a valid IP address.

A **controller** is a hardware and software device that controls data transfer from a computer to a peripheral device (examples are a monitor, a printer, or a keyboard) and vice versa.

A **multiplexer** is a hardware device that allows several nodes to share one communication channel.

In a **mesh topology** (also called "plex" or "interconnected"), every node (which can differ in size and configuration from the others) is connected to every other node.

Transmission Control Protocol/Internet Protocol (TCP/IP) is an industry-standard suite of communication protocols that enables interoperability.

6-6c Routing

To understand routing better, you will first examine packet switching, a network communication method that divides data into small packets and transmits them to an address, where they are reassembled. A **packet** is a collection of binary digits—including message data and control characters for formatting and transmitting—sent from computer to computer over a network. Packets are transmitted along the best route available between sender and receiver (see Exhibit 6.6). Any packet-switching network can handle multimedia data, such as text, graphics, audio, and video.

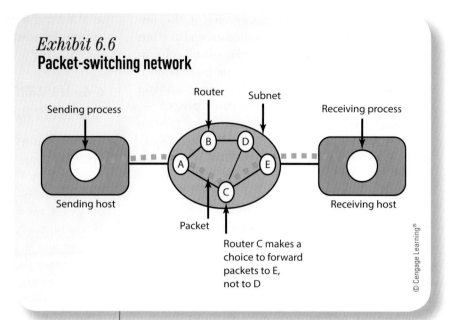

Exhibit 6.6
Packet-switching network

The path or route that data takes on a network is determined by the type of network and the software used to transmit data. The process of deciding which path that data takes is called **routing**. Routing is similar to the path you take from home to work. Although you probably take the same path most of the time, sometimes you have to change your path, depending on road and weather conditions, traffic, and time of day. Similarly, a packet's route can change each time a connection is made, based on the amount of traffic and the availability of the circuit. The decision about which route to follow is done in one of two ways: at a central location (centralized routing) or at each node along the route (distributed routing). In most cases, a **routing table**, generated automatically by software, is used to determine the best possible route for the packet. The routing table lists nodes on a network and the path to each node, along with alternate routes and the speed of existing routes.

In **centralized routing**, one node is in charge of selecting the path for all packets. This node, considered the network routing manager, stores the routing table, and any changes to a route must be made at this node. All network nodes periodically forward status information on the number of inbound, outbound, and processed messages to the network routing manager. The network routing manager, therefore, has an overview of the network and can determine whether any part of it is underused or overused. As with all centralized configurations, there are disadvantages to having control at one node. For example, if the network routing manager is at a point far from the network's center, many links and paths that make up the network are far from the central node. Status information sent by other nodes to initiate changes to the routing table have to travel a long distance to the central node, causing a delay in routing some data and reducing network performance. In addition, if the controlling node fails, no routing information is available.

Distributed routing relies on each node to calculate the best possible route. Each node contains its own routing table with current information on the status of adjacent nodes so packets can follow the

---

A **packet** is a collection of binary digits—including message data and control characters for formatting and transmitting—sent from computer to computer over a network.

Routing is the process of deciding which path to take on a network. This is determined by the type of network and the software used to transmit data.

A **routing table**, generated automatically by software, is used to determine the best possible route for a packet.

In **centralized routing**, one node is in charge of selecting the path for all packets. This node, considered the network routing manager, stores the routing table, and any changes to a route must be made at this node.

Distributed routing relies on each node to calculate its own best possible route. Each node contains its own routing table with current information on the status of adjacent nodes so packets can follow the best possible route.

best possible route. Each node also sends status messages periodically so adjacent nodes can update their tables. Distributed routing eliminates the problems caused by having the routing table at a centralized site. If one node is not operational, routing tables at other nodes are updated, and the packet is sent along a different path.

6-6d Routers

A **router** is a network connection device containing software that connects network systems and controls traffic flow between them. The networks being connected can be operating on different protocols, but they must use a common routing protocol. Routers operate at the Network layer of the OSI model and handle routing packets on a network. Cisco Systems and Juniper Networks are two major router vendors.

A router performs the same functions as a bridge but is a more sophisticated device. A bridge connects two LANs using the same protocol, and the communication medium does not have to be the same on both LANs.

Routers can also choose the best possible path for packets based on distance or cost. They can prevent network jams that delay packet delivery, and they can handle packets of different sizes. A router can also be used to isolate a portion of the LAN from the rest of the network; this process is called "segmenting." For example, you might want to keep information about new product development or payroll information isolated from the rest of the network, for confidentiality reasons.

There are two types of routers: static and dynamic. A **static router** requires the network routing manager to give it information about which addresses are on which network. A **dynamic router** can build tables that identify addresses on each network. Dynamic routers are used more often now, particularly on the Internet.

6-6e Client/Server Model

In the **client/server model**, software runs on the local computer (the client) and communicates with the remote server to request information or services. A server is a remote computer on the network that provides information or services in response to client requests. For example, on your client computer, you make this request: "Display the names of all marketing majors with a GPA greater than 3.8." The database server receives your request, processes it, and returns

the following names: Alan Bidgoli, Moury Jones, and Jasmine Thomas.

In the most basic client/server configuration, the following events usually take place:

1. The user runs client software to create a query.
2. The client accepts the request and formats it so the server can understand it.
3. The client sends the request to the server over the network.
4. The server receives and processes the query.
5. The results are sent to the client.
6. The results are formatted and displayed to the user in an understandable format.

The main advantage of the client/server architecture is its scalability, meaning its ability to grow. Client/server architectures can be scaled horizontally or vertically. Horizontal scaling means adding more workstations (clients), and vertical scaling means migrating the network to larger, faster servers.

To understand client/server architecture better, you can think of it in terms of these three levels of logic:

- Presentation logic
- Application logic
- Data management logic

Presentation logic, the top level, is concerned with how data is returned to the client. The Windows graphical user interface (GUI) is an example of presentation software. An interface's main function is to translate tasks and convert them to something users can understand. Application logic is concerned with the software processing requests for users. Data management logic is concerned with data management and storage operations. The real challenge in a client/server architecture

A **router** is a network connection device containing software that connects network systems and controls traffic flow between them.

A **static router** requires the network routing manager to give it information about which addresses are on which network.

A **dynamic router** can build tables that identify addresses on each network.

In the **client/server model**, software runs on the local computer (the client) and communicates with the remote server to request information or services. A server is a remote computer on the network that provides information or services in response to client requests.

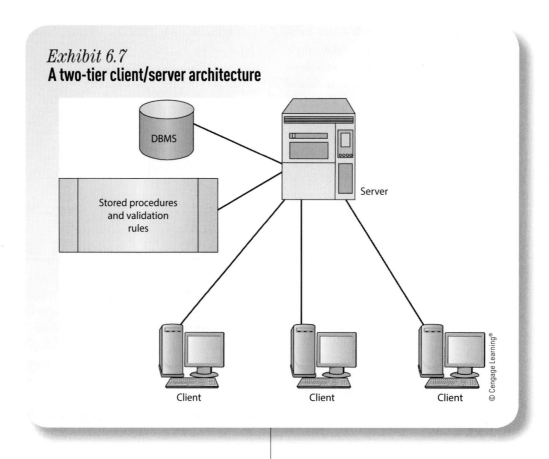

Exhibit 6.7
A two-tier client/server architecture

DBMS

Stored procedures and validation rules

Server

Client

Client

Client

© Cengage Learning®

is how to divide these three logics between the client and server. The following sections describe some typical architectures used for this purpose.

Two-Tier Architecture

In the **two-tier architecture**, a client (tier one) communicates directly with the server (tier two), as shown in Exhibit 6.7. The presentation logic is always on the client, and the data management logic is on the server. The application logic can be on the client, on the server, or split between them, although it is usually on the client side.

This architecture is effective in small workgroups (that is, groups of 50 clients or less). Because application logic is usually on the client side, a two-tier architecture has the advantages of application development speed, simplicity, and power. On the downside, any changes

in application logic, such as stored procedures and validation rules for databases, require major modifications of clients, resulting in upgrade and modification costs. However, this depends on the application.

N-Tier Architectures

In a two-tier architecture, if the application logic is modified, it can affect the processing workload. For example, if application software is placed on the client, changing the data management software requires modifying the software on all clients. An **n-tier architecture** attempts to balance the workload between client and server by removing application processing from both the client and server and placing it on a middle-tier server, as shown in Exhibit 6.8. The most common n-tier architecture is the three-tier architecture. This arrangement leaves the presentation logic on the client and the data management logic on the server (see Exhibit 6.9).

Improving network performance is a major advantage of this architecture. However, network management is more challenging because there is more network traffic, and testing software is more difficult in an n-tier architecture because more devices must communicate to respond to a user request.

In the **two-tier architecture** (the most common type), a client (tier one) communicates directly with the server (tier two).

An **n-tier architecture** attempts to balance the workload between client and server by removing application processing from both the client and server and placing it on a middle-tier server.

Exhibit 6.8
An n-tier architecture

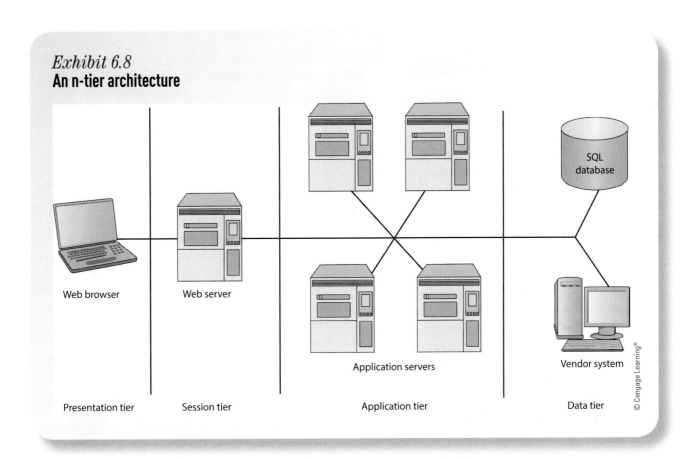

| Web browser | Web server | Application servers | SQL database / Vendor system |

| Presentation tier | Session tier | Application tier | Data tier |

Exhibit 6.9
A three-tier architecture

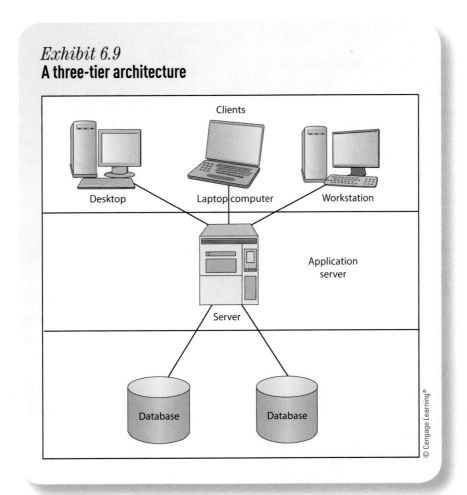

Clients

Desktop   Laptop computer   Workstation

Application server

Server

Database   Database

© Cengage Learning®

WIRELESS AND MOBILE NETWORKS

A **wireless network** is a network that uses wireless instead of wired technology. A **mobile network** (also called a "cellular network") is a network operating on a radio frequency (RF), consisting of radio cells, each served by a fixed transmitter, known as a "cell site" or "base station" (discussed later in this chapter). These cells are used to provide radio coverage over a wider area.

Wireless and mobile networks have the advantages of mobility, flexibility, ease of installation, and low cost. These systems are particularly effective when no infrastructure (such as communication lines or established wired networks) is in place, which is common in many developing nations and in old buildings that do not have the necessary wiring for a network. Drawbacks of mobile and wireless networks include the following:

- *Limited throughput*—**Throughput** is similar to bandwidth. It is the amount of data transferred or processed in a specified time, usually one second. Unlike the other disadvantages of wireless, throughput is expected to become a bigger problem in the future.

- *Limited range*—The distance a signal can travel without losing strength is more limited in mobile and wireless networks. For example, a Wi-Fi (Wireless Fidelity) network can have a range of 120 feet indoors and 300 feet outdoors. Chapter 14 covers range specifications for mobile and wireless networks in more detail.

- *In-building penetration problems*—Wireless signals might not be able to pass through certain building materials or might have difficulty passing through walls.

- *Vulnerability to frequency noise*—Interference from other signals, usually called "noise," can cause transmission problems. Common sources of noise include thunderstorms and lightning, which create radio waves (the same waves used by wireless networks), transformers, high-voltage cables, and fluorescent lights.

- *Security*—Wireless network traffic can be captured with sniffers. (Security is discussed in more detail later in this chapter.)

There are various definitions of mobile and wireless computing. Mobile computing might simply mean using a laptop away from the office or using a modem to access the corporate network from a client's office. Neither activity requires wireless technology. Wireless LANs usually refer to proprietary LANs, meaning they use a certain vendor's specifications, such as Apple Computer's LAN protocol for linking Macintosh devices. The term *wireless LAN* is also used to describe any wireless network. Wireless LANs have the same characteristics as wired LANs, except they use wireless media, such as infrared (IR) light or RF.

Exhibit 6.10 shows a wireless notebook connecting to a wired LAN. The transceiver on the laptop establishes radio contact with the wired LAN (although the figure does not show the entire wired network). The transceiver/receiver can be built in, attached to the notebook, mounted on a wall, or placed on a desk next to the notebook.

Wireless networks have many advantages. For example, healthcare workers who use handheld, notebook computers or tablets (such as the iPad) with wireless capabilities are able to get patient information quickly. Instead of writing notes on paper regarding the patient's condition, then transcribing the notes into an electronic form, they can enter information directly

A **wireless network** is a network that uses wireless instead of wired technology.

A **mobile network** (also called a cellular network) is a network operating on a radio frequency (RF), consisting of radio cells, each served by a fixed transmitter, known as a cell site or base station.

Throughput is similar to bandwidth. It is the amount of data transferred or processed in a specified time, usually one second.

Exhibit 6.10
A wireless notebook connecting to a wired LAN

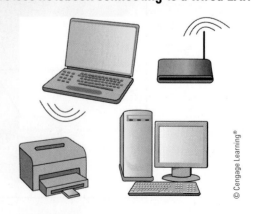

© Cengage Learning®

Mobile Computing and Mobile Apps

Mobile computing has become a familiar term because of the popularity of mobile apps. An app is designed to perform a specific task and can run on mobile devices, such as smartphones, tablets, and other handheld devices. Today, there is an app available for just about any task or application that has a general audience, including such areas as games, social media, retail, banking, finance, and medicine. Many of the apps are free; some must be purchased.

Apple started its App Store in 2008. In early 2013, the company announced that 50 billion apps had been downloaded from the App Store. At the same time, Google announced that 48 billion apps had been downloaded from Google Play, its own app store.[4]

Flurry, a company that provides apps developers with tools to measure how much people are using their apps, announced in 2013 that there were more than 300,000 apps from 100,000 developers in use worldwide. The company also indicated that it collects over 1.3 trillion different in-app events each month.[5]

Here are some examples of popular apps in use:

- *EPOCRATES (Epocrates Inc.)*—Gives doctors basic information about drugs, the right dosing for adults and children, and warnings about harmful interactions.[6]
- *ALIVECOR (AliveCor Inc.)*—Produces electrocardiograms when patients place their fingers over the monitor's sensors.
- *Google Shopper*—Provides information such as prices, reviews, and videos for millions of products.[7]
- *RBS (Citizens Bank)*—Performs all sorts of online banking tasks.[8]
- *Spotify*—Streams millions of songs from the Internet.[9]

into the handheld devices. Because the information can be sent to and saved on a centralized database, it is available to other workers instantly. In addition, entering notes directly prevents errors that are common during the transcription process, which improves the quality of information. The information box above discusses mobile apps, which have further enhanced the popularity of mobile computing.

6-7a Wireless Technologies

The use of wireless and mobile devices is on the rise. In October 2011, CTIA, the international association for the wireless telecommunications industry, reported that, for the first time, the number of wireless devices in the United States had surpassed the number of people.[10]

In a wireless environment, portable computers use small antennas to communicate with radio towers in the surrounding area. Satellites in near-Earth orbit pick up low-powered signals from mobile and portable network devices. The wireless communication industry has many vendors and is changing rapidly, but

> In October 2011, the number of wireless devices in the United States surpassed the number of people.

wireless technologies generally fall into two groups:

- *Wireless LANs (WLANs)*—These networks are becoming an important alternative to wired LANs in many companies. Like their wired counterparts, WLANs are characterized by having one owner and covering a limited area.
- *Wireless WANs (WWANs)*—These networks cover a broader area than WLANs and include the following devices: cellular networks, cellular digital packet data (CDPD), paging networks, personal communication systems (PCS), packet radio networks, broadband personal communications systems (BPCS), microwave networks, and satellite networks.

WLANs and WWANs rely on the RF spectrum as the communication medium, but they differ in the ways that are outlined in Table 6.2.

Note that 802.11a and 802.11b have been largely replaced by the current wireless standards: 802.11g and 802.11n. The 802.11g standard uses the 2.4 GHz frequency and has a data transfer rate of 54 Mbps; 802.11n

TABLE 6.2 WLNs Vs. WWANs

	WLANs	WWANs
Coverage	About 100 meters	Much wider area than for WLANs; capable of a regional, nation-wide, or international range
Speed	With the 802.11b wireless standard, data transfer rate up to 11 Mbps; with 802.11a, up to 54 Mbps; with 802.11n, up to 100 Mbps	Varies from 115 Kbps to 14 Mbps, depending on the technology
Data security	Usually lower than for WWANs	Usually higher than for WLANs

© Cengage Learning®

uses the same frequency but increases the data transfer rate to 100 Mbps.

6-7b Mobile Networks

Mobile networks have a three-part architecture, shown in Exhibit 6.11:

- Base stations send and receive transmissions to and from subscribers.
- Mobile telephone switching offices (MTSOs) transfer calls between national or global phone networks and base stations.
- Subscribers (users) connect to base stations by using mobile communication devices.

Mobile devices register by subscribing to a carrier service (provider) licensed for certain geographic areas. When a mobile unit is outside its provider's coverage area, roaming occurs. Roaming is using a cellular phone outside of a carrier's limited service area. By doing this, users are extending the connectivity service in a location that is different from the home location where the service was first registered.

To improve the efficiency and quality of digital communications, two technologies have been developed:

Exhibit 6.11
Mobile network architecture

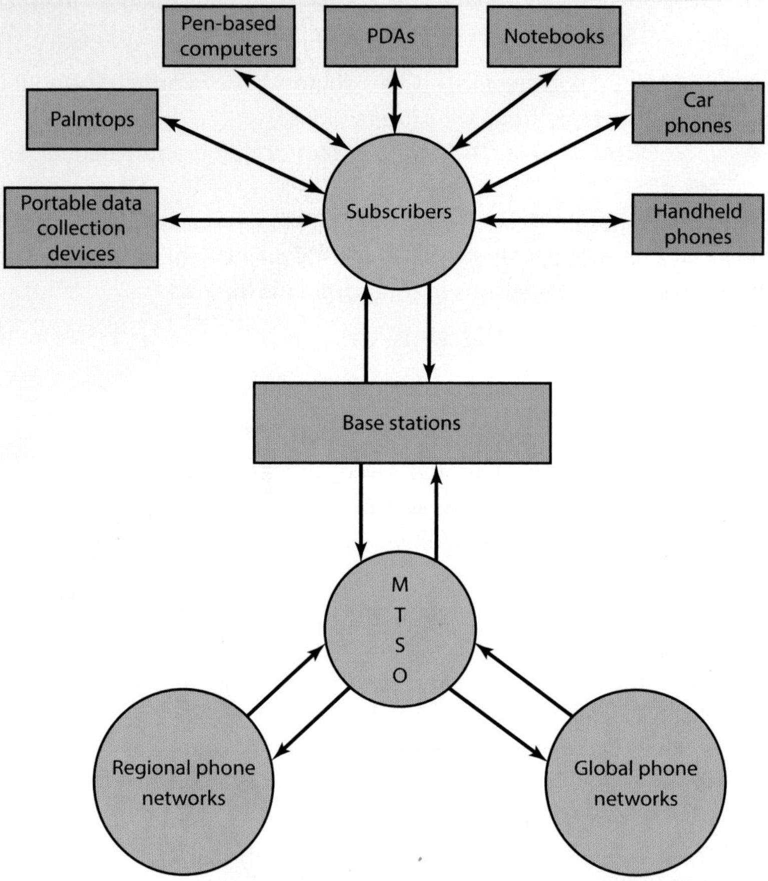

© Cengage Learning®

To improve the efficiency and quality of digital communications, **Time Division Multiple Access (TDMA)** divides each channel into six time slots. Each user is allocated two slots: one for transmission and one for reception. This method increases efficiency by 300 percent, as it allows carrying three calls on one channel.

To improve the efficiency and quality of digital communications, **Code Division Multiple Access (CDMA)** transmits multiple encoded messages over a wide frequency and then decodes them at the receiving end.

Time Division Multiple Access and Code Division Multiple Access. **Time Division Multiple Access (TDMA)** divides each channel into six time slots. Each user is allocated two slots: one for transmission and one for reception. This method increases efficiency by 300 percent, as it allows carrying three calls on one channel. **Code Division Multiple Access (CDMA)** transmits multiple encoded messages over a wide frequency and then decodes them at the receiving end.

TABLE 6.3 GENERATIONS OF CELLULAR NETWORKS

Generation	Description
1G	Analog transmission with limited bandwidth
2G	Support for voice, data, paging, and fax services added
2.5G	Added packet-switching technology, which transmits data packets over radio signals; different from the phone system's circuit-switching technology, which transmits data as a continuous stream of bits
3G	Supports transmission of high-quality multimedia data, including data, voice, and video
4G	Advanced version of 3G that provides broadband, large-capacity, high-speed data transmission, and high-quality interactive multimedia services

© Cengage Learning®

MOBILE COMPUTING IN ACTION: THE APPLE IPHONE

Apple released its first version of the iPhone in 2007. In 2008, the 3G version, which allowed fast 3G networking, was introduced. In 2009, the iPhone 3GS, which was much faster than the earlier versions and also allowed video recording, was introduced. In 2010, the iPhone 4 was released; it enabled video conferencing, enhanced e-mail, multitasking among all apps, and much more. In 2011, the iPhone 4S was released with advanced voice capabilities and multitasking features. In 2012, iPhone 5 was introduced with a slim design, 4-inch Retina display, and a faster chip. In 2013, iPhone 5s and 5c were introduced. The iPhone 5s is faster, has a new camera, and comes with a fingerprint reader. (For more information, go to apple.com.[11]) The iPhone 5c is basically a repackaged iPhone 5 that comes in different colors. (For more information, go to apple.com.[12]) There are numerous business applications for the iPhone. Here are some of the business benefits of using the iPhone[13]:

- It integrates with Microsoft Exchange so you can check your e-mails, your contacts, and your calendar, staying up to date no matter where you are.
- The Safari browser gives you access to your company's resources anytime and anywhere.
- It includes innovative apps like Maps, Voice Memos, and Voice Control.
- You can send SMS messages to multiple recipients.
- You can check stocks from anywhere or get a quick weather report before heading off on your next business trip.

Advanced Mobile Phone System (AMPS) is the analog mobile phone standard developed by Bell Labs and introduced in 1983. Digital technologies, however, are more widely used because of higher data capacities, improved voice quality, encryption capabilities, and integration with other digital networks. As of February 18, 2008, U.S. carriers were no longer required to support AMPS. Companies such as AT&T and Verizon have discontinued this service permanently. Table 6.3 describes the various generations of cellular networks. (Bluetooth, a wireless networking technology, is discussed in Chapter 14.)

Many businesses use wireless and mobile networks to improve customer service and reduce operational costs. The information box above gives an overview of the Apple iPhone and its business applications.

6-8 WIRELESS SECURITY

Security is important in any type of network, but it is especially important in a wireless network, because anyone walking or driving within the range of an access point (AP), even if outside the home or office, can use the network. An AP is the part of a WLAN that connects it to other networks. Finding WLANs is an easy task. A user can simply walk or drive around office buildings or homes with a WLAN-equipped computer and try to pick up a signal. Wireless signals can also be intercepted, and they are susceptible to the same DoS attacks (discussed in Chapter 5) to which wired networks are susceptible.

There are several techniques for improving the security of a wireless network:

- *SSID (Service Set Identifier)*—All client computers that try to access the AP are required to include an SSID in all their packets. A packet without an SSID is not processed by the AP. The major weakness of using an SSID is that it can be picked up by other devices within the range, given the right software.

- *WEP (Wired Equivalent Privacy)*—A key must be manually entered into the AP and the client computer. The key encrypts the message before transmission. Because this manual process is complex and time consuming, the WEP technique is not suitable for large networks.

- *EAP (Extensible Authentication Protocol)*—EAP keys are dynamically generated based on the user's ID and password. When the user logs out of the system, the key is discarded. A new key is generated when the user logs back into the network.

- *WPA (Wi-Fi Protected Access)*—This technique combines the strongest features of WEP and EAP. Keys are fixed, as in WEP, or dynamically changed, as in EAP. However, the WPA key is longer than the WEP key; therefore, it is more difficult to break. Also, the key is changed for each frame (a distinct and identifiable data set) before transmission.

- *WPA2 or 802.11i*—This technique uses EAP to obtain a master key. With this master key, a user's computer and the AP negotiate for a key that will be used for a session. After the session is terminated, the key is discarded. This technique uses Advanced Encryption Standard, which is more complex than WPA and much harder to break.

The adjoining information box highlights privacy and ethical issues related to wireless devices.

In data communication, **convergence** refers to integrating voice, video, and data so that multimedia information can be used for decision making.

PRIVACY AND ETHICAL ISSUES OF WIRELESS DEVICES

Some wireless devices, including iPhones, iPads, and Android devices, track the locations of their users. The data points are saved either on the device itself or on other computers and networks. Although the users of these devices are able to turn off the GPS feature, many do not do that. In fact, having GPS capability is probably one of the reasons many people buy these devices in the first place. However, location tracking can involve privacy concerns for users and ethical issues for IT personnel. Today, many corporations issue these devices to their employees to help them perform their jobs more efficiently. Of course, the employers have a legal right to monitor how these devices are being used by their employees. This can include monitoring the Web sites their employees visit, the e-mails they send or receive, their telephone conversations, and so forth. Many corporations require their employees to carry company-issued devices at all times, which means that the employees' off-hours movements are being tracked. Is it ethical for an employer to monitor its employees' movement in this way?[14,15]

6-9   CONVERGENCE OF VOICE, VIDEO, AND DATA

In data communication, **convergence** refers to integrating voice, video, and data so that multimedia information can be used for decision making. In the past, separate networks were used to transmit voice, video, and data, but as the demand for integrated services

In the past, separate networks were used to transmit data, voice, and video, but as the demand for integrated services increased, technology has developed to meet this demand.

TELEPRESENCE: A NEW USE OF DATA COMMUNICATION AND CONVERGENCE

Telepresence has attracted a lot of attention recently, particularly because of the economic downturn; companies are finding they can reduce business travel and business meeting expenses. This technology integrates audio and video conferencing into a single platform, and recent improvements have resulted in higher quality, greater ease of use, and better reliability.

Telepresence systems can be used to record meetings for later use or to incorporate multimedia technologies into presentations. They can also offer plug-and-play collaboration applications. Some products offer a telepresence room, with customized lighting and acoustics as well as large high-density (HD) screens that can be configured for up to 20 users. Others are on a smaller scale, with a single HD screen. Major vendors of telepresence products include Cisco, Polycom, Tandberg, Teliris, and HP.[16]

has increased, technology has developed to meet this demand.

Convergence requires major network upgrades, because video requires much more bandwidth. This has changed, however, with the availability of high-speed technologies, such as Asynchronous Transfer Mode (ATM), Gigabit Ethernet, 3G and 4G networks, and more demand for applications using these technologies. Gigabit Ethernet is a LAN transmission standard capable of 1 Gbps and 10 Gbps data transfer speeds. The ATM is a packet-switching service that operates at 25 Mbps and 622 Mbps, with maximum speed of up to 10 Gbps. As mentioned earlier, the 3G network is the third generation of mobile networking and telecommunications. It features a wider range of services and a more advanced network capacity than the 2G network. The 3G network has increased the rate of information transfer, its quality, video and broadband wireless data transfers, and the quality of Internet telephony or Voice over Internet Protocol (VoIP). It has also made streaming video possible as well as much faster uploads and downloads. The 4G network will further enhance all these features.

More content providers, network operators, telecommunication companies, and broadcasting networks, among others, have moved toward convergence. Even smaller companies are now taking advantage of this fast-growing technology by offering

multimedia product demonstrations and using the Internet for multimedia presentations and collaboration. Convergence is possible now because of a combination of technological innovation, changes in market structure, and regulatory reform. Common applications of convergence include the following:

- E-commerce
- More entertainment options as the number of TV channels substantially increases and movies and videos on demand become more available
- Increased availability and affordability of video and computer conferencing
- Consumer products and services, such as virtual classrooms, telecommuting, and virtual reality

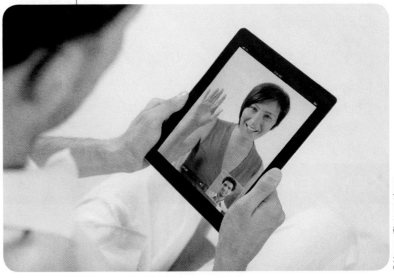

© bloomua/Shutterstock.com

As a tool for delivering services, the Internet is an important contributor to the convergence phenomenon. Advances in digital technologies are helping to move convergence technologies forward, and when standards in data collection, processing, and transmission become more available and acceptable, their use should increase even further. See the information box on the previous page on telepresence.

The Industry Connection box highlights Cisco Systems, Inc., which offers many products and services used in data communication systems.

Industry Connection: Cisco Systems, Inc.[17]

The main goal of Cisco Systems, Inc., the largest vendor of networking equipment, is to make it easier to connect different computers. Cisco offers a wide variety of products, including routers, switches, network management tools, optical networking, security software, VPNs, firewalls, and collaboration and telepresence products. The variety of products makes it possible for organizations to get everything they need for networking solutions from one vendor. Cisco's products and services include the following:

PIX Firewall Series: Allows corporations to protect their internal networks from outside intruders.

Network Management Tools: Allows network managers to automate, simplify, and integrate their networks to reduce operational costs and improve productivity.

Identity Management Tools: Protect information resources through identity policies, access control, and compliance features.

TelePresence Network Management: Integrates audio, high-definition video, and interactive features to deliver face-to-face collaboration capabilities.

Cisco 800 Series routers: Offer built-in security, including content filtering, WAN connection with multiple access options, four 10/100 Mbps Fast Ethernet managed switch ports, and more. The routing system released in 2010, CRS-3 (Cisco Carrier Routing System-3), delivers Internet speeds of up to 322 terabits per second, which can offer video and other content 12 times faster than its rivals' systems.

STUDY TOOLS 6

LOCATED AT BACK OF THE TEXTBOOK
☐ Rip out Chapter Review Card

LOCATED AT WWW.CENGAGE.COM/LOGIN
☐ Review Key Terms Flash Cards (Print or Online)

☐ Download audio and visual summaries to review

☐ Complete both Practice Quizzes to prepare for tests

☐ Play "Beat the Clock" and "Quizbowl" to master concepts

☐ Complete "Crossword Puzzle" to review key terms

☐ Watch video on "Method" for a real company example

KEY TERMS

attenuation 113
bandwidth 113
broadband 113
bus topology 120

centralized processing 115
centralized routing 122
client/server model 123
Code Division Multiple Access (CDMA) 128

REVIEWS AND DISCUSSIONS

1. What are three business applications of data communication?

2. Among the five media (channels) discussed in the chapter, which one has the highest throughput? Which one is the most secure?

3. What are two advantages of the bus topology?

4. Why is TCP/IP so popular?

5. Among the three types of networks (LAN, WAN, and MAN), which is the fastest?

6. Define the client/server model. Which architecture in this model is the most common?

7. What are three advantages of mobile and wireless networks compared to their wired counterparts?

8. What are three examples of security threats to a smartphone?

PROJECTS

1. A newly established Internet company with 40 employees needs your advice. They are looking for a collaboration tool and have narrowed their choices to GoToMeeting, WebEx, and My Web Conferences. After reading the information presented in this chapter and other sources, prepare a two-page document that includes two advantages and two disadvantages of each tool. Which one is your final recommendation? Why did you choose that tool over the other two?

2. Cisco and Polycom are two major vendors of telepresence products. After reading the information presented in this chapter and other sources, write a one-page paper that identifies one top-of-the-line product from each company. Which product would you recommend to the company mentioned in Project 1? What are you basing your recommendation on?

3. Mobile and wireless devices are being increasingly used in the health care industry. After reading the information presented in this chapter and other sources, write a two-page paper that outlines five applications of these devices in this industry. Also, identify three mobile apps that could be used by medical personnel to increase their productivity.

4. After reading the information presented in this chapter and other sources, write a two-page paper that identifies five mobile apps that could be used in the banking industry. How do these apps increase the productivity of the bankers and their customers? Do you see any drawbacks involved in using these apps?

5. After reading the information presented in this chapter and other sources, write a one-page paper that includes a six-item bulleted list for improving the privacy and security of your smartphone. Generally speaking, are iOS devices more or less secure than Android devices?

6. After reading the information presented in this chapter and other sources, write a two-page paper that describes five business applications of convergence. Which industries are expected to gain the most from the convergence trend?

ARE YOU READY TO MOVE ON?

1. In broadband data transmission, multiple pieces of data are sent simultaneously to increase the transmission rate. True or False?

2. All Internet connections require a modem. True or False?

3. Wireless and mobile networks have the advantages of mobility, flexibility, ease of installation, and low cost. True or False?

4. Which of the following components is not included in a typical data communication system?

   a. Sender and receiver devices

   b. Modems and routers

   c. Network interface card (NIC)

   d. Communication medium (channel)

5. Which of the following layers is not part of the Open Systems Interconnection (OSI) model?

   a. Distribution

   b. Application

   c. Session

   d. Transport

6. Which of the following is not a network topology?

   a. Star

   b. Ring

   c. Mesh

   d. Client/server

CASE STUDY 6-1

Data Communication at Walmart

Walmart has made several changes in its data communication systems to improve its suppliers' access to sales and inventory data. For example, the company added a customized Web site for its suppliers, such as Mattel, Procter & Gamble, and Warner-Lambert. Walmart's goal is to improve efficiency in order to keep prices low and maintain a high level of customer service. With Walmart's network, suppliers can access sales, inventory, and forecasting data over extremely fast

connections. To ensure confidentiality of data, a sophisticated security system has been implemented to prevent suppliers from accessing data about one another's products.

Walmart has also added Web-based access to its RetailLink system so suppliers can use information in the database. Other data communication applications at Walmart include automated distribution, computerized routing, and electronic data interchange (EDI).[18,19]

Walmart uses the latest in wireless technology

© Anton Gvozdikov/Shutterstock.com

in its operations for warehouse management systems (WMS) to track and manage the flow of goods through its distribution centers. Another application of wireless technology is for controlling and monitoring forklifts and industrial vehicles that move merchandise inside its distribution centers. The Vehicle Management System (VMS) is the latest application of data communication in Walmart. Among other features, the VMS includes a two-way text-messaging system that enables management to effectively divert material-handling resources to where they are needed the most. (The VMS works effectively with RFID systems.) According to Walmart, the VMS has improved safety and has also significantly improved the productivity of its operations.[20]

Answer the following questions:

1. How has Walmart improved its data communication systems for suppliers?
2. What are some typical data communication applications at Walmart?
3. What are some of the applications of wireless technology at Walmart?
4. What are some of the features and capabilities of the VMS?

CASE STUDY 6-2

Protecting the Security and Privacy of Mobile Devices

As the number of smartphone and tablet devices increases, so does the risk that hackers and computer criminals will target these devices. Since they were first introduced, certain mobile devices have been vulnerable to eavesdropping and fraudulent charges. In addition, many of them automatically track a user's location.[21]

The major security risks associated with mobile devices include:[22]

Malware—Android apps, in particular, are vulnerable to malware because of the platform's openness. To guard against malware threats, users have to upload the most recent versions of the operating system and use mobile security tools.

Premium SMS billing—With their devices' vulnerability to malware, smartphone users face an added risk of subscribing to premium text-messaging services that charge every time users interact with them. Most cell phone carriers allow subscribers to block premium SMS messaging, however.

E-mail and SMS phishing—Because it is more difficult to establish a link's legitimacy on a mobile device, mobile users are more likely to click on them, which is a phisher's dream come true. Mobile users should therefore use a lot of caution when using e-mail on these devices.

Spyware—Commercially available software can be used by intruders to track and control the user's mobile activities.

Malicious Web sites—These could pose a threat in the future, given that many smartphone browsers are based on a browser engine with vulnerabilities.

Many of the recommendations discussed in Chapters 4 and 5 apply here. Never leave your mobile device unlocked. Never leave it unattended. Always protect it with a password.

© Goodluz/Shutterstock.com

Answer the following questions:

1. What are some examples of security risks associated with a mobile device?
2. How can these devices automatically track a user's location?
3. What are a couple of recommendations for protecting your mobile device against these threats?

7 | The Internet, Intranets, and Extranets

LEARNING OUTCOMES

After studying this chapter, you should be able to:

7-1 Describe the makeup of the Internet and the World Wide Web.

7-2 Discuss navigational tools, search engines, and directories.

7-3 Describe common Internet services.

7-4 Summarize widely used Web applications.

7-5 Explain the purpose of intranets.

7-6 Explain the purpose of extranets.

7-7 Summarize the trends of the Web 2.0 and Web 3.0 eras and Internet2.

7-8 Describe the Internet of Everything.

After you finish this chapter, go to **PAGE 159** for the **STUDY TOOLS**

© Ostaf/Shutterstock.com

This chapter introduces you to the Internet and Web technologies. It provides an overview of the Domain Name System and various types of Internet connections. You learn how navigational tools, search engines, and directories are used on the Internet, and there is a brief survey of common Internet services and Web applications. This chapter also explains intranets and extranets and how they are used. Finally, you learn about the trends of the Web 2.0 and Web 3.0 eras, Internet2, and the Internet of Everything.

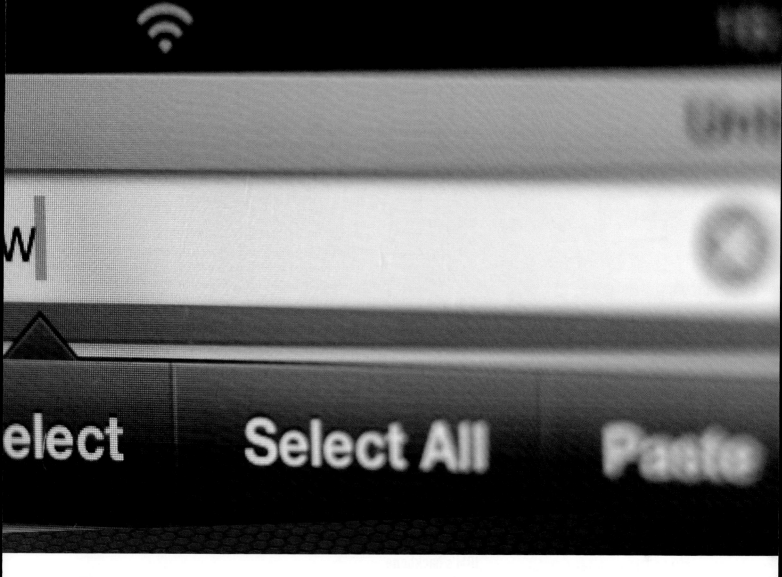

7-1 THE INTERNET AND THE WORLD WIDE WEB

The **Internet** is a worldwide collection of millions of computers and networks of all sizes. The term *Internet* is derived from the term *internetworking*, which means connecting networks. Simply put, the Internet is a network of networks. No one actually owns or runs the Internet, and each network is administered and funded locally.

In an international survey of people in their 20s, conducted by Cisco Systems, a third of the respondents said they consider Internet access one of life's necessities. Among college students, 55 percent said they could not function without the Internet. Among recent graduates with jobs, 62 percent said the same thing. In both age

> **Among college students, 55 percent said they could not function without the Internet.**

categories, over 30 percent said they could survive, but it would be a struggle.[1]

The Internet started in 1969 as a U.S. Department of Defense project called the **Advanced Research Projects Agency Network (ARPANET)** that connected four nodes: University of California

The **Internet** is a worldwide collection of millions of computers and networks of all sizes. It is a network of networks.

The **Advanced Research Projects Agency Network (ARPANET)**, a project started in 1969 by the U.S. Department of Defense, was the beginning of the Internet.

at Los Angeles, University of California at Santa Barbara, Stanford Research Institute at Stanford University in California, and University of Utah at Salt Lake City. Other nodes composed of computer networks from universities and government laboratories were added to the network later. These connections were linked in a three-level hierarchical structure: backbones, regional networks, and local area networks.

The Internet started in 1969 as a U.S. Department of Defense project.

In 1987, ARPANET evolved into the National Science Foundation Network (NSFNET), which is considered the first Internet backbone. The NSF initially restricted Internet use to research and to educational institutions; commercial use was not allowed. Eventually, because of increased demand, other backbones were allowed to connect to NSFNET.

The **Internet backbone** is a foundation network linked with fiber-optic cables that can support very high bandwidth. It is called a *backbone* because it supports all the other networks that form the Internet, just as the human backbone is the foundation of the nervous system. The Internet backbone is made up of many interconnected government, academic, commercial, and other high-capacity data routers.

Several private companies operate their own Internet backbones that interconnect at network access points (NAPs). You can find a list of Internet backbones based in the United States at *http://www.nthelp.com/maps.htm*. Exhibit 7.1 shows IBM's backbone. NAPs determine how traffic is routed over the Internet. As you learned in Chapter 6, local area networks (LANs) serve as localized Internet connections, and they use NAPs to connect to the Internet backbone.

The World Wide Web (WWW, or "the Web") changed the Internet in 1989 by introducing a graphical interface to the largely text-based Internet. The Web was proposed by Tim Berners-Lee at the European Organization for Nuclear Research (CERN), the world's largest particle physics center. (*CERN* stands for *Conseil Europeen pour la Recherche Nucleaire*.)

The Web organizes information by using **hypermedia**, meaning documents that include embedded references to audio, text, images, video, or other documents. Composed of billions of hypermedia documents, the Web constitutes a large portion of the Internet. The embedded references in hypermedia documents are called **hypertext**; they consist of links users can click to follow a particular thread (topic). By using hypertext links, users can access files, applications, and

---

The **Internet backbone** is a foundation network linked with fiber-optic cables that can support very high bandwidth. It is made up of many interconnected government, academic, commercial, and other high-capacity data routers.

With **hypermedia**, documents include embedded references to audio, text, images, video, and other documents.

The embedded references in hypermedia documents are called **hypertext**; they consist of links users can click to follow a particular thread (topic).

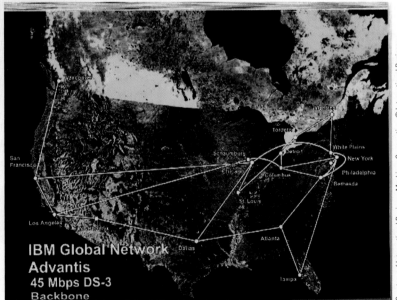

Exhibit 7.1
IBM's backbone

IBM Global Network
Advantis
45 Mbps DS-3
Backbone

other computers in any order they like (unlike in paper documents) and retrieve information with the click of a button. In essence, hypertext is an approach to data management, in which data is stored in a network of nodes connected by links. Data in these nodes is accessed with an interactive browsing system, meaning the user determines the order in which information is accessed.

Any computer that stores hypermedia documents and makes them available to other computers on the Internet is called a server or Web server, and computers requesting these documents are called clients. A client can be a home computer or a node in an organization's LAN. The most exciting feature of the Web is that hypermedia documents can be stored anywhere in the world, so users can jump from a site in the United States to a site in Paris, France, in just a few milliseconds. The information box below summarizes major events in the development of the Internet.

7-1a The Domain Name System

Domain names, such as IBM.com or whitehouse.gov, are unique identifiers of computer or network addresses on the Internet. Each computer or network also has an Internet Protocol (IP) address, such as 208.77.188.166, which is assigned by the Internet Corporation for Assigned Names and Numbers (ICANN). These numbers are difficult to remember, however, so language-based domain names are used more often to access Web sites. When information is transferred from one

MAJOR EVENTS IN THE DEVELOPMENT OF THE INTERNET

September 1969: ARPANET is born.

1971: Ray Tomlinson of BBN invents an e-mail program to send messages across a network.

January 1983: Transition from Network Control Protocol (NCP) to Transmission Control Protocol/Internet Protocol (TCP/IP), the protocol for sending and receiving packets.

1987: The National Science Foundation creates a backbone to the National Research and Education Network called NSFNET; it signifies the birth of the Internet.

November 1988: A worm attacks more than 6,000 computers, including those at the Department of Defense.

1989: The World Wide Web is developed at CERN.

February 1991: The Bush administration approves Senator Al Gore's idea to develop a high-speed national network, and the term "information superhighway" is coined.

November 1993: Pacific Bell announces a plan to spend $16 billion on the information superhighway.

January 1994: MCI announces a 6-year plan to spend $20 billion on an international communication network.

April 1995: Netscape becomes the most popular graphical navigator for surfing the Web.

August 1995: Microsoft releases the first version of Internet Explorer.

April 1996: Yahoo! goes public.

June 1998: A U.S. appellate court rules that Microsoft can integrate its browser with its operating system, allowing the company to integrate almost any application into the Microsoft OS.

February 2000: A denial-of-service attack (DoS) shuts down several Web sites (including Yahoo!, Ameritrade, and Amazon.com) for several hours.

2004: A worm variously called MyDoom or Novarg spreads through Web servers. About 1 of every 12 e-mail messages is infected.

2005: YouTube.com is launched.

2008: Microsoft offers to buy Yahoo! for $44.6 billion, and San Francisco federal judge Jeffrey S. White orders the disabling of Wikileaks.org, a Web site that discloses confidential information.

2009: The number of worldwide Internet users surpasses one billion per month.[2]

2013: The number of worldwide Facebook users tops 1.11 billion.

network to another, domain names are converted to IP addresses by the **Domain Name System (DNS)** protocol. Servers using this protocol (called DNS servers) maintain lists of computers' and Web sites' addresses and their associated IP addresses. DNS servers translate all domain names into IP addresses.

You see domain names used in **uniform resource locators (URLs)**, also called *universal resource locators*, to identify a Web page. A URL is the address of a document or site on the Internet. For example, in the URL *http://www.csub.edu*, the domain name is *csub .edu*. Every domain name has a suffix indicating the top-level domain (TLD) it belongs to. In this example, the suffix is .edu, which stands for educational institutions. Combinations of letters, the numerals 0 through 9, and hyphens can be used in domain names, too. Spaces are not allowed.

The TLD denotes the type of organization or country the address specifies. TLDs are divided into organizational domains (generic top-level domains, gTLDs) and geographic domains (country code top-level domains, ccTLDs). Table 7.1 lists common gTLDs.

Many new gTLDs have been proposed, including .aero (aviation industry), .museum, .law, and .store.

Some are already in use, such as .info for organizations providing information services, .biz for businesses, and .news for news-related sites.

In addition, most countries have geographic domains. These ccTLDs include .au for Australia, .ca for Canada, .fr for France, .jp for Japan, and .uk for the United Kingdom. You can find a complete list of ccTLDs at *www.thrall.org/domains.htm*.

Starting in January 2012, corporations and individuals were able to choose from an expanded list of TLDs that was approved by ICANN in June 2011. Before that, there were only 22 generic TLDs, such as .com, .edu, and so forth. Now, Internet address names can end with almost any word in any language, giving organizations around the world the opportunity to market their brands, products, or causes in new and creative ways. The TLDs can be in any language or character set and can contain any phrase, including a company or brand name. The new gTLDs fall into four categories[3,4]:

- Generic word TLDs (e.g., .company or .TV)
- Corporate TLDs (e.g., .Microsoft or .Intel), owned by corporations to control use of their brands
- Community TLDs (e.g., .Spanish or .Persian), limited to members of a defined community
- Geographic TLDs (e.g., .London or .Madrid), owned by cities and geographic regions and used to promote business and tourism

Here are brief explanations of each part of a URL, using *http://www.csub.edu/~hbidgoli/books.html* as an example:

- *http*—Stands for Hypertext Transfer Protocol, the protocol used for accessing most Web sites.
- *www.csub.edu*—The *www* stands for *WWW, World Wide Web*, or *the Web*. The *csub* stands for California State University at Bakersfield. And the *.edu* is the suffix for educational institutions. Together, *csub.edu* uniquely identifies this Web site.

TABLE 7.1 GENERIC TOP-LEVEL DOMAINS

gTLD	Purpose
.com	Commercial organizations (such as Microsoft)
.edu	Educational institutions (such as California State University)
.int	International organizations (such as the United Nations)
.mil	U.S. military organizations (such as the U.S. Army)
.gov	U.S. government organizations (such as the Internal Revenue Service)
.net	Backbone, regional, and commercial networks (e.g., the National Science Foundation's Internet Network Information Center)
.org	Other organizations, such as research and nonprofit organizations (e.g., the Internet Town Hall)

© Cengage Learning®

When information is transferred from one network to another, domain names are converted to IP addresses by the **Domain Name System (DNS)** protocol. Servers using this protocol (called DNS servers) maintain lists of computers' and Web sites' addresses and their associated IP addresses.

uniform resource locators (URLs), also called *universal resource locators*, identify a Web page. A URL is the address of a document or site on the Internet.

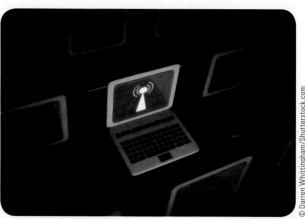

WHAT IS HTML?

Hypertext Markup Language (HTML) is the language used to create Web pages. It defines a page's layout and appearance by using tags and attributes. A tag delineates a section of the page, such as the header or body; an attribute specifies a value, such as a font color, for a page component. (Note that HTML codes are not case sensitive.) The most recent version of HTML is HTML6, which competes with Adobe Flash. Flash is a multimedia platform used to add animation, video, and interactivity to Web pages. A typical structure for an HTML document is as follows:

```
<HTML>

<HEAD>

(Enter the page's description.)

</HEAD>

<BODY>

(Enter the page's content.)

</BODY>

</HTML>
```

- *~/hbidgoli*—This part is the name of the directory in which files pertaining to the books the author has written are stored. A server can be divided into directories for better organization.
- *books.html*—This part is the document itself. The .html extension means it is a Hypertext Markup Language (HTML) document. (See the information box above for more on HTML.) Servers that do not support long extensions display just *.htm*; other servers display *.html*.

7-1b Types of Internet Connections

As you learned in Chapter 6, there are several methods for connecting to a network, including the Internet. These methods include dial-up and cable modems as well as Digital Subscriber Line (DSL). Several types of DSL services are available:

- *Symmetric DSL (SDSL)*—SDSL has the same data transmission rate to and from the phone network (called upstream and downstream), usually up to 1.5 Mbps (million bits per second) in both directions.
- *Asymmetric DSL (ADSL)*—ADSL has a lower transmission rate upstream (3.5 Mbps) than downstream

(typically 24 Mbps)—for example, the ITU G.992.5 Annex M standard.

- *Very High-Speed DSL (VDSL)*—VDSL has a downstream/upstream transmission rate of up to 100 Mbps over short distances—for example, the ITU G.993.2 standard.

Organizations often use T1 or T3 lines. These are provided by the telephone company and are capable of transporting the equivalent of 24 conventional telephone lines using only two pairs of copper wires. T1 uses two pairs of copper wires to carry up to 24 simultaneous conversations (called channels) and has a transmission rate of 1.544 Mbps; it is more widely used than T3. (In other countries, T1 is called E1 and has a transmission rate of 2.048 Mbps.) A T3 line is a digital communication link that supports transmission rates of 43–45 Mbps. A T3 line actually consists of 672 channels, each supporting rates of 64 Kbps. The 10 countries with the fastest Internet connections are Hong Kong, South Korea, Romania, Ireland, Czech Republic, Belgium, Switzerland, Bulgaria, Israel, and Singapore.[5]

7-2 NAVIGATIONAL TOOLS, SEARCH ENGINES, AND DIRECTORIES

Now that you know what the Internet is and how to connect to it, you will need tools to get around it and find what you are looking for. These tools can be divided into three categories:

- **Navigational tools**—These are used to travel from Web site to Web site (i.e., "surf" the Internet).
- **Search engines**—These allow you to look up information on the Internet by entering keywords related to your topic of interest.

Hypertext Markup Language (HTML) is the language used to create Web pages. It defines a page's layout and appearance by using tags and attributes. A tag delineates a section of the page, such as the header or body; an attribute specifies a value, such as a font color, for a page component.

Navigational tools are used to travel from Web site to Web site—as in "surf" the Internet.

A **search engine**, such as Google.com or Ask.com, is an information system that enables users to retrieve data from the Web by using search terms.

- **Directories**—These are indexes of information, based on keywords embedded in documents, that allow search engines to find what you are looking for. Some Web sites (such as Yahoo!) also use directories to organize content into categories.

Originally, Internet users used text-based commands for simple tasks, such as downloading files or sending e-mails. However, it was tedious to type commands at the command line, and users also had to have certain programming skills to use these systems. The graphical browsers changed all this by providing menus and graphics-based tools that allowed users to point and click. These systems make the user-system interface more easy to use, and the graphical browsers also support multimedia information, such as images and sound.

These three tools for getting around the Internet and finding information are described in more detail in the following sections.

7-2a Navigational Tools

Many graphical Web browsers are available, such as Microsoft Internet Explorer (IE), Mozilla Firefox, Google Chrome, Apple Safari, and Opera. Typically, these browsers have menu options you have seen in word-processing programs, such as File, Edit, and Help. They also include options for viewing your browsing history, bookmarking favorite Web sites, and setting viewing preferences, as well as navigation buttons to move backward and forward in Web pages you have visited. With some browsers, you can also set up specialized toolbars for accessing frequently visited sites or conducting searches.

7-2b Search Engines and Directories

A search engine, such as Google, Bing, DuckDuckGo, or Ask.com, is an information system that enables users to retrieve data from the Web by using search terms. All search engines follow a three-step process:

1. *Crawling the Web*—Search engines use software called *crawlers*, *spiders*, *bots*, and other similar names. These automated modules search the Web continuously for new data. When you post a new Web page, crawlers find it (if it is public), and when you update it, crawlers find the new data. Crawlers also check to see what links are on your page and make sure they work; if a link is broken, crawlers identify it and include this information as part of the data about that page. In addition, crawlers can go through the other pages that are part of your Web site, as long as there are links to those pages. All the gathered data is sent back to the search engine's data center so the search engine always has the most current information on the Web.

2. *Indexing*—Housed at server farms, search engines use keywords to index data coming in from crawlers. Each keyword has an index entry that is linked to all Web pages containing that keyword. For example, a company selling picture frames includes the term *picture frame* on its Web site several times. The indexing process recognizes the frequency of use and creates an index entry for the term *picture frame*. This index entry is linked to the company's Web site, along with all other sites containing the term *picture frame*. Indexing makes it possible for search engines to retrieve all related Web pages when you enter a search term.

3. *Searching*—When you enter a search term, the search engine uses the index created in Step 2 to look up the term. If the term exists in the index, the search engine identifies all Web pages linked to the term. However, it needs some way of prioritizing Web pages based on how close each one is to the search term. For example, say your Aunt Emma makes picture-frame cookies and has a Web site for selling them. Someone searching on the term *picture frame* might see Aunt Emma's site listed, too. Because search engines are programmed to try to differentiate different types of search requests, they can use other terms, such as *posters*, *photos*, and *images*, to give a higher priority to Web pages containing these additional terms, along with the search term *picture frame*, and a lower priority to Web pages containing terms such as *cookies* or *baked goods* along with *picture frame*. Search engines vary in intelligence, which is why you can use the same search term and get different results with two different search engines.

In recent years, two new types of searches are gaining in popularity: Graph Search, which Facebook introduced in 2013, and Knowledge Graph, which Google introduced in 2012. Graph Search allows Facebook users to find people, photos, places, and interests that

Directories are indexes of information based on keywords embedded in documents, which make it possible for search engines to find what you are looking for.

are relevant to them. Example searches using this method include[6]:

- People who like sausage and live in the southwest part of the city I live in
- Photos of a particular friend of mine that were taken after 2005
- Photos of all my friends who live in Los Angeles
- Chinese restaurants in Boston that my friends have liked
- Tourist attractions in Southern France visited by my friends

In a traditional Google search, users type in key words and Google displays the most relevant Web pages based mostly on the number of times those Web pages have been linked to. With Facebook's Graph Search, the objects that users are searching for are not Web pages; instead, they're searching for the virtual representations of real-world objects: people, places, photos, and so forth. And what determines the results of the search are Facebook Likes.[7]

According to Google, Knowledge Graph provides answers, not just links. So, alongside the usual results, there are now panels containing important facts about people, places, and things. The goal is to return pages that match the entities or concepts that the key words evoke. Google's database includes over 3.5 billion facts about 500 million objects. Categories include cities, countries, sports teams, and so forth.[8]

Directories organize information into categories. There are two kinds of directories on the Web. The first is the automated, or crawler-based, directory that search engines use; it creates indexes of search terms and collects these terms automatically by using crawlers. Google, Yahoo!, Ask.com, and others fall into this category. When your Web page changes, for example, these directories update their indexes and databases automatically to reflect the change. The second type of directory is the human-powered directory. If you want your Web page to be listed in a search engine's results, you have to manually submit keywords to a human-powered directory. It does not use crawlers to collect data; instead, it relies on users to supply the data. After key words are submitted, they are indexed with search engines and can then be listed in search results. The main difference is that if your Web page changes, the directory does not have the updated content until you submit changes to the directory. Open Directory is an example of a human-powered directory. However, Google has made many directories obsolete,

and directories in general are not as relevant as they used to be.

Crawler-based directories are based on index terms, just as the phone book's white pages are based on the last names and first names of people. Some search engines, in addition to their index-term-based directory, offer directories based on popular categories, such as business, sports, entertainment, travel, and dining. Each category can have subcategories; for example, an Entertainment category might contain Movies, Music, and Theater subcategories. Yahoo! Travel, Yahoo! Business, and Yahoo! RealEstate are some categories in the Yahoo! directory, considered top of the line by both users and experts.

7-3 INTERNET SERVICES

Many services are available via the Internet, and most are made possible by the TCP suite of protocols in the Application layer (introduced in Chapter 6). For instance, TCP/IP provides several useful e-mail protocols, such as Simple Message Transfer Protocol (SMTP) for sending e-mails and Post Office Protocol (POP) for retrieving messages. Popular services include e-mail, newsgroups, and discussion groups; Internet Relay Chat (IRC) and instant messaging; and Internet telephony. All are discussed in the following sections.

7-3a E-Mail

E-mail is one of the most widely used services on the Internet. In addition to personal use, many companies use e-mail for product announcements, payment confirmations, and newsletters. Recently, businesses have also begun to send receipts for items purchased by customers through e-mail. This helps reduce costs and is more environmentally friendly than using paper.

There are two main types of e-mail. Web-based e-mail enables you to access your e-mail account from any computer and, in some cases, store your e-mails on a Web server. MSN Hotmail and Google Gmail are two examples of free Web-based e-mail services. The other type of e-mail is client-based e-mail, which consists of an e-mail program you install on your computer; e-mail is downloaded and stored locally on your computer. Examples of client-based e-mail programs include Microsoft Outlook, Mozilla Thunderbird, and Apple Mail.

Most e-mail programs include a folder system for organizing your e-mails and an address book in which to

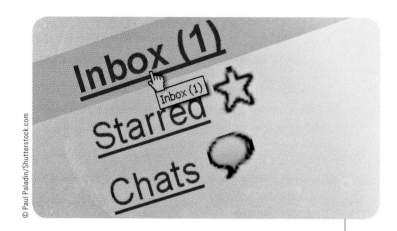

store e-mail addresses. Many address books include an autocompletion feature so you can just type a recipient's name and the e-mail address is filled in automatically. You can also set up distribution groups for sending an e-mail to several people at the same time. Other commonly available features are spell checkers and delivery notifications. You can also attach documents and multimedia files to e-mails.

7-3b Newsgroups and Discussion Groups

The Internet serves millions of people with diverse backgrounds and interests. Discussion groups and newsgroups are a great way for people with similar interests to find one another. Although newsgroups and discussion groups are alike in many ways, **discussion groups** are usually for exchanging opinions and ideas on a specific topic, usually of a technical or scholarly nature. Group members post messages or articles that others in the group can read. **Newsgroups** are typically more general in nature and can cover any topic; they allow people to get together for fun or for business purposes. For example, you could join a newsgroup for people interested in ancient civilizations or one that is used to help

people write and debug computer programs. Newsgroups can also serve as an effective advertising medium in e-commerce.

7-3c Instant Messaging

Internet Relay Chat (IRC) enables users in chat rooms to exchange text messages with people in other locations in real time. Think of it as a coffee shop where people sit around a table and chat, except each person could be in a different country. In addition, the entire conversation is "recorded," so you can scroll back to see something you missed, for example. You can find chat rooms on a variety of topics, such as gardening, video games, and relationships.

Instant messaging (IM) is a service for communicating with others via a private "chat room" on the Internet. Many IM applications are available, such as Windows Messenger, Yahoo! Messenger, and Google Chat, and the capabilities and features vary depending on the application. For example, some IM applications notify you when someone on your chat list comes online; others have features for audio or video conversations.

A new type of messaging is offered by a mobile app called Snapchat. Here, users combine pictures, videos, text, and drawings into "Snaps" that are sent to other Snapchat users. These Snaps self-destruct in a matter of seconds, seeming to not leave a trace. However, while they are on the screen they can be captured or somebody can take a picture of them. They can also be undeleted and brought back to life, using recently developed tools. Some people have been using Snapchat for sexting and/or sending nude pictures of themselves to others, which may cause them legal issues.[9] Also, the Snapchat site was hacked in late December 2013, after which the hackers posted on

Discussion groups are usually for exchanging opinions and ideas on a specific topic, usually of a technical or scholarly nature. Group members post messages or articles that others in the group can read.

Newsgroups are typically more general in nature and can cover any topic; they allow people to get together for fun or for business purposes.

Internet Relay Chat (IRC) enables users in chat rooms to exchange text messages with people in other locations in real time.

Instant messaging (IM) is a service for communicating with others via a private "chat room" on the Internet.

the Web the phone numbers and other information of more than 4 million users.[10]

7-3d Internet Telephony

Internet telephony is using the Internet rather than the telephone network to exchange spoken conversations. The protocol used for this capability is **Voice over Internet Protocol (VoIP)**. To use VoIP, you need a high-speed Internet connection and usually a microphone or headset. Some companies have special adapters that connect to your high-speed modem and allow you to use your regular phone. Because access to the Internet is available at local phone connection rates, international and other long-distance calls are much less expensive. Many businesses use VoIP to offer hotlines, help desks, and other services at far lower cost than with telephone networks.

WEB APPLICATIONS

Several service industries use the Internet and its supporting technologies to offer services and products to a wide range of customers at more competitive prices and with increased convenience. In the current economic downturn, the Internet is playing an important role in helping organizations reduce expenses, because Web applications can be used with minimum costs. The following sections describe how a variety of service industries use Web applications.

7-4a Tourism and Travel

The tourism and travel industry has benefited from e-commerce Web applications. For example, the Tropical Island Vacation (*www.tropicalislandvacation.com*) home page directs prospective vacationers to online brochures

> Many businesses use VoIP to offer hotlines, help desks, and other services at far lower cost than with telephone networks.

VoIP is also used to route traffic starting and ending at conventional public switched telephone network (PSTN) phones. The only drawback is the call quality, which is not as good as with regular phone lines. However, the quality has been improving steadily. In addition to cost savings, VoIP offers the following advantages:

- Users do not experience busy lines.
- Voicemails can be received on the computer.
- Users can screen callers, even if the caller has caller ID blocked.
- Users can have calls forwarded from anywhere in the world.
- Users can direct calls to the correct departments and take automated orders.

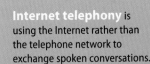

after users respond to questions about the type of vacation they want to take. They can click appealing photographs or phrases to explore further. Many travel Web sites allow customers to book tickets for plane trips and cruises as well as make reservations for hotels and rental cars. On some sites, such as InfoHub.com, specialty travel adventures are offered covering activities such as artist workshops, yoga retreats, hunting, scuba diving, and more. Expedia.com, Travel.com, Travelocity.com, Priceline.com, Hotels.com, Google.com/flights/, and

Internet telephony is using the Internet rather than the telephone network to exchange spoken conversations.

Voice over Internet Protocol (VoIP) is the protocol used for Internet telephony.

Yahoo! Travel are other examples of sites that offer all types of tourism and travel services. Yet another example, TripAdvisor (*www.tripadvisor.com*), features reviews and advice on hotels, resorts, flights, vacation rentals, vacation packages, travel guides, and much more.

7-4b Publishing

Many major publishers in the United States and Europe have Web sites that offer descriptions of forthcoming books, sample chapters, online ordering, and search features for looking up books on certain topics or by specific authors. Some publishers even offer books that can be read online free for 90 days or allow you to buy e-book versions or even selected chapters. Exhibit 7.2 shows the home page of this book's publisher, Cengage Learning (*www.cengage.com*).

7-4c Higher Education

Most universities have Web sites with information about departments, programs, faculty, and academic resources. Some even offer virtual tours of the campus for prospective students, and more universities are creating virtual divisions that offer entire degree programs via the Internet. Online degree programs help colleges and universities facing an enrollment decline, because they make it possible for students who could not attend school otherwise to enroll in classes. With online classes, universities can also have renowned experts give lectures or seminars, usually at a reduced cost, because travel expenses are not a factor. In addition, many professional certification programs are offered through the Internet, which is convenient for people who live in remote areas or cannot attend regular classes.

7-4d Real Estate

Real-estate Web sites provide millions of up-to-date listings of homes for sale. Buyers can review neighborhoods, schools, and local real-estate prices, and customers can use these sites to find realtors and brokerage firms and learn home-buying tips. Some sites have virtual tours of houses for sale, which is convenient for buyers moving to another state. (Chapter 14 covers virtual reality technologies.) Other services include appraisals, neighborhood and school profiles, financing options, and home-improvement advice. Major real-estate Web sites include Remax (*www.remax.com*), Century 21 (*www.century21.com*), Prudential (*www.prudential.com*), ERA (*www.era.com*), and Zillow (*www.zillow.com*). Zillow is a home and real-estate marketplace that assists its users in making home-related decisions. There are also apps available for both iPhone and Android devices that can simplify real-estate decisions. One popular example is the free iPhone app called HomeSnap, which uses the Multiple Listing Service (MLS) and public record data to offer home prices and other information about houses both on and off the market.[11]

7-4e Employment

Employment services are widely available on the Internet. You might be familiar with Monster.com, for example. These sites offer comprehensive services to job seekers, including the following:

- Expert advice and tools for managing your career
- Resume assistance, including tools for creating professional-looking resumes
- Job search tutorials
- Resume posting and distribution
- Searches by company, industry, region, or category
- Announcements of job fairs
- Career tests to see what career is right for you
- Salary calculators

Exhibit 7.2
Cengage Learning home page

© Cengage Learning®

7-4f Financial Institutions

Almost all U.S. and Canadian banks and credit unions, and many others worldwide, offer online banking services and use e-mail to communicate with customers and send account statements and financial reports. E-mail helps banks reduce the time and costs of communicating via phone (particularly long-distance calls) and postal mail. Customers can get more up-to-date account information and check balances at any time of the day or night. Despite all these advantages, consumer acceptance has been slow. Measures are being taken, however, to ensure that a secure nationwide electronic banking system is in place, which should help ease consumers' concerns. For example, digital signatures (discussed in Chapter 5) are a key technology because they provide an electronic means of guaranteeing the authenticity of involved parties and verifying that encrypted documents have not been changed during transmission.

The following list describes some banking services available via the Internet:

- Accessing customer service by e-mail around the clock
- Viewing current and old transactions
- Online mortgage applications
- Interactive tools for designing a savings plan, choosing a mortgage, or getting insurance quotes online
- Finding loan status and credit card account information online
- Paying bills and credit card accounts
- Transferring funds
- Viewing digital copies of checks

7-4g Software Distribution

Many vendors distribute software on the Internet as well as drivers and patches. For example, most antivirus vendors make updates available for download to keep up with new viruses and worms. Typically, patches, updates, and small programs such as new browser versions are fast and easy to download. Trying to download large programs, such as Microsoft Office Suite, takes too long, so these types of programs are not usually distributed via the Internet.

Developing online copyright-protection schemes continues to be a challenge. If users need an encryption code to "unlock" software they have downloaded, making backups might not be possible. Despite these challenges, online software distribution provides an inexpensive, convenient, and fast way to sell software.[12]

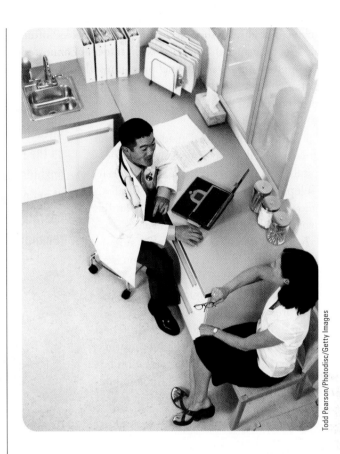

Todd Pearson/Photodisc/Getty Images

7-4h Health Care

With patient records stored on the Internet, healthcare workers can order lab tests and prescriptions, admit patients to hospitals, and refer patients to other physicians more easily; also, test and consultation results can be directed to the right patient records automatically. All patient information can be accessible from one central location; finding critical health information is faster and more efficient, especially if a patient falls ill while away from home. However, these systems have potential problems involving information privacy, accuracy, and currency. Popular health-related Web sites include:

- Yahoo! Health (*http://health.yahoo.net*)
- National Institutes of Health (NIH) (*www.nih.gov*)
- WebMD (*www.webmd.com*)
- MedicineNet (*www.medicinenet.com/script/main/hp.asp*)

There are other uses for healthcare Web sites. Telemedicine, for example, enables medical professionals to conduct remote consultation, diagnosis, and conferencing, which can save on office overhead and travel costs. In addition, personal health information systems (PHISs) can make interactive medical tools available to the public. These systems use public kiosks (often in shopping malls) equipped with Internet-connected computers and a diagnostic procedure that prompts patients with a series of questions. These systems can be useful in detecting early onset of diseases.[13]

In addition, virtual medicine on the Internet enables specialists at major hospitals to operate on patients remotely. Telepresence surgery, as it is called, allows surgeons to operate all over the world without physically traveling anywhere. A robot performs the surgery based on the digitized information sent by the surgeon via the Internet. These robots have stereoscopic cameras to create three-dimensional images for the surgeon's virtual reality goggles and tactical sensors that provide position information to the surgeon. The information box below discuses a successful implementation of electronic health records at Kaiser Permanente.

> Virtual medicine on the Internet enables specialists at major hospitals to operate on patients remotely.

7-4i Politics

Most political candidates now make use of Web sites in campaigns. The sites are a helpful tool for announcing candidates' platforms, publicizing their voting records, posting notices of upcoming appearances and debates,

and even raising campaign funds. President Barack Obama's fund-raising effort in his first and second presidential campaigns is a good example of how successful Web sites can be in this area.

Some claim the Internet has helped empower voters and revitalize the democratic process. Being well informed about candidates' stances on political issues is much easier with Web sites, for example, and online voting may make voting easier for people who in the past could not make it to polling sites. In addition, there is the possibility of legislators being able to remain in their home states, close to their constituents, and voting on bills via an online system. However, a stringent ID system would have to be in place, one that most likely would use biometric security measures. Currently, the U.S. House of Representatives is attempting to put all pending legislation online; and presidential documents, executive orders, and other materials are available on the White House's Web site. You can also find full-text versions of speeches, proclamations, press briefings, daily schedules, the proposed federal budget, healthcare reform documents, and the Economic Report of the President. The information box on the next page highlights some of the possible features of the Internet in 2020.

ELECTRONIC HEALTH RECORDS PAYS OFF FOR KAISER PERMANENTE

According to an article in InfoWorld, only 9 percent of U.S. hospitals were using even a basic form of electronic health records (EHR) in 2009, but that has changed. For example, Kaiser Permanente, a managed care consortium with over nine million members, implemented an EHR system that became fully operational in 2010. It took Kaiser 10 years to implement its system, with a total cost of $4 billion ($444 per member).

The Kaiser system, which covers every department and every patient, is called HealthConnect, and it uses data mining and analytics to improve patient service and reduce their bills. For example, the system reminds a patient about a particular test or a need to refill a prescription. According to Kaiser, duplicate testing is approximately 15 to 17 percent of total healthcare costs, and HealthConnect eliminates this cost by integrating all the patient's medical records, including X-rays and ER visits. The system also allows patients to access their medical records (including lab results) from iOS and Android devices, and they can e-mail their doctors directly through the system. Kaiser has already launched the next phase of its EHR system, Health360, which embraces the latest in mobile technology.[14]

THE INTERNET IN 2020

There are some radical predictions for the future of the Internet having to do with architectural changes, IP addresses, the DNS, and routing tables. However, most experts agree that the Internet in 2020 will have the following characteristics[15]:

1. More people will use it—as many as 5 billion, compared to 2.2 billion at the end of 2011.[16]
2. It will be more geographically dispersed, stretching to more remote corners of the globe. (It will also support more languages.)
3. It will be more of a network of devices—for example, sensors on buildings to monitor security—than a network of computers.
4. It will carry much more content—a veritable "exaflood," which is the term coined by researchers for the onrushing flow of exabytes.
5. It will be wireless. By 2014, as many as 2.5 billion people worldwide will subscribe to mobile broadband, according to one estimation.
6. More Internet services will be offered in the cloud, given that cloud computing will continue to grow, generating an estimated $45.5 billion in annual revenue as early as 2015.
7. It will be greener, thanks to more energy-efficient technologies.
8. Network management will be more automated.
9. Connectivity will become less important, thanks to techniques that will tolerate delays or forward communications from one user to another.
10. There will be more hackers.

7-5 INTRANETS

Many of the applications and services made possible with the Internet can be offered to an organization's users by establishing an intranet. An **intranet** is a network within an organization that uses Internet protocols and technologies (e.g., TCP/IP, which includes File Transfer Protocol [FTP], SMTP, and others) for collecting, storing, and disseminating useful information that supports business activities, such as sales, customer service, human resources, and marketing. Intranets are also called *corporate portals*. You might wonder what the difference is between a company's Web site and its intranet. The main difference is that the company Web site is usually public; an intranet is for internal use by employees. However, many companies also allow trusted business partners to access their intranets, usually with a password or another authentication method to protect confidential information.

An intranet uses Internet technologies to solve organizational problems that have been solved in the past by proprietary databases, groupware, scheduling, and workflow applications. An intranet is different from a LAN, although it uses the same physical connections. An intranet is an application or service that uses an organization's computer network. Although intranets are physically located in an organization, they can span the globe, allowing remote users to access the intranet's information. However, carefully defining and limiting access is important for security reasons, so intranets are typically set up behind a firewall.

In a typical intranet configuration (Exhibit 7.3), users in the organization can access all Web servers, but the system administrator must define each user's level of access. Employees can communicate with one another and post information on their departmental Web servers.

Departmental Web servers can be used to host Web sites. For example, the Human Resources Department

> An **intranet** is a network within an organization that uses Internet protocols and technologies (e.g., TCP/IP, which includes File Transfer Protocol [FTP], SMTP, and others) for collecting, storing, and disseminating useful information that supports business activities, such as sales, customer service, human resources, and marketing.

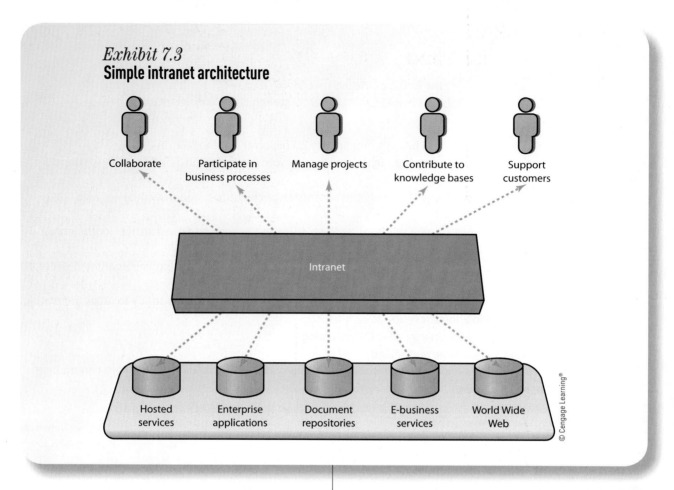

Exhibit 7.3
Simple intranet architecture

Collaborate · Participate in business processes · Manage projects · Contribute to knowledge bases · Support customers

Intranet

Hosted services · Enterprise applications · Document repositories · E-business services · World Wide Web

© Cengage Learning®

might have a separate Web site containing information that employees need to access frequently, such as benefits information or 401K records. Similarly, the Marketing Department could have a Web site with the latest product information. Employees can also bookmark important sites in the intranet.

7-5a The Internet vs. Intranets

The Internet is a public network; an intranet is a private network. Any user can access the Internet, but access to an intranet is only for certain users and must be approved. Table 7.2 summarizes the major differences between the Internet and intranets.

Despite these differences, both use the same protocol, TCP/IP, and both use browsers for accessing information. Typically, they use similar languages for developing applications, such as Java, and offer files in similar formats.

An advantage of an intranet is that because the organization can control which browser is used, it can specify a browser that supports the technologies the organization uses, such as Internet telephony or video conferencing. In addition, the organization knows documents will be displayed the same way in all users' browsers; on the Internet, there is no assurance that a Web page will be displayed the same way for every user who views it. Intranets also enable organizations to share software, such as an office suite or a DBMS.

Key feature	Internet	Intranet
TABLE 7.2 THE INTERNET VS. INTRANETS		
User	Anybody	Approved users only
Geographical scope	Unlimited	Limited or unlimited
Speed	Slower than an intranet	Faster than the Internet
Security	Less than an intranet's	More than the Internet's; user access more restricted

© Cengage Learning®

7-5b Applications of an Intranet

A well-designed intranet can make the following types of information, among others, available to the entire organization in a timely manner to improve an organization's efficiency and effectiveness[17]:

- *Human resources management*—401K plans, upcoming events, the company's mission statement and policies, job postings, medical benefits, orientation materials, online training sessions and materials, meeting minutes, vacation time
- *Sales and marketing*—Call tracking, information on competitors, customer information, order tracking and placement, product information
- *Production and operations*—Equipment inventory, facilities management, industry news, product catalog, project information
- *Accounting and finance*—Budget planning, expense reports

Intranets can also help organizations move from a calendar or schedule-based document-publishing strategy to one that is based on events or needs. In the past, for example, a company usually published an employee handbook once a year, and it was not updated until the following year, even if major changes happened that required updating, such as organizational restructuring. The company might occasionally have sent single pages as updates, and employees would have to insert these pages in a binder. Needless to say, these binders were often out of date and difficult to use. With an intranet, however, a company can make updates as soon as they are needed, in response to company events rather than a set schedule.

Intranets reduce the costs and time of document production, too. In the past, document production went through several steps, such as creating content, producing and revising drafts, migrating content to desktop publishing, duplicating, and distributing. Intranets eliminate the duplication and distribution steps, and often the step of migrating to a publishing application can be streamlined or eliminated.

7-6 EXTRANETS

An **extranet** is a secure network that uses the Internet and Web technologies to connect intranets of business partners so communication between organizations or between consumers is possible. Extranets are considered a type of interorganizational system (IOS). These systems facilitate information exchange among business partners. Some of these systems, such as electronic funds transfer (EFT) and e-mail, have been used in traditional businesses as well as in e-commerce. Electronic data interchange (EDI) is another common IOS.

As mentioned, some organizations allow customers and business partners to access their intranets for specific purposes. For example, a supplier might want to check the inventory status, or a customer might want to check an account balance. Often, an organization makes a portion of its *intranet* accessible to external parties as its *extranet*. Comprehensive security measures must ensure that access is granted only to authorized users and trusted business partners. Exhibit 7.4 shows a simple extranet. In Exhibit 7.4, DMZ refers to the demilitarized zone, an area of the network that is separate from the organization's LAN, where the extranet server is placed. Table 7.3 compares the Internet, intranets, and extranets.[18]

There are numerous applications of extranets. For example, Toshiba America, Inc., has designed an extranet for timely order-entry processing. Using this extranet, more than 300 dealers can place orders for parts until 5 p.m. for next-day delivery. Dealers can also check accounts receivable balances and pricing arrangements, read press releases, and more. This secure system has decreased costs and improved customer service.[19]

Another example of an extranet is the Federal Express Tracking System (*www.fedex.com*). Federal Express uses its extranet to collect information and make it available to customers over the Internet. Customers can enter a package's tracking number and locate any package still in the system as well as prepare and print shipping forms, get tracking numbers, and schedule pickups.

Extranets not only allow companies to reduce internetworking costs, they give companies a competitive advantage, which can lead to increased profits. A successful extranet requires a comprehensive security system and management control, however. The security system should include access control, user-based authentication, encryption, and auditing and reporting capabilities.

An **extranet** is a secure network that uses the Internet and Web technologies to connect intranets of business partners so communication between organizations or between consumers is possible.

Exhibit 7.4
Simple extranet architecture

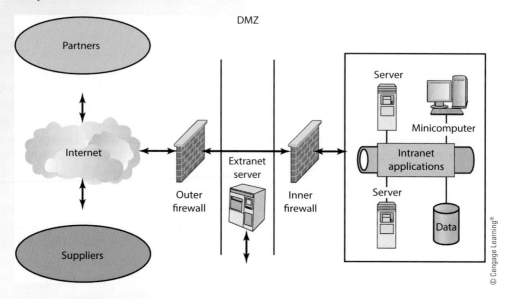

TABLE 7.3 COMPARISON OF THE INTERNET, INTRANETS, AND EXTRANETS			
	Internet	**Intranet**	**Extranet**
Access	Public	Private	Private
Information	General	Typically confidential	Typically confidential
Users	Everybody	Members of an organization	Groups of closely related companies, users, or organizations

An extranet offers an organization the same benefits as an intranet but provides other advantages, such as the following[20]:

- *Coordination*—Improves coordination between business partners, such as suppliers, distributors, and customers. Critical information can be made available quickly so decisions can be made without delays. For example, a manufacturer can coordinate production by checking the inventory status.

- *Feedback*—Provides instant feedback from customers and other business partners to an organization, and gives consumers an opportunity to express opinions on new products or services before they are introduced to the market.

Extranets not only allow companies to reduce internetworking costs, they give companies a competitive advantage, which can lead to increased profits.

- *Customer satisfaction*—Links customers to an organization so they can get more information on products and services. Customers can also order products online, expediting B2B (business-to-business). E-commerce is a major beneficiary of an extranet.

- *Cost reduction*—Reduces inventory costs by providing information to participants in a supply network program. For example, Mobil Corporation designed an extranet that allows its distributors to submit purchase orders, which has increased efficiency and expedited delivery of goods and services.

- *Expedited communication*—Improves communication by linking intranets for access to critical information. A traveling salesperson can get the latest product information remotely before going to a sales meeting, for example.

7-7 NEW TRENDS: THE WEB 2.0 AND WEB 3.0 ERAS

Web 2.0 refers to the trend toward Web applications that are more interactive than traditional Web applications. Collaboration or e-collaboration is one of the key components of Web 2.0. Table 7.4 compares various Web 1.0 and Web 2.0 applications,[21] and the following sections describe some Web 2.0 applications in more detail.

Most experts agree that Web 3.0, also known as "the Semantic Web," provides personalization that allows users to access the Web more intelligently. Computers, not their users, will perform the tedious work involved in finding, combining, and acting upon information on the Web. For example, if a user searches for the word *spring* in a sentence, the search engine figures out if the user is searching for the season, a coil, or *jump*, and then it provides the most relevant search results.

The main focus of Web 2.0 has been on social networking and collaboration. Web 3.0, on the other hand, focuses on "intelligent" Web applications using various artificial intelligent technologies (discussed in Chapter 13). These include natural language processing, artificial neural networks, and intelligent agents. The goal is to tailor online searching and requests to users' specific search patterns, preferences, and needs.

One part of Web 3.0 could be the semantic Web proposed by World Wide Web inventor Tim Berners-Lee. According to Berners-Lee, the Web can be made more useful by using methods (such as content tags) that will enable computers to understand what they are displaying and communicate more effectively with one another. Using the semantic Web, computers will be able to read Web sites as easily as humans read them. Nova Spivack's Twine (*www.slideshare.net/novaspivack/web-evolution-nova-spivack-twine*) is one of the first online services to use Web 3.0 technologies. Twine automatically organizes information, learns about users' specific interests and search patterns, and makes recommendations based on this information.[22]

Bing and Google Search already offer many Web 3.0 features.

> **Web 2.0** refers to the trend toward Web applications that are more interactive than traditional Web applications. Collaboration or e-collaboration is one of its key components.

TABLE 7.4 WEB 1.0 vs. WEB 2.0	
Web 1.0	**Web 2.0**
DoubleClick (used for online marketing)	Google AdSense
Ofoto (sharing digital photos)	Flickr
Akamai (streaming media services)	BitTorrent
mp3.com	iTunes
Britannica Online	Wikipedia
Personal Web sites	Blogging
eVite (type of wiki for event planning)	Upcoming.org and Events and Venues Database (EVBD)
Domain name speculation	Search engine optimization
Page views	Cost per click
Content management systems	Wikis
ERoom and Groove (collaboration software)	Collaboration portals, such as IBM Quickr and Microsoft Sharepoint
Posting a movie file on a personal Web page	YouTube

7-7a Blogs

A **blog** (short for *Weblog*) is a journal or newsletter that is updated frequently and intended for the general public. Blogs reflect their authors' personalities and often include philosophical reflections and opinions on social or political issues. Sometimes, they are simply used as a way for families or groups of friends to keep in touch with one another. Automated tools have made creating and maintaining blogs easy, so even people with very little technical background can have blogs. In addition, many Web sites, such as Blogger.com, offer free space for blogs and even allow bloggers to post photos. One popular blogging tool is Tumblr (*www.tumblr.com*), which allows users to post anything—text, photos, quotes, links, music, and videos—from their browsers, phones, desktops, e-mail programs, and so forth. Another popular site is Pinterest (*http://pinterest.com*).

There are also blogs on Web sites that are dedicated to particular topics or organizations; these are periodically updated with the latest news and views. For example, on the CNN Web site, you can find blogs written by Anderson Cooper and Wolf Blitzer, among others. Blogs are becoming a popular source of online publication, too, especially for political information, opinions, and alternative news coverage; some examples can be found at *www.huffingtonpost.com* and *www.slate.com*. Microblogs, a newer version of traditional blogs, enable users to create smaller versions of blog posts (known as *microposts*); these can take the form of short sentences or individual images or links.

7-7b Wikis

A **wiki** is a type of Web site that allows users to add, delete, and sometimes modify content. One of the best-known examples is the online encyclopedia Wikipedia. What is unique about wikis is that an information user can also be an information provider. The most serious problem with wikis is the quality of information, because allowing anyone to modify content affects the content's accuracy. Wikipedia is currently working on methods to verify the credentials of users contributing to the site because of past problems with contributors falsifying credentials.

Wikis have caught on at many companies, too. For example, an Intel employee developed Intelpedia as a way for employees around the world to share information on company history, project progress, and more. However, some employees do not like their content being edited by others. For this reason, "corporate wikis" were developed; these include tighter security and access controls. Corporate wikis are used for a variety of purposes, such as posting news about product development. Many open-source software packages for creating wikis are available, such as MediaWiki and TWiki. Companies are also creating wikis to give customers information. For example, Motorola and T-Mobile have set up wikis about their products that function as continually updated user guides, and eBay has formed eBay Wiki, where buyers and sellers can share information on a wide range of topics.[23]

7-7c Social Networking Sites

Social networking refers to a broad class of Web sites and services that allow users to connect with friends, family, and colleagues online as well as meet

LINKEDIN: A PROFESSIONAL SOCIAL NETWORKING SITE

LinkedIn is the world's largest professional networking site, with over 259 million members from around the world. LinkedIn allows its members to connect with professional partners to exchange ideas, opportunities, and knowledge.[25] As a future knowledge worker, you can use LinkedIn to network with industry professionals, maintain current professional relationships, create new relationships, build your personal network of industry experts, gain technical information, search for jobs, bookmark jobs for which you would like to apply, and enhance your career opportunities. You can use LinkedIn's search tool to look for jobs and professional expertise, you can join LinkedIn groups for a particular interest, and much more. Over 2 million companies have a company profile on LinkedIn. Businesses can use LinkedIn to acquire new customers, keep in touch with existing customers, find the right vendors, build their industry networks, list jobs and search for potential candidates, get answers to technical questions, do fund-raising, and so forth. This book also has a LinkedIn page.[26]

Twitter: Real-time Networking with Your Followers

Twitter is a social network that connects you to the latest stories and to people whom you like to follow. As of January 31, 2013, there were an estimated 485 million Twitter users around the world, and it is considered the fastest growing social network.[28] Each tweet is no more than 140 characters long; you can also see photos, videos, and conversations directly in tweets. This gives you a complete picture in one place. Businesses use Twitter to connect with their customers, and individuals use it to connect with their followers in real time. Using Twitter, businesses can gather real-time market information and feedback as well as create relationships with their customers.[29]

The following list highlights some of the organizations that use Twitter to stay connected with their customers and promote their products and services[30]:

- The U.S. Chamber of Commerce uses Twitter to reach an influential audience.
- MTV incorporates Twitter into its Video Music Awards.
- The Fox television network integrates a real-time Twitter feed into its TV ads.
- The professional soccer team LA Galaxy uses Twitter to guide conversations and build awareness.
- The Red Cross uses Twitter for donor support and to mobilize communities in times of need.
- The UN Foundation extends the reach of its events globally with Twitter.

people with similar interests or hobbies. According to a study conducted by Experian Marketing Services, 16 minutes of every hour that American spend online takes place on social networks.[24] More than 100 of these social networks are available on the Internet. Two of the most popular are Facebook and Twitter. In addition, LinkedIn is a professional networking site where you can connect with professional contacts and exchange ideas and job opportunities with a large network of professionals. Many people now use both LinkedIn and Facebook to keep their professional and social contacts separate. The information box on the previous page discusses some additional features of LinkedIn.

Social networking sites are also popular for business use. For example, many companies use Twitter to keep track of customer opinions about their products. (See the information box above.) Comcast, Dell, General Motors, and H&R Block are some additional examples.[27] Companies also use social networking sites for advertising; they might include links to their company Web sites or use pay-per-click (PPC) features. PPC is an Internet advertising method used on Web sites, in which advertisers pay their host only when their ad is clicked.

Twitter is extremely popular, and the term *tweet* is often used for a response or comment no longer than 140 characters, the maximum length allowed for a Twitter post. Even this book has a Twitter account! Follow it at *twitter.com/4LTRPress_MIS*.

7-7d RSS Feeds

RSS (Really Simple Syndication) feeds are a fast, easy way to distribute Web content in Extensible Markup Language (XML) format. RSS is a subscription service, and new content from Web sites you have selected is delivered via a feed reader. The content all goes to one convenient spot where you can read "headlines." With this service, you do not have to keep checking a site for updates.

XML, a subset of the Standard Generalized Markup Language (SGML), is a flexible method for creating common formats for information. Unlike HTML tags that specify layout and appearance, XML tags represent the kind of content being posted and transmitted. Although HTML contains some layout and appearance features, these "presentational attributes" are deprecated by the W3C, which suggests that HTML only be used for creating structured documents through markup. Layout and appearance should be handled by CSS (Cascading Style Sheets).

Data can be meaningless without a context for understanding it. For example, consider the following:

RSS (Really Simple Syndication) feeds are a fast, easy way to distribute Web content in Extensible Markup Language (XML) format. It is a subscription service, and new content from Web sites you have selected is delivered via a feed reader to one convenient spot.

"Information Systems, Smith, John, 357, 2014, Cengage, 45.00, 02-139-4467-X." From this string of data, you might make an educated guess that this refers to a book, including the title, author name, number of pages, year of publication, publisher, price, and ISBN. A computer, however, might interpret this same data as indicating that Smith John spoke about information systems for 357 minutes at a conference organized by Cengage in 2014 and received 45 Euros as compensation, with a transaction ID of 02-139-4467-X for the payment.

XML prevents this kind of confusion by defining data with a context. So, you would format the preceding data string as follows:

```
<book>
<title>Information Systems</title>
<authorlastname>Smith</authorlastname>
<authorfirstname>John</authorfirstname>
<pages>357</pages>
<yearofpub>2014</yearofpub>
<publisher>Cengage</publisher>
<pricein$>45.00</pricein$>
<isbn>02-139-4467-X</isbn>
</book>
```

As you can see, each piece of data is defined with its context by using tags, which makes the data much easier to interpret. Although both HTML and XML are tag-based languages, they have different purposes. XML was designed to improve interoperability and data sharing between different systems, which is why RSS feeds are in XML. Any system can interpret the data in an RSS feed the correct way because it is based on the data's meaning, not its format and layout.

7-7e Podcasting

A **podcast** is an electronic audio file, such as an MP3 file, that is posted on the Web for users to download to their mobile devices or even their computers. Users can

also listen to it over the Web. A podcast has a specific URL and is defined with an XML item tag.

Podcasts are usually collected by an "aggregator," such as iTunes or iPodder. You can also subscribe to various podcasts; NPR, *The Economist*, and ESPN all offer podcast subscriptions. What differentiates a podcast from a regular audio file is that users can subscribe to it. Each time a new podcast is available, the aggregator collects it automatically, using the URL, and makes it available for subscribers. Subscribers can then "sync" the podcast with their mobile devices and listen to it whenever they want.

> Financial institutions offer podcasts to inform customers about investment strategies, market performance, and trading.

This subscription model makes podcasts more useful and popular and increases their accessibility. Syndication feeds are one way of announcing a podcast's availability. Organizations use podcasts to update people on their products and services, new trends, changes in organizational structure, and merger/acquisition news. Financial institutions, for example, offer podcasts to inform customers about investment strategies, market performance, and trading. When multimedia information is involved, the terms *video podcast, vodcast,* or *vidcast* are sometimes used.

7-7f The Internet2

Another recent development is **Internet2 (I2)**, a collaborative effort involving more than 200 U.S. universities and corporations (including AT&T, IBM, Microsoft, and Cisco Systems) to develop advanced Internet technologies and applications for higher education and academic research.

The I2 project started in 1987 and was planned as a decentralized network in which universities located in the same geographic region form alliances to create a local connection point-of-presence called a gigapop. **Gigapops** connect a variety of high-performance networks, and a gigapop's main function is the exchange of I2 traffic with a specified bandwidth. One major objective of the I2 project is to develop new applications that can enhance researchers' ability to collaborate and conduct scientific experiments.

I2 relies on the NSFNET and MCI's very high-speed backbone network service (vBNS), which was designed in 1995 as a high-bandwidth network for research applications. This nationwide network operates at 622 Mbps, using MCI's advanced switching and fiber-optic transmission technologies. Applications of I2 include:

- *Learningware*—This suite of applications is intended to make education more accessible, targeting distance learning and self-education. The proposed Instructional Management System (IMS) provides an environment that enables students to learn in an "anytime, anywhere" fashion. This technology also gives instructors access to a broad range of teaching materials for online classes. Some software, such as WebEx and Elluminate Live, is already available.

- *Digital Library*—This initiative, started in the 1990s, aimed to create an electronic repository of educational resources, such as textbooks and journals. The goal was to include rare books and documents, such as the Dead Sea Scrolls and the Magna Carta, although this goal has not been achieved yet. The bottom line is that researchers can have access to everything they need without leaving their offices, including access to experts who can guide them in their work. Imagine learning the theory of relativity directly from Albert Einstein, for example.

- *Teleimmersion*—A teleimmersion system allows people in different locations to share a virtual environment created on the Web. Virtual reality has important applications in education, science, manufacturing, and collaborative decision making. For instance, a pediatric cardiologist in Los Angeles can work with a surgeon in India to explain how to perform an operation transposing the pulmonary artery and vein in an infant born with an abnormality. The cardiologist can train the surgeon in a virtual reality environment with an imaginary infant. This technology can be used in many other training settings, such as putting out oil-rig fires, repairing complex machinery, and even co-teaching a virtual class. Oil companies have used this technology to conduct an exploratory dig and evaluate situations such as high-pressure natural gas in the vicinity or too much water in the dig. In this case, a company builds its own teleimmersion center, and all participants are in the same location.

- *Virtual laboratories*—These are environments designed specifically for scientific and engineering applications, and they allow a group of researchers connected to I2 to work on joint projects, such as large-scale simulations and global databases.

 ## 7-8 THE INTERNET OF EVERYTHING: THE NEXT BIG NETWORK

The **Internet of Everything (IoE)** refers to a Web-based development in which people, processes, data, and things are interconnected via the Internet using various means, such as RFID devices, barcodes, wireless systems (using Bluetooth and WI-FI), and QR codes. It is a network of networks in which billions of connections create unparalleled opportunities and challenges.

Whereas "IoE" refers to all the connections that would be made, the **Internet of Things (IoT)** refers to the physical objects that are connected to the Internet and, therefore, to all the other physical objects. As an example, consider a smart refrigerator that signals when your milk's consumption period has expired. By 2020, over 50 billion objects will be connected to the Internet, according to one estimate.[31,32] And they will likely be smart devices that are uniquely identified though IP addresses, RFIDs, QR codes, or sensors.

The technology behind the Internet of Everything will facilitate, among other things, automated inventory systems in the retail industry, automated and programmable appliances in domestic households, and road and bridge systems that will be able to detect a problem as soon as it occurs and notify the authorities. Consumers and businesses will save money by preserving energy when they control their room temperatures onsite or remotely through smart devices, while companies will save time and money on labor due to automation.

A **gigapop** is a local connection point-of-presence that connects a variety of high-performance networks, and its main function is the exchange of I2 traffic with a specified bandwidth.

The **Internet of Everything (IoE)** refers to a Web-based development in which people, processes, data, and things are interconnected via the Internet using various means, such as RFID devices, barcodes, wireless systems (using Bluetooth and WI-FI), and QR codes.

The **Internet of things (IoT)** refers to the physical objects that are connected to the Internet and, therefore, to all the other physical objects.

In general, the Internet of Everything could help solve many 21st-century social problems, such as hunger, water pollution, adverse climate change, and increasing energy costs. More specific benefits could include people being able to take more effective preventive measures regarding their health by wearing sensor-embedded clothing that measures vital signs; the resulting data can be securely and quickly transmitted to doctors.[33]

Many smart objects are currently on the market, such as: 94Fifty's Bluetooth basketball, which can sense dribble force and shot angle; Nike Hyperdunk Plus basketball shoes, which can tell the users how fast they are running and how high they are jumping; and the Under Armour Armour39 shirt, which can tell the user's heart rate and lung capacity. Currently, these smart objects cannot be connected to one another, however; thus, if all three of them were on one individual's body, they would not be aware of each one another's existence. The Internet of Everything should change this by creating connectivity and communication.[34]

Individuals, businesses, and governments around the globe will benefit from IoE technology; security, privacy, and reliability will play a major role in the success of this technology, as it does for any network. There needs to be close coordination and communication among these three key players to protect the privacy and integrity of the information that is being shared on this global network. The information box below highlights two real-life examples of IoE as it exists today.

The Industry Connection for this chapter highlights the role of Google, Inc., as a leader in search technology and Web advertising platforms.

THE INTERNET OF EVERYTHING IN ACTION

In parts of Africa, people used to go from well to well in order to get some drinking water. After hiking miles to get to a particular well, they often found out that the well was not operational. German pump manufacturer Grundfos installed sensors in these wells and developed an app so that people, using their smartphones, can check the well's condition before hiking miles to get there.[35]

Elsewhere, the City24/7 project is designed to vitalize large cities by gathering and sharing useful information accessible to everyone, everywhere though multiple devices. In collaboration with Cisco IBSG and the City of New York, the interactive platform was launched. The gathered information is displayed on large screens placed at public places such as bus stations, subways, and telephone booths. The information is location based and will be displayed as soon as it becomes available, so that citizens can access it. First responders, including the police and fire departments, can use this information and respond accordingly. The next phase is the expansion of the project into other large cities.[36]

Industry Connection: Google, Inc.[37]

Google, founded in 1998, offers one of the most widely used search engines in the world. It also offers products and services in the following categories:

Search: There are more than 20 search categories, including Web, blog (by blog name, posts on a certain topic, a specific date range, and more), catalogs (for mail-order catalogs, many previously unavailable online), images, books (searches the full text of books and magazines that Google has scanned), and maps. There is also Google Earth, an exciting feature that combines searching with the viewing of satellite images, terrain maps, and 3-D models.

Ads: AdSense, AdWords, and Analytics are used for displaying ads on your Web site or with Google search results. For example, with AdWords, you create an ad and choose key words related to your business. When people search Google with one of your key words, your ad is displayed next to the search results, so you are reaching an audience that is already interested in your product or service.

Applications: Google has many applications for account holders, such as Gmail, Google Talk (instant messaging and voice calls), Google Groups, YouTube, Blogger, Google Checkout (for shopping online with just a single Google sign-in), and Google Docs, a free, Web-based word-processing and spreadsheet program.

YouTube: A video-sharing Web site on which users can upload, share, and view videos.

Enterprise: Google has applications for organizations, such as Google Maps for Enterprise, which is particularly useful for planning operations and logistics, and SketchUp Pro, a 3-D modeling tool for architects, city planners, game developers, and others that can be used in presentations and documents.

Mobile: Many of Google's services and applications, such as Blogger, YouTube, and Gmail, are available on mobile devices. You can also use text messaging to get real-time information from Google on a variety of topics, including weather reports, sports scores, and flight updates. Google also offers Android OS for smartphone and mobile devices.

Google Wave: This is a collaboration and communication tool designed to consolidate features from e-mail, instant messaging, blogging, wikis, and document sharing while offering a variety of social networking features.

Google Plus: A social networking feature to compete with Facebook, Google Plus is based on the concept of circles, so a user may create a circle called "college friends," a circle called "work friends," and so forth. This may give a user more control over who can see what about a person.

Google Now: Available on smartphones and tablets, this serves as a personal assistant by providing relevant information on weather, sports, traffic, and much more.

STUDY TOOLS 7

LOCATED AT BACK OF THE TEXTBOOK
☐ Rip out Chapter Review Card

LOCATED AT WWW.CENGAGE.COM/LOGIN
☐ Review Key Terms Flash Cards (Print or Online)

☐ Download audio and visual summaries to review

☐ Complete both Practice Quizzes to prepare for tests

☐ Play "Beat the Clock" and "Quizbowl" to master concepts

☐ Complete "Crossword Puzzle" to review key terms

☐ Watch video on "Method" for a real company example

KEY TERMS

Advanced Research Projects Agency Network (ARPANET) 137
blog 154
directories 142
discussion groups 144
Domain Name System (DNS) 140
extranet 151
gigapop 157
hypermedia 138
hypertext 138
Hypertext Markup Language (HTML) 141
instant messaging (IM) 144
Internet 137
Internet backbone 138
Internet Relay Chat (IRC) 144

Internet telephony 145
Internet2 (I2) 156
intranet 149
navigational tools 141
newsgroups 144
podcast 156
RSS (Really Simple Syndication) feeds 155
search engine 141
social networking 154
the Internet of Everything (IoE) 157
the Internet of Things (IoT) 157
uniform resource locators (URLs) 140
Voice over Internet Protocol (VoIP) 145
Web 2.0 153
wiki 154

REVIEWS AND DISCUSSIONS

1. What is a domain name system?
2. What is the format of the majority of the documents on the Web?
3. What are three examples of popular search engines?
4. What are three business applications of e-mail?
5. What are three applications of the Internet in the healthcare industry?
6. What are two business applications of extranets?
7. What are three key features of Web 2.0?
8. What is the Internet of Everything?

PROJECTS

1. Snapchat is a popular messaging system used by over four million teenagers and young adults. After reading the information presented in this chapter and other sources, write a one-page paper that summarizes some of the possible legal and social issues related to this messaging system. Why are some parents concerned when their teens use this system?

2. AskMD is a popular medical app. After reading the information presented in this chapter and other sources, write a one-page paper that summarizes some of the features of this app. What are three other examples of medical apps for iOS and Android devices?

3. The human resources departments of many organizations are creating intranets to improve the efficiency and effectiveness of the departments' operations. After reading the information presented in this chapter and other sources, write a one-page paper that summarizes the applications of intranets in HR operations. What are some of the challenges in designing and using such an application?

4. Small as well as large businesses could use social networks to improve their efficiency and help their bottom lines. After reading the information presented in this chapter and other sources, write a two-page paper that identifies two applications of Twitter, Facebook, and Yelp in running a small business.

5. Graph Search by Facebook and Knowledge Graph by Google offer features that are not available in traditional search engines. After reading the information presented in this chapter and other sources, write a one-page paper that outlines two business applications of each search method. How could these new search methods help a business generate more revenue?

6. The Internet of Everything has created a lot of excitement in the business world. After reading the information presented in this chapter and other sources, write a one-page paper that lists five business applications of this new platform. What are some of the legal and social issues related to the Internet of Everything?

ARE YOU READY TO MOVE ON?

1. ARPANET started in 1991. True or False?
2. Google Chrome is an example of a search engine. True or False?
3. A wiki is a type of Web site that allows users to add, delete, and sometimes modify content. True or False?
4. Which of the following is not an example of a Web browser?
   a. Microsoft Internet Explorer (IE)
   b. Mozilla Firefox
   c. Apple Safari
   d. DuckDuckGo
5. Which of the following is the most popular professional social network?
   a. Google+
   b. LinkedIn
   c. Facebook
   d. Twitter
6. Which of the following is not among the features of Web 2.0?
   a. Blogs
   b. Wikis
   c. Internet2
   d. RSS feeds

Scotts Miracle-Gro's Intranet: The Garden

Scotts Miracle-Gro is a major provider of lawn, garden, and outdoor-living products and services. Its headquarters is in Marysville, Ohio, and it has over 8,000 employees.[38] In 2012, the company's internal network, called "The Garden," was chosen by the Nielsen Norman Group as one of the Ten Best Intranets of 2012. Since it was launched in 2011, The Garden has served as an effective internal communication tool for Scotts employees. On any given day, some 2,250 employees visit the site in order to get the latest news about the company, among other things.

In fact, the system was developed based on employee needs and through close communication between the employees and the IT department.[39] According to Tyler Kerr, Manager of Electronic Communications at Scotts, the company conducted surveys, engaging its stakeholders in discussions about their needs and opinions. As a result, the Garden is very user friendly, and employees are constantly asked about features they would like to be added.

The Garden gives Scotts employees fast access to important and frequently used documents, such as travel expense and HR forms. Employees can also organize their important links into "My Favorites" lists. In October 2011, Scotts further enhanced The Garden by designing "The Vine," an internal social media site. Using the Vine, employees are able to communicate with one another through their profile categories, such as job title, location, years of experience, and other relevant information.[40]

Answer the following questions:

1. When was The Garden launched?
2. What are some key features of The Garden?
3. Why has The Garden been so successful?
4. What is the purpose of The Vine?

Social Networking in Support of Small Businesses

The Internet puts small businesses on the same footing as large organizations by providing an inexpensive platform for interacting with customers and selling products and services. With their global reach, social networking sites are a good example of how the Internet can level the playing field. Here are five ways that small businesses can take advantage of these sites:

- *Creating local social networks*—Small businesses can use sites such as Yelp (*www.yelp.com*), which help people find local restaurants, dentists, hair stylists, mechanics, and so on. People also use these sites to find out about upcoming events, take advantage of special offers, and talk to other customers.
- *Creating a blog or social hub*—This allows small businesses to keep customers engaged by creating useful content such as how-to-do lists or industry insights.
- *Using Twitter*—This allows small businesses to connect with their consumers in real time.
- *Creating a Facebook fan page*—This allows small business to visualize and build a community that can be customized by adding maps, coupons, and so forth.
- *Using a custom wiki*—A wiki can be used as a public forum that alerts customers to problems and concerns related to the company's products and services, and resolves issues or answers questions to keep customers engaged.[41]

Answer the following questions:

1. How does the Internet put small businesses on the same footing as large organizations?
2. What are two ways social networking sites such as Twitter can help a small business?
3. How can a site such as Yelp help small businesses?

8 | E-Commerce

LEARNING OUTCOMES

After studying this chapter, you should be able to:

8-1 Define e-commerce and describe its advantages, disadvantages, and business models.

8-2 Explain the major categories of e-commerce.

8-3 Describe the business-to-consumer e-commerce cycle.

8-4 Summarize the major models of business-to-business e-commerce.

8-5 Describe mobile-based and voice-based e-commerce.

8-6 Explain three supporting technologies for e-commerce.

After you finish this chapter, go to **PAGE 178** for the **STUDY TOOLS**

This chapter provides an overview of e-commerce and value chain analysis, then compares e-commerce with traditional commerce. It explains e-commerce business models and the major categories of e-commerce. Along the way, it shows what the major activities are in the business-to-consumer e-commerce cycle and the major models of business-to-business e-commerce as well as mobile-based and voice-based e-commerce. Finally, it provides an overview of electronic payment systems, Web marketing, and search engine optimization—three supporting technologies for e-commerce operations.

8-1 DEFINING E-COMMERCE

E-commerce and e-business differ slightly. **E-business** encompasses all the activities a company performs in selling and buying products and services using computers and communication technologies. In broad terms, e-business includes several related activities, such as online shopping, sales force automation, supply chain management, electronic procurement (e-procurement), electronic payment systems, Web advertising, and order management. **E-commerce** is buying and selling goods and services over the Internet. In other words, e-commerce is part of e-business. However, the two terms are often used interchangeably.

E-business includes not only transactions that center on buying and selling goods and services to generate revenue but also transactions that generate demand for goods and services, offer sales support and customer service, and facilitate communication between business partners.

E-commerce builds on traditional commerce by adding the flexibility that networks offer and the availability

> **E-business** encompasses all the activities a company performs in selling and buying products and services using computers and communication technologies.
>
> **E-commerce** is buying and selling goods and services over the Internet.

of the Internet. The following are common business applications that use the Internet:

- Buying and selling products and services
- Collaborating with other companies
- Communicating with business partners
- Gathering business intelligence on customers and competitors
- Providing customer service
- Making software updates and patches available
- Offering vendor support
- Publishing and disseminating information

8-1a The Value Chain and E-Commerce

One way to examine e-commerce and its role in the business world is through value chain analysis. Michael Porter introduced the **value chain** concept in 1985.[1] It consists of a series of activities designed to meet business needs by adding value (or cost) in each phase of the process. Typically, a division within a business designs, produces, markets, delivers, and supports its products or services. Each activity adds cost and value to the product or service delivered to the customer (see Exhibit 8.1).

In Exhibit 8.1, the top four components—organizational infrastructure, human resource management, technological development, and procurement (gathering input)—are supporting activities. The "margin" represents the value added by supporting primary activities (the components at the bottom). The following list describes primary activities:

- *Inbound logistics*—Movement of materials and parts from suppliers and vendors to production or storage facilities; includes tasks associated with receiving, storing, and disseminating incoming goods or materials.
- *Operations*—Processing raw materials into finished goods and services.
- *Outbound logistics*—Moving and storing products, from the end of the production line to end users or distribution centers.

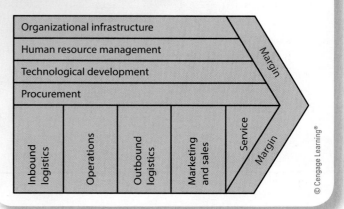

Exhibit 8.1
Michael Porter's value chain

Organizational infrastructure					Margin
Human resource management					
Technological development					
Procurement					
Inbound logistics	Operations	Outbound logistics	Marketing and sales	Service	Margin

© Cengage Learning®

- *Marketing and sales*—Activities for identifying customer needs and generating sales.
- *Service*—Activities to support customers after the sale of products and services.

For instance, by having superior relationships with suppliers (through prompt payments, electronic ordering, loyalty, and so forth), the company can ensure timely delivery and a high quality of raw materials. These, in turn, add value for customers by providing a high-quality product at a lower cost. If good quality and lower costs are top priorities for customers, the company knows on which parts of the value chain to focus (e.g., better suppliers to ensure quality and reduce costs, superior operations to ensure quality, better distribution to reduce costs, better after-sales service to ensure quality with warranties). So the value chain is really about understanding what aspects of an organization's business add value for customers and then maximizing those aspects.

A furniture manufacturing company, for example, buys raw materials (wood) from a logging company, converts the raw materials into finished products (chairs), ships the finished products to retailers, distributors, or customers, and markets these products. The company can continue the value chain after delivering the chairs to furniture stores by offering other products and services. Value chain analysis can also help the furniture company

> The value chain is really about understanding what aspects of an organization's business add value for customers and then maximizing those aspects.

A **value chain** is a series of activities designed to meet business needs by adding value (or cost) in each phase of the e-commerce process.

spot manufacturing opportunities. For example, it could cut costs of raw materials if it owned or formed a partnership with a lumber company. In any industry, a company is part of a value chain when it buys goods or services from suppliers, adds features to increase value, and sells the goods or services to customers. As another example, a computer vendor can buy components from different vendors and then assemble them into complete PCs. The vendor has added value by assembling the components and, therefore, can charge a higher price than the combined cost of the components.

E-commerce, its applications, and its supporting technologies, as discussed in this chapter, are an example of using Porter's value chain concept. The Internet can increase the speed and accuracy of communication between suppliers, distributors, and customers. Moreover, the Internet's low cost means companies of any size can take advantage of value chain integration, which is the process of multiple companies in a shared market working together to plan and manage the flow of goods, services, and information from manufacturers to consumers. This process optimizes the value chain's efficiency, thus creating a competitive advantage for all companies involved.

E-commerce can enhance a value chain by offering new ways to reduce costs or improve operations, including the following:

- Using e-mail rather than regular mail to notify customers of upcoming sales can reduce costs.

- Selling to customers via the company Web site can generate new sources of revenue, particularly from customers who live far away from the company's headquarters or physical store.

- Offering online customer service can make products or services more appealing to customers.

As you learn throughout this book, many companies have taken advantage of the Web and e-commerce to reduce costs, increase revenue, and improve customer service. For example, Dell Computer generates a large portion of its revenue through the Web and eliminates the middleman in the process. Similarly, Cisco Systems sells networking hardware and software over the Web, and customers can track packages on the Web with United Parcel Service (UPS) and FedEx. The information box below describers how Twitter, a popular social networking site, can help businesses.

8-1b E-Commerce vs. Traditional Commerce

Although the goal of e-commerce and traditional commerce is the same—selling products and services to generate profit—they do it quite differently. In e-commerce, the Web and telecommunication technologies play a major role. Often, there is no physical store, and the buyer and seller do not see each other. Many companies now operate as a mix of traditional commerce and e-commerce, however, and have some kind of e-commerce presence. These companies, referred to as **click-and-brick e-commerce**, capitalize on the advantages of online interaction with their customers yet retain the benefits of having a physical store location. For example, customers can buy items from the company's Web site but take them to the physical store if they need to return items. Table 8.1 compares e-commerce and traditional commerce.

> **Click-and-brick e-commerce** mixes traditional commerce and e-commerce. It capitalizes on the advantages of online interaction with customers yet retains the benefits of having a physical store location.

TWITTER HELPS BUSINESSES FIND CUSTOMERS

In addition to its personal and social applications, Twitter can be used by businesses as a promotional tool and as a way to find sales leads. John Pepper, CEO and cofounder of Boloco, a burrito restaurant chain, posted a photo of a coupon on Twitter and invited customers to bring in any image of the coupon—a photocopy, a printout, or an image on a mobile phone—to get the discount. The promotion was a big success, given that 900 customers redeemed the coupon (including by bringing their mobile phones) as opposed to the usual 350. Similarly, Dell Computer announced that it had generated $3 million in sales on Twitter during a 2-year period by posting the coupon numbers for discounts on Dell Outlet items. Before long, the company's Twitter account, @DellOutlet, had attracted over 700,000 followers. Twitter also helps businesses find customers who are looking for their products or services, and it allows businesses to deal with unhappy customers quickly.[2]

TABLE 8.1 E-COMMERCE VS. TRADITIONAL COMMERCE

Activity	Traditional Commerce	E-Commerce
Product information	Magazines, flyers	Web sites, online catalogs
Business communication	Regular mail, phone calls	E-mail
Check product availability	Phone calls, faxes, and letters	E-mail, Web sites, and extranets
Order generation	Printed forms	E-mail, Web sites
Product acknowledgments	Phone calls, faxes	E-mail, Web sites, and electronic data interchange (EDI)
Invoice generation	Printed forms	Web sites

© Cengage Learning®

8-1c Advantages and Disadvantages of E-Commerce

Businesses of all sizes use the Internet and e-commerce applications to gain a competitive edge. For example, IBM does business with more than 12,000 suppliers over the Web and uses the Web for sending purchase orders, receiving invoices, and paying suppliers. In addition, IBM uses the Internet and Web technologies for its transaction-processing network.

Similar to traditional business, e-commerce has many advantages and disadvantages. If e-commerce is based on a sound business model (discussed in the next section), its advantages outweigh its disadvantages. Advantages of e-commerce include the following:

- Creating better relationships with suppliers, customers, and business partners
- Creating "price transparency," meaning all market participants can trade at the same price
- Being able to operate around the clock and around the globe
- Gathering more information about potential customers
- Increasing customer involvement (e.g., offering a feedback section on the company Web site)
- Improving customer service
- Increasing flexibility and ease of shopping
- Increasing the number of customers

- Increasing opportunities for collaboration with business partners
- Increasing return on investment because inventory needs are reduced
- Offering personalized services and product customization
- Reducing administrative and transaction costs

E-commerce also has the following disadvantages, although many of these should be eliminated or reduced in the near future:

- Bandwidth capacity problems (in certain parts of the world)
- Security and privacy issues
- Accessibility (not everybody is connected to the Web yet)
- Acceptance (not everybody accepts this technology)

8-1d E-Commerce Business Models

The Internet has improved productivity in many organizations, but this improvement must be converted to profitability. As the fall of many dot.com companies in 2000 and 2001 shows, just improving productivity is not enough. The companies that survive have a sound business model governing how they plan to make a profit and sustain a business for future growth.

To achieve profitability, e-commerce companies focus their operations in different parts of the value chain,

The Internet has improved productivity in many organizations, but this improvement must be converted to profitability.

© iStockphoto.com/Micah Young

discussed earlier. To generate revenue, for example, an e-commerce company might decide to sell only products or services or cut out the middleman in the link between suppliers and consumers. Many business-to-consumer (B2C) business models do the latter by using the Web to deliver products and services to customers, which helps reduce prices and improve customer service. As you learned in the discussion of Michael Porter's differentiation strategies in Chapter 1, by differentiating themselves from their competitors in this fashion, these companies can increase their market shares as well as their customer loyalty.

The products that e-commerce companies sell could be traditional products, such as books and apparel, or digital products, such as songs, software, and e-books. Similarly, e-commerce models can be traditional or "digital." Traditional e-commerce models are usually an extension or revision of traditional business models, such as advertising or being a merchant, or a new type of model suitable for Web implementation, such as an infomediary (described in the following list). The following are the most widely used business models in e-commerce[3]:

- *Merchant*—The **merchant model** transfers the old retail model to the e-commerce world by using the medium of the Internet. In the most common type of merchant model, an e-commerce company uses Internet technologies and Web services to sell goods and services over the Web. Companies following this model offer good customer service and lower prices to establish a presence on the Web. Amazon uses this model, but traditional businesses, such as Dell, Cisco, and Hewlett-Packard, have adopted this model to eliminate the middleman and reach new customers.

- *Brokerage*—Using the **brokerage model** brings sellers and buyers together on the Web and collects commissions on transactions between these parties. The best example of this model is an online auction site, such as eBay (*www.ebay.com*), DealDash (*www.dealdash.com*), or QuiBids (*www.quibids.com*).

Auction sites can generate additional revenue by selling banner advertisements. Other examples of the brokerage model are online stockbrokers, such as TDAmeritrade.com and Schwab.com, which generate revenue by collecting commissions from buyers and sellers of securities.

- *Advertising*—The **advertising model** is an extension of traditional advertising media, such as radio and television. Directories such as Yahoo! provide content (similar to radio and TV) to users for free. By creating more traffic with this free content, they can charge companies for placing banner ads or leasing spots on their sites. Google, for example, generates revenue from AdWords, which offers pay-per-click (PPC) advertising and site-targeted advertising for both text and banner ads.

- *Mixed*—The **mixed model** refers to generating revenue from more than one source. For example, ISPs such as AOL generate revenue from advertising and from subscription fees for Internet access. An auction site can also generate revenue from commissions collected from buyers and sellers and from advertising.

- *Informediary*—E-commerce sites that use the **informediary model** collect information on consumers and businesses and then sell this information to other companies for marketing purposes. For example, Bizrate (*bizrate.com*) collects information about the performance of other sites and sells this information to advertisers.

- *Subscription*—Using the **subscription model**, e-commerce sites sell digital products or services to customers. For example, the *Wall Street Journal* and *Consumer Reports* offer online subscriptions, and antivirus vendors use this model to distribute their software and updates. The information box on the next page predicts where e-commence will be in 2020.

The **merchant model** transfers the old retail model to the e-commerce world by using the medium of the Internet.

Using the **brokerage model** brings sellers and buyers together on the Web and collects commissions on transactions between these parties.

The **advertising model** is an extension of traditional advertising media, such as radio and television. Directories such as Yahoo! provide content (similar to radio and TV) to users for free. By creating more traffic with this free content, they can charge companies for placing banner ads or leasing spots on their sites.

The **mixed model** refers to generating revenue from more than one source.

Under the **informediary model**, e-commerce sites collect information on consumers and businesses and then sell this information to other companies for marketing purposes.

Under the **subscription model**, e-commerce sites sell digital products or services to customers.

E-COMMERCE IN 2020

According to Goldman Sachs, a multinational investment banking firm based in the United States, e-commerce in the United States will reach $624.17 billion in 2020. That means it will account for about 17 percent of U.S. retail sales. In 2011, by contrast, e-commerce accounted for 5–6 percent of U.S. retail sales. Peer reviews and the opinions of other shoppers who post on social media will have an increasing impact on customers' buying decisions. Smartphones will play a major role in this trend as well. Social networking sites such as Facebook, Twitter, Foursquare, and Groupon will continue to influence shopping decisions. Google searches through mobile devices will continue to influence online shoppers.[3]

Another retail research firm, Conlumino, has predicted that, by 2020, m-commerce (mobile commerce) will play a role in 25 percent of all online sales, either through price checking or through direct purchases.[4]

And yet another research group, this one based at the Alibaba Group, China's largest e-commerce company, has predicted that the total sales for e-commerce in China will reach $8.15 trillion by 2020.[5]

It also looks like new online shopping models, such as "flash sales" and "daily deal" services will become more popular. Flash sales, which last only a brief period of time, are invite-only events that offer name-brand merchandise.[6]

MAJOR CATEGORIES OF E-COMMERCE

E-commerce transactions occur among consumers, businesses, and government, resulting in nine major categories, which are summarized in Table 8.2. These categories are described in the following sections.

8-2a Business-to-Consumer E-Commerce

Business-to-consumer (B2C) companies—such as Amazon, Barnesandnoble.com, and Onsale.com—sell directly to consumers. As discussed in this chapter's Industry Connection, Amazon and its business partners sell a wide array of products and services, including books, DVDs, prescription drugs, clothing, and household products. Amazon is also an example of a pure-play company, which means that it relies exclusively on the Web to distribute its products. In recent years, companies that used to have only physical stores—called brick-and-mortar companies—have entered the virtual marketplace by establishing comprehensive Web sites and virtual storefronts. Walmart, the Gap, and Staples are examples of this. These companies supplement their traditional commerce with e-commerce. Some experts believe that these companies may become more successful than pure-play companies because of the advantages a physical space can offer, such as customers being able to visit a store to see merchandise and make returns.

© Boumen Japet/Shutterstock.com

TABLE 8.2 MAJOR E-COMMERCE CATEGORIES

	Consumer	Business	Government
Consumer	C2C	C2B	C2G
Business	B2C	B2B	B2G
Government	G2C	G2B	G2G

© Cengage Learning®

Business-to-consumer (B2C) companies sell directly to consumers.

8-2b Business-to-Business E-Commerce

Business-to-business (B2B) e-commerce involves electronic transactions between businesses. These transactions have been around for many years in the form of electronic data interchange (EDI) and electronic funds transfer (EFT). In recent years, the Internet has increased the number of B2B transactions and made B2B the fastest-growing segment of e-commerce. As discussed in Chapter 7, extranets have been used effectively for B2B operations, as companies rely on other companies for supplies, utilities, and services. Companies using B2B applications for purchase orders, invoices, inventory status, shipping logistics, business contracts, and other operations report millions of dollars in savings by increasing transaction speed, reducing errors, and eliminating manual tasks. Walmart is a major player in B2B e-commerce. Its suppliers—Procter & Gamble, Johnson & Johnson, and others—sell their products to Walmart electronically, which allows them to check the inventory status in each store and replenish products in a timely manner.

8-2c Consumer-to-Consumer E-Commerce

Consumer-to-consumer (C2C) e-commerce involves business transactions between users, such as when consumers sell to other consumers via the Internet. When people use online classified ads (e.g., Craigslist) or online auction sites (e.g., eBay), that is C2C e-commerce. People can also advertise products and services on organizations' intranets (discussed in Chapter 7) and sell them to other employees.

8-2d Consumer-to-Business E-Commerce

Consumer-to-business (C2B) e-commerce involves people selling products or services to businesses, such as when a consumer creates online surveys for a company to use. Another example is when businesses use crowdsourcing by asking consumers to perform services—such as contributing to a Web site—for a fee.

8-2e Government and Nonbusiness E-Commerce

Many government and other nonbusiness organizations use e-commerce applications, including the Department of Defense, the Internal Revenue Service,

and the Department of Treasury. These applications are broadly called **e-government** (or just "e-gov") applications and are divided into these categories:

- *Government-to-citizen (G2C)*—Tax filing and payments; completing, submitting, and downloading forms; requests for records; online voter registration
- *Government-to-business (G2B)*—Sales of federal assets, license applications and renewals
- *Government-to-government (G2G)*—Disaster assistance and crisis response
- *Government-to-employee (G2E)*—E-training

See the information box on the next page on e-gov. Exhibit 8.2 shows the home page of USA.gov, a Web site for delivering government-related information to users.

Exhibit 8.2
USA.gov home page

United States Government

Universities are an example of nonbusiness organizations that use e-commerce applications; for example, many universities use Web technologies for online classes, registration, and grade reporting. In addition, nonprofit, political, and social organizations use e-commerce applications for activities such as fund-raising, political forums, and purchasing.

8-2f Organizational or Intrabusiness E-Commerce

Organizational (intrabusiness e-commerce) involves e-commerce activities that take place inside an organization, typically via the organization's intranet (discussed in Chapter 7). These e-commerce activities are not a part of the nine major categories we just discussed. However, they do support overall e-business activities. They can include the exchange of goods, services, or information among employees (such as the C2C e-commerce discussed previously). Other examples include conducting training programs and offering human resource services. Some of these activities, though not specifically selling and buying, are considered supporting

> **Organizational (intrabusiness e-commerce)** involves e-commerce activities that take place inside an organization, typically via the organization's intranet. These activities can include the exchange of goods, services, or information among employees.

activities in Porter's value chain. For example, a human resources department supports the personnel involved in producing and distributing a company's products.

 B2C E-COMMERCE CYCLE

As Exhibit 8.3 shows, five major activities are involved in conducting B2C e-commerce:

1. *Information sharing*—A B2C e-commerce company can use a variety of methods to share information with its customers, such as company Web sites, online catalogs, e-mail, online advertisements, video conferencing, message boards, and newsgroups.

2. *Ordering*—Customers can use electronic forms or e-mail to order products from a B2C site.

3. *Payment*—Customers have a variety of payment options, such as credit cards, e-checks, and e-wallets. Electronic payment systems are discussed in the section titled "E-Commerce Supporting Technologies."

4. *Fulfillment*—Delivering products or services to customers varies, depending on whether physical products (books, videos, CDs) or digital products (software, music, electronic documents) are being delivered. For example, delivery of physical products can take place via air, sea, or ground at varying costs and with different options. Delivery of digital

products is more straightforward, usually involving just downloading, although products are typically verified with digital signatures (see Chapter 5). Fulfillment also varies depending on whether the company handles its own fulfillment operations or outsources them. Fulfillment often includes delivery address verification and digital warehousing, which maintains digital products on storage media until they are delivered. Several third-party companies are available to handle fulfillment functions for e-commerce sites.

5. *Service and support*—Service and support are even more important in e-commerce than in traditional commerce, given that e-commerce companies do not have a physical location to help maintain current customers. Because maintaining current customers is less expensive than attracting new customers, e-commerce companies should make an effort to improve customer service and support by using some of the following methods: e-mail confirmations and product updates, online surveys, help desks, and guaranteed secure transactions.

The activities listed in Exhibit 8.3 are the same in traditional commerce and probably occur in the same sequence, too. However, each stage has been transformed by Web technologies and the Internet.

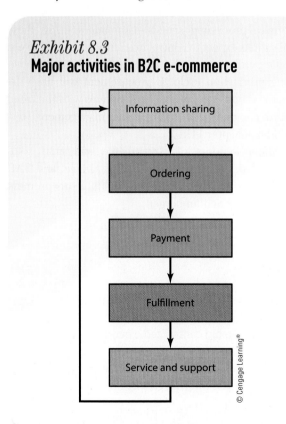

Exhibit 8.3
Major activities in B2C e-commerce

Information sharing → Ordering → Payment → Fulfillment → Service and support

© Cengage Learning®

8-4 B2B E-COMMERCE: A SECOND LOOK

B2B e-commerce may include the activities shown in Exhibit 8.3; however, the following technologies are used extensively: intranets, extranets, virtual private networks, EDI, and EFT. B2B e-commerce reduces delivery time, inventory requirements, and prices, and it helps business partners share relevant, accurate, and timely information. The end result is improved supply chain management among business partners.

B2B e-commerce also lowers production costs and improves accuracy by eliminating many labor-intensive tasks, such as creating invoices and tracking payments manually. In addition, the information flow among business partners is improved by creating a direct online connection in the supply chain network, which reduces delivery time. In other words, raw materials are received faster and information related to customers' demands is transferred faster. Improved electronic communication between business partners improves overall communication, which results in better inventory management and control.

8-4a Major Models of B2B E-Commerce

There are three major models of B2B e-commerce, based on who controls the marketplace—seller, buyer, or intermediary (third party). This results in three marketplace models: seller-side marketplace, buyer-side marketplace, and third-party exchange marketplace. A fourth marketplace model, trading partner agreements, which facilitate contracts and negotiations among business partners, is also gaining popularity. The following sections discuss these models.

Seller-Side Marketplace

The **seller-side marketplace** model occurs most often. In this model, sellers who cater to specialized markets, such as chemicals, electronics, and auto components, come together to create a common marketplace for buyers—sort of a one-stop shopping model. Sellers can pool their market power, and buyers' search for alternative sources is simplified.

The **seller-side marketplace** model occurs most often. In this model, sellers who cater to specialized markets, such as chemicals, electronics, and auto components, come together to create a common marketplace for buyers—sort of a one-stop shopping model.

© Alexey Pustoshilov/Shutterstock.com

E-procurement applications often have purchase-approval procedures that allow users to only connect to company-approved e-catalogs that give employees prenegotiated prices. The main objectives of e-procurement are to prevent purchases from suppliers that are not on the approved list of sellers and eliminate the processing costs of purchases. Not following this process can be costly for the receiving partner, because it can result in paying higher prices for supplies. E-procurement can also qualify customers for volume discounts or special offers.

> The main objective of e-procurement is to avoid buying from suppliers not on the approved list of sellers and to eliminate the processing costs of purchases.

A popular application of this model is **e-procurement**, which enables employees in an organization to order and receive supplies and services directly from suppliers. Typically, a company negotiates reduced prices with suppliers ahead of time. E-procurement streamlines the traditional procurement process, which reduces costs, saves time, and improves relationships between suppliers and participating organizations.

E-procurement enables employees in an organization to order and receive supplies and services directly from suppliers.

E-procurement applications can automate some buying and selling activities, which reduces costs and improves processing speeds. Companies using these applications expect to control inventory more effectively, reduce purchasing overhead, and improve the manufacturing production cycle. E-procurement will likely be integrated into standard business systems with the trend toward computerized supply-chain management (discussed in Chapter 11).

Major vendors of e-commerce and B2B solutions include I2 Technologies, IBM, Oracle, and SAP. The information box below highlights e-procurement applications at Schlumberger.

E-PROCUREMENT AT SCHLUMBERGER

Schlumberger, an oil field services provider, developed an e-procurement system for order processing that has reduced the cost per order. The new system reduces costs by streamlining the paperwork that was required both to route purchase orders for approval and for other administrative tasks. The old centralized electronic data interchange (EDI) procurement system was replaced with a Web-based system that enables employees to contact any approved supplier directly from their workstations. The system has an easy-to-use, flexible interface that has allowed Schlumberger to conduct business with a more diverse group of suppliers. The Internet connection for the system is inexpensive and fast, and the system's open platform has been an advantage.[9]

Buyer-Side Marketplace

Large corporations (such as General Electric or Boeing) as well as consortiums of large companies use the **buyer-side marketplace** model. Here is how it works: A buyer, or a group of buyers, opens an electronic marketplace and invites sellers to bid on announced products or make a request for quotation (RFQ). Using this model, buyers can manage the procurement process more efficiently, lower administrative costs, and implement uniform pricing. Companies invest in buyer-side marketplaces with the goal of establishing new sales channels that increase their market presence and lower the cost of each sale. By participating in buyer-side marketplaces, sellers can do the following:

- Conduct sales transactions.
- Automate the order management process.
- Conduct post-sales analysis.
- Automate the fulfillment function.
- Improve understanding of buying behaviors.
- Provide an alternative sales channel.
- Reduce order placement and delivery time.

Third-Party Exchange Marketplace

The **third-party exchange market-place** model is not controlled by sellers or buyers. Instead, it is controlled by a third party, and the marketplace generates revenue from the fees charged for matching buyers and sellers. A **vertical market** concentrates on a specific industry or market, such as the utilities industry, the beef and dairy industries, and the sale of medical products. A **horizontal market** concentrates on a specific function or business process and automates this function or process for different industries. Employee-benefits administration and media buying are examples of horizontal markets.

This model offers suppliers a direct channel of communication to buyers through online storefronts. The interactive procedures in the marketplace have features such as product catalogs, requests for information (RFI), rebates and promotions, broker contacts, and product sample requests.

Trading Partner Agreements

The main objectives of **trading partner agreements** are to automate negotiating processes and enforce contracts between participating businesses. Using this model, business partners can send and receive bids, contracts, and other information needed when offering and purchasing products and services. This model will become more common with the development of electronic business Extensible Markup Language (ebXML), a worldwide project working on using XML to standardize the exchange of e-commerce data, including electronic contracts and trading partner agreements.

Using this model enables customers to submit, via the Internet, electronic documents that previously required hard copies with signatures. The Digital Signature Act of 1999 gives digital signatures the same legal validity as handwritten signatures. Accepting an electronic trading agreement binds the parties to all its terms and conditions.

With ebXML, contracts can be transmitted electronically, and many processes between trading partners can be performed electronically, including inventory status, shipping logistics, purchase orders, reservation systems, and electronic payments. The main advantage of ebXML or XML over HTML is that you can assign data-type definitions to information on a page so Web browsers select only the data requested in a search. This feature makes data transfer easier because not all data is transferred, just the data needed in a particular situation. It is particularly useful in m-commerce (mobile commerce), because loading only the necessary data in a browser makes searches more efficient. This process reduces traffic on the Internet and helps prevent delays during peak usage hours.

In a **buyer-side marketplace** model, a buyer, or a group of buyers, opens an electronic marketplace and invites sellers to bid on announced products or make a request for quotation (RFQ). Using this model, buyers can manage the procurement process more efficiently, lower administrative costs, and implement uniform pricing.

The **third-party exchange marketplace** model is not controlled by sellers or buyers. Instead, it is controlled by a third party, and the marketplace generates revenue from the fees charged for matching buyers and sellers.

A **vertical market** concentrates on a specific industry or market. The utilities industry, the beef and dairy industries, and the sale of medical products are examples of vertical markets.

A **horizontal market** concentrates on a specific function or business process and automates this function or process for different industries.

Trading partner agreements automate negotiating processes and enforce contracts between participating businesses.

8-5 MOBILE AND VOICE-BASED E-COMMERCE

Mobile commerce (m-commerce), based on the Wireless Application Protocol (WAP), has been around for several years, particularly in European countries. M-commerce is the use of handheld devices, such as smartphones or PDAs, to conduct business transactions, such as making stock trades with an online brokerage firm. Supporting technologies for m-commerce applications include wireless wide area networks (WWANs) and 3G and 4G networks (discussed in Chapter 6) as well as short-range wireless communication technologies, such as Wi-Fi, WiMAX, Bluetooth, and RFID (discussed in Chapter 14).

Many telecommunication companies offer Web-ready cell phones. In addition, a wide variety of m-commerce applications are available, among the most popular being iPhone apps, which include games, entertainment, news, and travel information. Some iPhone apps are free; others must be purchased via iTunes.

Microsoft has a wireless version of Internet Explorer called Internet Explorer Mobile. And many e-commerce companies are developing the simple, text-based interface required by current cell phones. For example, Google offers mobile Internet users features such as search, news, map, and Gmail (*www.google.com/mobile*). MSN Mobile (*http://mobile.msn.com*) provides a special browser for accessing e-mail, news, sports, entertainment, maps, and Windows Live services, such as Hotmail and Windows Live Messenger. Other applications of m-commerce include banking, traffic updates, tourism services, shopping, and video conferencing. Mobile user-to-user applications, such as sharing games and pictures, are also popular.

You can already use a mobile phone to access a Web site and order a product. The next step is **voice-based e-commerce**, which relies on voice recognition and text-to-speech technologies

that have improved dramatically in the past decade.[10,11,12] For example, you will soon be able to simply speak the name of the Web site or service you want to access and use voice commands to search a database by product name and find the merchant with the most competitive prices. Voice-based e-commerce is suitable for such applications as making stock trades, looking up sports scores, reserving movie tickets, and getting directions to a restaurant.

Currently, the iPhone 3GS and beyond support voice-based Google searches. The iPhone 4S and beyond are equipped with Siri, which lets users send messages, make calls, set reminders, and perform other tasks using their voices. (Android devices use Google Voice for this purpose.) One method of conducting voice-based e-commerce is using e-wallets (virtual wallets), which are discussed further in the next section. In addition to storing financial information, e-wallets can store such information as the customer's address and driver's license number. Security features for voice-based e-commerce are expected to include the following:

- Call recognition, so that calls have to be placed from specific mobile devices
- Voice recognition, so that authorizations have to match a specific voice pattern
- Shipping to a set address that cannot be changed by voice commands

Several voice portals are already available, including Nuance (*nuance.com*), Internet Speech (*internetspeech.com*), and Google Voice (*www.google.com/googlevoice/about.html*). You can visit these sites to discover the latest in voice-based e-commerce.

Mobile commerce (m-commerce) is using handheld devices, such as smartphones or PDAs, to conduct business transactions.

Voice-based e-commerce relies on voice recognition and text-to-speech technologies.

iStockphoto.com/Ismail Akin Bostanci

8-6 E-COMMERCE SUPPORTING TECHNOLOGIES

A number of technologies and applications support e-commerce activities: electronic payment systems, Web marketing, and search engine optimization. They are discussed in the following sections.

8-6a Electronic Payment Systems

An **electronic payment** is a transaction in which money or scrip is exchanged only electronically. This usually involves the use of the Internet, other computer networks, and digitally stored value systems. Payment cards—credit cards, debit cards, charge cards, and smart cards—are the most popular instruments for electronic payment transactions. **Smart cards** have been used in Europe, Asia, and Australia for many years and are slowly gaining acceptance in the United States because of their multiple functions. A smart card is about the size of a credit card and contains an embedded microprocessor chip for storing important financial and personal information. The chip can be loaded with information and updated periodically.

E-cash, a secure and convenient alternative to bills and coins, complements credit, debit, and charge cards and adds convenience and control to everyday cash transactions. E-cash usually works with a smart card, and the amount of cash stored on the chip can be "recharged" electronically.

An **e-check**, the electronic version of a paper check, offers security, speed, and convenience for online transactions. Many utility companies offer customers the opportunity to use e-checks to make their payments, and most banks accept e-checks for online bill paying. E-checks are a good solution when other electronic payment systems are too risky or otherwise not appropriate.

E-wallets (virtual wallets), which are available for most handheld devices, offer a secure, convenient, and portable tool for online shopping. They store personal and financial information, such as credit card numbers, passwords, and PINs. E-wallets can be used for micropayments (discussed later in this section), and online shoppers find them useful because personal and financial information does not have to be reentered each time they place an order.

> Payment cards—credit cards, debit cards, charge cards, and smart cards—are the most popular instrument for electronic payment transactions.

You are probably familiar with **PayPal**, a popular online payment system used on many online auction sites. Users with valid e-mail addresses can set up PayPal accounts and make secure payments for online transactions using their credit cards or bank accounts.

Micropayments are transactions on the Web involving very small amounts of money. They began as a way for advertisers to pay for cost per view or cost per click, which is typically one-tenth of a cent. Such fractional amounts are difficult to handle with traditional currency methods, and electronic micropayments reduce the cost of handling them for financial institutions. Payment amounts are accumulated for customers until they are large enough to offset the transaction fee, and then the account deduction or charge is submitted to the bank. Of course, micropayment systems charge a fee for tracking and processing the transactions. However, the World Wide Web Consortium (W3C; *www.w3c.org*), which defines standards for Web-related technologies, has canceled support for

Electronic payment refers to money or script that is exchanged electronically. It usually involves use of the Internet, other computer networks, and digitally stored value systems. It includes credit cards, debit cards, charge cards, and smart cards.

A **smart card** is about the size of a credit card and contains an embedded microprocessor chip for storing important financial and personal information. The chip can be loaded with information and updated periodically.

E-cash, a secure and convenient alternative to bills and coins, complements credit, debit, and charge cards and adds convenience and control to everyday cash transactions.

An **e-check**, the electronic version of a paper check, offers security, speed, and convenience for online transactions.

E-wallets, which are available for most handheld devices, offer a secure, convenient, and portable tool for online shopping. They store personal and financial information, such as credit card numbers, passwords, and PINs.

PayPal is a popular online payment system used for many online transactions. Users with valid e-mail addresses can set up accounts and make secure payments for online transactions using their credit cards or bank accounts.

Micropayments are transactions on the Web involving very small amounts of money. They began as a method for advertisers to pay for cost per view or cost per click.

CHALLENGES IN USING MOBILE PAYMENT SYSTEMS

Credit and debit cards have some serious shortcomings. They can be lost, hacked, and easily copied. And they are difficult to replace. Also, each bank account has its own credit card. Thus, if you have 12 accounts, you will need 12 separate credit cards. Despite these shortcomings, credit cards remain very popular and are used in both e-commerce and in traditional commerce. That's because everyone in the credit card industry—banks, stores, and users—agrees on the standards and the interface.

There are many players in the mobile payment environment, and their numbers are increasing. Major players include PayPal, Google, Apple, Square, Facebook, Groupon, Isis, Dwolla, LevelUp, and Venmo. Mobile carriers and credit card companies are also involved in mobile payments. However, most customers are still using credit cards as the main method of payment because mobile payments are still not accepted by many merchants and the interface is not universal either.

Due to the lack of standards for mobile payment systems, new solutions have arisen to enhance the usability of traditional credit cards. One such solution is offered by Coin, a San Francisco-based startup company. Coin combines all of a customer's credit cards into one credit card, which is enhanced by a mobile app that allows the user to load information from other credit cards the user has or swipe each of them one-by-one (using a device that comes with Coin) to duplicate the information from those cards into Coin. This eliminates having to carry multiple cards in a wallet. Still, Coin faces the challenge of being accepted by most merchants.[14]

micropayments and is no longer working on standards for them. In 2012, Google launched a micropayments feature for Google Wallet that allows site owners to charge small amounts of money for Web content.[13]

The information box above highlights some of the challenges in using mobile payment systems.

8-6b Web Marketing

Web marketing uses the Web and its supporting technologies to promote goods and services. Although traditional media, such as radio and TV, are still used for marketing, the Web offers many unique capabilities, such as message boards for customers to post questions and newsgroups for sending information to customers. To better understand Web marketing, review the following list of terms:

- *Ad impression*—One user viewing one ad
- *Banner ads*—Usually placed on frequently visited Web sites, these ads are rather small (around 468 × 60 pixels) and employ simple animation. Clicking a banner ad displays a short marketing message or transfers the user to another Web site.

- *Click*—When users click URLs or banner ads, they are transferred to other Web sites or shown marketing messages, and this is recorded by the Web server. For example, each time a certain keyword used for a search takes a user to a particular Web page, the advertiser who owns that site pays the search engine a cost per click (discussed later in this list). A consortium of Yahoo!, Microsoft, Google, and the Interactive Advertising Bureau has formed the Click Measurement Working Group, which is trying to define what a "legal and valid" click ought to be.

- *Cost per thousand (CPM)*—Most Web and e-mail advertising is priced based on the cost per thousand ad impressions. (*M* stands for *mille*, which means thousand.) For example, a $125 CPM means it costs $125 for 1,000 ad impressions.

- *Cost per click (CPC)*—This refers to the cost of each click on an ad. For example, $1.25 CPC means that for every click an advertiser gets, the advertiser pays $1.25 to the sponsoring Web site. MIVA (*www.miva.com*) is an example of a cost-per-click network.

- *Click-through rate (CTR)*—This is computed by dividing the number of clicks an ad gets by the total impressions bought. For example, if an advertiser buys 100,000 impressions and gets 20,000 clicks, the CTR is 20 percent (20,000/100,000).

- *Cookie*—This is information a Web site stores on the user's hard drive so it can be used for a later visit, such as for greeting a visitor by name. This information is also used to record user preferences and browsing habits.

Web marketing uses the Web and its supporting technologies to promote goods and services.

- *Hit*—Any element of a Web page (including text, graphics, and interactive items) that is clicked counts as a hit to a server. Hits are not the preferred unit of measurement for site traffic because the number of hits per page can vary widely, depending on the number of graphics, type of browser used, and page size.

- *Meta tag*—This HTML tag does not affect how a Web page is displayed; it simply provides information about a Web page, such as keywords that represent the page content, the Web designer, and frequency of page updates. Search engines use this information (particularly the keywords) to create indexes.

- *Page view (PV)*—This refers to one user viewing one Web page.

- *Pop-up ads*—These display ads appear on top of a browser window, blocking the user's view.

- *Pop-under ads*—These display ads appear underneath a browser window. They are less intrusive than pop-up ads.

- *Splash screen*—A Web page displayed when the user first visits the site; it is designed to capture the user's attention and motivate the user to browse the site. The splash screen may display the company's logo as well as a message about any requirements for viewing the site, such as the need to install plug-ins.

- *Spot leasing*—Search engines and directories offer space that companies can purchase for advertising purposes. Spots have an advantage over banner ads because their placement is permanent; banner ad placement can change from visit to visit. However, spots can be more expensive than banner ads, especially on high-traffic sites, such as Yahoo!.

Intelligent agents (discussed in Chapter 13) and push technology (discussed in Chapter 14) are also used as Web-marketing tools. Intelligent agents are an artificial intelligence application that can be used for Web marketing. For example, product-brokering agents can alert customers about a new product. Push technology is the opposite of pull technology, in which users search the Web to find (pull) information. With push technology, information is sent to users based on their previous inquiries, interests, or specifications. This technology can be used to send and update marketing information, product and price lists, and product updates. The information box below discusses some of the challenges in using digital ads.

8-6c Search Engine Optimization

Search engine optimization (SEO) is a method for improving the volume or quality of traffic to a Web site. It helps a Web site receive a high ranking in the search results, which tends to generate more revenue for the Web site. For the average keyword used in a search, the search engine might list hundreds or thousands of Web sites, but most people visit only the top 5 or 10 sites and ignore the rest. Therefore, it is important to be among the top sites.

A comprehensive Web marketing campaign should use a variety of methods, and SEO is another method that

> **Search engine optimization (SEO)** is a method for improving the volume or quality of traffic to a Web site. A higher ranking in search results should generate more revenue for a Web site.

CHALLENGES IN USING DIGITAL ADS

Advertisers lose billions of dollars every year as a result of phony traffic to Web sites. Phony traffic is created by bots that are controlled by hijacked PCs around the world. Solve Media estimates that up to 29 percent of current display-advertising traffic is driven by these bots.

Songsrpeople.com is an example of a Web site that generates most of its traffic by bots, according to the security firm White Ops. Interestingly enough, major corporations such as Target, Amazon, and State Farm have display ads on the Web site. But most of the visitors to the Web site are not human; they are computer-generated. Bots mislead advertisers, creating the impression that a Web site receives lots of traffic, which allows the Web site to charge advertisers more for putting ads on these sites. Some bots are so sophisticated that they can mimic a human shopper by clicking from one site to another, "watching" videos, or even adding items to shopping carts. There are thousands of these sites on the Internet, according to government authorities and Internet security experts, who are trying to identify the sites and put them on a blacklist. In the meantime, advertisers need to distinguish legitimate traffic from phony traffic before investing in digital ads.[15]

can help improve business. Some companies offer SEO services. Unlike Web-marketing methods that involve paying for listings on search engines, SEO aims at increasing a Web site's performance on search engines in a natural (and free) fashion. As you learned in Chapter 7, a typical search engine, such as Google or Bing, uses a crawler or spider to find a Web site and then, based on the site's contents and relevance, indexes it and gives it a ranking. Optimizing a Web site involves editing a site's contents and HTML code to increase its relevance to specific keywords. SEO includes techniques that make it easier for search engines to find and index a site for certain keywords. The following are some common ways to optimize a Web site's traffic:

- *Keywords*—Decide on a few keywords that best describe the Web site and use them consistently throughout the site's contents.
- *Page title*—Make sure the page title reflects the site and its contents accurately.
- *Inbound links*—Get people to comment on your Web site, using one of your top keywords.
- *Content—Update your Web content regularly.*
- *Links to others—Develop relationships with other Web sites.*

This chapter's Industry Connection focuses on Amazon, one of the leaders in e-commerce.

Industry Connection: Amazon[16]

Amazon, a leader in B2C e-commerce, offers a variety of products and services, including books, CDs, videos, games, free e-cards, online auctions, and other shopping services and partnership opportunities. By using customer accounts, shopping carts, and its 1-Click feature, Amazon makes shopping fast and convenient, and it uses e-mail for order confirmation and customer notifications of new products tailored to customers' shopping habits. In addition, Amazon has created an open forum with customers by posting customers' book and product reviews and allowing customers to rate products on a scale of one to five stars. Most recently, Amazon has become one of the major players in the cloud computing environment. Here are some of the activities customers can do on Amazon:

- Search for books, music, and many other products and services.
- Browse in hundreds of product categories, from audio books, jazz, and video documentaries, to coins and stamps available for auction.
- Get personalized recommendations based on previous purchases.
- Sign up for an e-mail subscription service to get the latest reviews of new titles in categories of interest.

- Create wish lists that can be saved for later viewing.
- Search book content on specific keywords, and view selected pages in some books.

Amazon is known for its personalization system, which is used to recommend goods, and its collaborative filtering, which is used to improve customer service. (Both features are discussed in Chapter 11.) Amazon also collaborates with business partners via Amazon Marketplace, where other merchants can sell their products through Amazon.

STUDY TOOLS 8

LOCATED AT BACK OF THE TEXTBOOK
- ☐ Rip out Chapter Review Card

LOCATED AT WWW.CENGAGE.COM/LOGIN
- ☐ Review Key Terms Flash Cards (Print or Online)
- ☐ Download audio and visual summaries to review

- ☐ Complete both Practice Quizzes to prepare for tests
- ☐ Play "Beat the Clock" and "Quizbowl" to master concepts
- ☐ Complete "Crossword Puzzle" to review key terms
- ☐ Watch video on "Method" for a real company example

KEY TERMS

advertising model 167
brokerage model 167
business-to-business (B2B) 169
business-to-consumer (B2C) 168
buyer-side marketplace 173
click-and-brick e-commerce 165
consumer-to-business (C2B) 169
consumer-to-consumer (C2C) 169
e-business 163
e-cash 175
e-check 175
e-commerce 163
e-government 169
e-procurement 172
e-wallets 175
electronic payment 175
horizontal market 173

informediary model 167
merchant model 167
micropayments 175
mixed model 167
mobile commerce (m-commerce) 174
organizational (intrabusiness e-commerce) 170
PayPal 175
search engine optimization (SEO) 177
seller-side marketplace 171
smart card 175
subscription model 167
third-party exchange marketplace 173
trading partner agreements 173
value chain 164
vertical market 173
voice-based e-commerce 174
Web marketing 176

REVIEWS AND DISCUSSIONS

1. What is the value chain and how is it used in the e-commerce environment?

2. What are two examples of B2C e-commerce business models?

3. What are the nine major categories of e-commerce?

4. What are two advantages of e-procurement compared to traditional procurement?

5. What are three examples of Web marketing?

6. What are two limitations of m-commerce?

7. What are two examples of popular voice portals?

8. Why have e-payment systems not taken off yet? What are two challenges that must be overcome first?

PROJECTS

1. Local, state, and federal government agencies are establishing e-gov sites in order to improve the efficiency and effectiveness of their operations. After reading the information presented in this chapter and other sources, write a one-page paper that summarizes six advantages of using e-gov applications for running a city. What are some of the challenges?

2. Yelp is a successful social networking site. After reading the information presented in this chapter and other sources, write a one-page paper that describes the business model used by Yelp. How does Yelp generate revenue? How can businesses benefit from this site?

3. Twitter has become a very popular personal as well as business social networking site. After reading the information presented in this chapter and other sources, write a one-page paper that summarizes five ways that a business can use Twitter in order to increase its revenue. What does "trending" mean on Twitter? What are promoted tweets?

4. After reading the information presented in this chapter and other sources, write a one-page paper that compares and contrasts traditional marketing with Web marketing. What are two advantages and two disadvantages of each?

5. The future of m-commerce appears to be very promising. After reading the information presented in this chapter and other sources, write a two-page paper that identifies four factors that will contribute to the future growth of m-commerce. What are some of the challenges that must be overcome?

6. Search engine optimization should be a part of any successful Web marketing strategy. After reading the information presented in this chapter and other sources, write a one-page paper that offers seven recommendations for improving SEO in an e-commerce site.

ARE YOU READY TO MOVE ON?

1. The value chain concept is only used in manufacturing organizations. It does not apply to service organizations. True or False?

2. Creating "price transparency" is one of the disadvantages of e-commerce. True or False?

3. An electronic payment is a transaction in which money or scrip is exchanged only electronically. True or False?

4. Which if the following is not an example of a B2C e-commerce business model?
   a. Merchant
   b. Brokerage
   c. SEO
   d. Advertising

5. Which of the following is not an example of a B2B e-commerce business model?
   a. Seller-side marketplace
   b. Informediary marketplace
   c. Buyer-side marketplace
   d. Third-party exchange marketplace

6. Which of the following is not an example of Web marketing?
   a. Banner ad
   b. Spot leasing
   c. TV ad
   d. Splash screen

CASE STUDY 8-1

E-Commerce Applications in Online Travel

People use online travel agents (OTAs) to book hotels, airline tickets, cruises, and more. Currently, Expedia, Travelocity, Priceline, and Orbitz are the top four OTAs, and they have transferred the traditional travel agency model to the Internet, using Web features and technologies to sell travel services. OTAs are considered the first generation of travel services. Most OTAs charge a fee for performing these services, but customers get confirmed travel arrangements as well as guarantees in case flights are canceled, among other things. Travel suppliers, such as airlines, hotels, and car rental companies, prefer that customers book through the suppliers' own Web sites, because they often have to pay high fees to be included on an OTA. JetBlue and InterContinental Hotels, for example, do not list their services on OTAs.

© arek malang/Shutterstock.com

The second generation of travel services is represented by travel search engines, which use Web technologies to help customers make travel arrangements quickly and efficiently; major ones include SideStep, Kayak, and Mobissimo. Customers can use travel search engines to search for the best deal and then click to directly book flights, hotels, and cars with the travel supplier. Using these services is less expensive than using OTAs, but it is also less convenient.[17]

Answer the following questions:

1. What are the differences between online travel agents, such as Orbitz, and travel search engines, such as Mobissimo?

2. What are the differences between Priceline and other OTA business models?

3. From the perspective of travel suppliers, what are the advantages of travel search engines?

CASE STUDY 8-2
Bridging the Gap Between E-Commerce and Traditional Commerce

Various tools and technologies are helping to bridge the gap between e-commerce and traditional commerce, including mobile devices and social media. Often, potential buyers see what their social networking friends have to say about products before making a purchase. Transparency is the key, given that buyers can now compare prices from nearly anywhere—including from a physical store.

New sets of technologies are being introduced that track the customer when the customer is inside the store. Apple's iBeacon, which competes with near field communication (NFC), is one such technology. It is a small wireless device that uses Bluetooth to detect and communicate with new generations of iPhones and iPads that are running iOS7.[18] A GPS guides a customer to a store; iBeacon then tracks the customer inside the store. (Qualcomm has developed its own version of iBeacon, Gimbal proximity beacons, which work with either iOS or Android.)

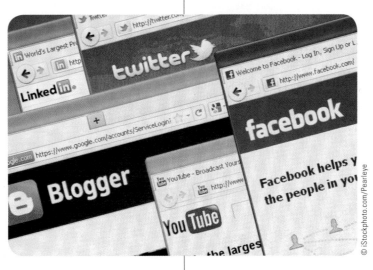

© iStockphoto.com/Pearleye

There are many potential applications of this technology. For example, a retailer can send a coupon as soon as it detects that a customer has walked to a particular aisle and is looking at a particular product. In Apple stores, for another example, as soon as the customer walks by the iPhone table, he/she will get a notification on upgrades. Major League Baseball has already announced that it will use iBeacon to customize fans' experiences at ballparks by using its "At The Ballpark" app. Some businesses are already using iPads as cash registers; these customers can potentially be tracked by Apple. Also, with Apple's huge user base, this technology has potential for tremendous growth.[19,20]

Privacy issues may be a concern as these tools and technologies roll out, given that customers will be tracked in stores that are equipped with this technology. However, customers may not mind this tracking as long as there is something in it for them.

Answer the following questions:

1. What is the function of Apple's iBeacon?
2. What are the differences between iBeacon and GPS?
3. How might a retail business benefit from iBeacon?
4. What are some of the concerns that customers may have about this technology?

9 Global Information Systems

LEARNING OUTCOMES

After studying this chapter, you should be able to:

9-1 Discuss the reasons for globalization and for using global information systems, including e-business and Internet growth.

9-2 Describe global information systems and their requirements and components.

9-3 Explain the types of organizational structures used with global information systems.

9-4 Discuss obstacles to using global information systems.

After you finish this chapter, go to **PAGE 197** for the **STUDY TOOLS**

Ian McKinnell/Getty Images

In this chapter, you review the reasons organizations should go global and adopt global information systems, including the rise in e-business and the growth of the Internet. Global information systems are a growing application of telecommunications and networking, and you learn the requirements and components of these systems as well as the types of organizational structures used with global information systems. Offshore outsourcing, as one beneficiary of global information systems, is also discussed. Finally, you explore some obstacles to using global information systems.

9-1 WHY GO GLOBAL?

The global economy is creating customers who demand integrated worldwide services, and the expansion of global markets is a major factor in developing global information systems to handle these integrated services. To understand the need for integrated worldwide services, consider the example of a U.S.-based shoe company that procures leather and has the upper parts of its shoes produced in Italy because of the high quality of leather and the expertise in shoe stitching available there. The uppers are then shipped to China, where they are attached to soles, thereby taking advantage of the inexpensive manufacturing labor available in that country. The shoes are then shipped to Ireland for testing because of Ireland's high concentration of high-tech facilities. Finally, the shoes are shipped to a variety of retail outlets in the United States, where they are sold. The entire supply-chain logistics—from Italy to China to Ireland—must be managed and coordinated from the U.S. headquarters. This example shows why companies choose other countries for different manufacturing processes and how important integration is in making sure all these processes are coordinated.

Many companies have become international. In 2010, for example, the Coca-Cola Company generated more than 75 percent of its revenue from outside the United States. Major corporations, such as Procter & Gamble, IBM, HP, McDonald's, Unilever, Nestle, and

Motorola, have been prime users of global information systems. Because today's multinational corporations operate in a variety of markets and cultures, a clear understanding of factors such as customs, laws, technological issues, and local business needs and practices is a prerequisite to the success of a global information system.

Airline reservation systems are considered the first large-scale interactive global system; hotels, rental car companies, and credit card services also now require worldwide databases to serve their customers more efficiently and effectively.[1] Global products, which are products or services that have been standardized for all markets, are becoming increasingly important in international marketing efforts. In addition, a manufacturer might "regionalize" operations—that is, move them to another country—because of advantages available in certain regions. For example, raw materials might be less expensive in Indonesia than in Singapore, and specialized skills needed for production might be available in India but not in Brazil.

The growing trend toward global customers and products means globalization has also become an important factor in purchasing and the supply chain. Worldwide purchasing gives suppliers the incentive to consider foreign competition as well as domestic competition. Furthermore, large global organizations can reduce costs in purchasing, manufacturing, and distribution because they have access to cheaper labor and can sell products and services locally as well as internationally.[2,3] The information box below highlights the use of global information systems at Rohm & Haas.

In 2010, the Coca-Cola Company generated more than 75 percent of its revenue from outside the United States.

9-1a E-Business: A Driving Force

E-business is a major factor in the widespread use of global information systems. As discussed in Chapter 8, e-business includes transactions that support revenue generation as well as those that focus on buying and selling goods and services. These revenue-generating transactions include generating demand for goods and services, offering sales support and customer service, and facilitating communication between business partners. An effective global information system can support all these activities.

E-business builds on the advantages and structures of traditional business by adding the flexibility that networks offer. By generating and delivering timely and relevant information supported by networks, e-business creates new opportunities for conducting commercial activities. For example, by using online information for commercial activities, e-business makes it easier for different groups to cooperate. Branches of a multinational company can share information to plan a new marketing campaign, different companies can work together to design new products or offer new services, and businesses can share information with customers to improve customer relations.

The Internet can simplify communication, improve business relationships, and offer new opportunities to both consumers and businesses. As e-business matures and more companies conduct business online, consumers can engage in comparison shopping more easily, for example. Even though direct buyer–seller communication has increased, there are still new opportunities for intermediaries. For example, some businesses

GLOBAL INFORMATION SYSTEMS AT ROHM & HAAS

Rohm & Haas, part of Dow Chemical, has production units in many countries. In the past, each country site operated independently and had its own inventory system. A major problem with this setup was that a country site might not be able to supply customers with the products they wanted. For example, if the site in France reported to customers that it was out of a certain product, it did not have an easy way to check whether the site in Germany, only 20 miles away, had a supply of this product. To solve these problems, Rohm & Haas overhauled its global information system by upgrading the order entry system and installing a company-wide materials management system. These systems were tied in with a global demand planning system. Rohm & Haas can now provide better service to its customers and ship products from other sites as quickly as needed. These improvements have given Rohm & Haas more of a competitive advantage in the global marketplace.[4]

MAKING A COMPANY WEB SITE GLOBAL[6]

Here are some suggestions for making a company Web site global:

- *Language*—Present your Web site in one of the seven most-used languages: English, French, Italian, German, Spanish, Japanese, or Chinese.
- *Font*—Make sure the Web site's main features are readable in different languages, depending on which font is selected.
- *Cultural differences*—Keep cultural variation in mind. For example, although white is the color of purity in the United States, it is the color of mourning in Eastern cultures. Also, a woman smiling without covering her mouth would be considered sexually suggestive in Japan.
- *Currency*—Make sure the Web site includes a currency conversion feature.
- *Date format*—This varies around the world. Many countries use day/month/year instead of month/day/year.
- *Spare use of graphics and flash features*—Because of bandwidth limitations in many parts of the world, you may want to keep the Web site rather simple.
- *E-mail*—Be prepared to send and receive e-mails in foreign languages.
- *Payments*—Not everybody uses credit cards. In Germany, for example, bank transfers are popular, whereas in Japan both COD and bank transfers are popular.
- *International logistics*—To ship internationally, you must address the various regulations that pertain to each country.
- *International listing*—List your Web site with international search engines and indexes.
- *Local involvement*—Invite local people to review the Web site before you launch it to the global market.
- *International marketing*—Promote your Web site in specific languages.

can become intermediaries or brokers to track special markets, notify clients of bargains or changes in market conditions, locate hard-to-find items, and even conduct searches for special products on clients' behalf.

Of course, it is the Internet that makes e-business possible. Small companies have discovered they can not only conduct business online just as large companies do, but they can lower costs by using the Internet to replace internal networks. The following section discusses the Internet's growth, which has contributed to the increase in e-business.

9-1b Growth of the Internet

Today, the Internet is a part of daily life in most parts of the world. According to the Miniwatts Marketing Group, which tracked the worldwide growth of the Internet from 2000 to 2012, the highest growth has occurred in Africa, the lowest in North America. As of June 30, 2012, there were approximately 2.4 billion worldwide Internet users, with Asia having 1.08 billion, Europe having 519 million, North America having 274 million, Latin America/Caribbean having 255 million, Africa having 167 million, the Middle East having 90 million, and Oceania/Australia having 24 million.[5]

With the explosive growth of the Internet and e-commerce, businesses that are active in the global market should make their Web sites more appealing to global customers. Some companies create separate Web

sites for each country in which they do business. This is called "localization of a Web site." When this is the case, the company's main Web site needs to make a clear and specific reference to these additional Web sites, preferably through drop-down menus. Still, some global customers may only use the main Web site. The adjoining information box offers some practical suggestions for making a company Web site global.

9-2 GLOBAL INFORMATION SYSTEMS: AN OVERVIEW

A **global information system (GIS)** is an information system that works across national borders, facilitates communication between headquarters and subsidiaries in other countries, and incorporates all the technologies and applications found in a typical

A **global information system (GIS)** is an information system that works across national borders, facilitates communication between headquarters and subsidiaries in other countries, and incorporates all the technologies and applications found in a typical information system to gather, store, manipulate, and transmit data across cultural and geographic boundaries.

information system to gather, store, manipulate, and transmit data across cultural and geographic boundaries.[7] In other words, a GIS is an information system for managing global operations, supporting an international company's decision-making processes, and dealing with complex variables in global operations and decision making.

With a GIS in place, an international company can increase its control over its subsidiaries and better coordinate their activities, thereby gaining access to new global markets.[8] Strategic planning is also a core function of a GIS. By being able to efficiently share information

High coordination has the following advantages[10]:

- Flexibility in responding to competitors in different countries and markets
- Ability to respond in one country to a change in another country
- Ability to maintain control of market needs around the world
- Ability to share and transfer knowledge between departments and international branches
- Increased efficiency and effectiveness in meeting customers' needs
- Reduced operational costs

> With a GIS in place, an international company can increase its control over its subsidiaries and better coordinate their activities, thereby gaining access to new global markets.

among subsidiaries, international companies can track performance, production schedules, shipping alternatives, and accounting items.

A GIS can be defined along two dimensions: control and coordination. Control consists of using managerial power to ensure adherence to the organization's goals. Coordination is the process of managing the interaction among activities in different, specialized parts of an organization. Control requires a centralized architecture for data, standardized definitions used across the organization, standard formats for reports, defined behaviors for different processes (such as how to respond when a customer has a complaint), and performance-tracking systems. Coordination requires a decentralized architecture for data, standardization within departments, and the ability to communicate these standards to other departments, collaboration systems, and technologies that support informal communication and socialization. The trade-off between the amount of control needed and the amount of coordination needed defines the organization's globalization strategy. Global organizations may use a combination of high control and high coordination, high control and low coordination, low control and high coordination, or low control and low coordination.[9]

9-2a Components of a Global Information System

Although a GIS can vary quite a bit depending on a company's size and business needs, most GISs have two basic components:

- A global database
- Information-sharing technologies

Global Database

Designing and implementing a global database is a technical challenge, mainly because of the different character sets required for the names of people and places and the different formats required for phone numbers and postal codes. Currency conversion is also a challenge in database development, although some software is available for this task. For example, SAP (originally called Systems Applications and Products in Data Processing) offers valuable features and capabilities for GISs.

Information-Sharing Technologies

International companies can use a variety of technologies for an integrated GIS. Small companies might outsource to take advantage of expertise that is not available inside the company. On the other hand, large companies with the resources and technical expertise might develop custom applications to be shared across borders. Depending on the system's use, a GIS might consist of a network for e-mail, remote data entry, audio, video, computer conferencing, and distributed databases. However, small companies might take advantage of existing public network providers, such as the Internet or value-added networks, for multicountry communication.[11] Value-added networks are private multipoint networks managed by a third party and used by organizations on a subscription basis. They offer electronic data interchange standards, encryption, secure e-mail, data synchronization, and other services. However, with the popularity of the Internet, they are not used as much anymore; today, businesses of all sizes typically use the Internet to conduct international business. No matter what the organization's size or scope, an integrated network for global control over the organization's resources is the foundation of any GIS. (See the information box below.)

An information system manager faces design and implementation issues when developing a global network. In addition to the usual components of a domestic network, a global network requires bridges, routers, and gateways that allow several networks to connect worldwide. In addition, a global network must have switching nodes to guide packets to their destinations.[13] (These components were discussed in Chapter 6.)

THE INTERNET AND GLOBALIZATION IN ACTION

The Internet allows entrepreneurs in developing countries to start and expand businesses without making large investments. For example, Muhammad Hassaan Khan, a young entrepreneur, was able to establish Zuha Innovation, a design and consulting business based in Faisalabad, Pakistan. The Internet has lowered the barriers for small businesses all over the world, and entrepreneurs have responded, including ones from India, China, Mexico, and Brazil. Online businesses can be created regardless of where the creator is located; all that is needed is Internet access. These businesses vary widely in the activities in which they are engaged, from e-commerce sites selling local products to bloggers making incomes from advertising revenue. As Adam Toren, president of YoungEntrepreneur.com, puts it, "The Internet brings all continents, races, cities, and villages together into a global network of trade and communications."[12]

An information system manager must determine the best communication media to meet global performance and traffic needs, such as fiber optics, satellite, microwave, or conventional phone lines. Factors to consider include bandwidth, range, noise, and cost. You learned about bandwidth and range in Chapter 6. Global providers such as SprintLink, AT&T, and MCI can supply information on the range specifications for companies. The noise factor involves how immune a medium is to outside electronic interference. As always, component, installation, and leasing costs must be balanced with these other factors.

In addition, an information system manager must choose the best transmission technology for the global network's needs. Without reliable transmission, a network has no value. Current transmission technologies are synchronous, asynchronous, multiplexing, digital (baseband), and analog (broadband). With synchronous transmission, both parties are connected at the same time, as in a phone call. With asynchronous transmission, they do not have to be connected at the same time, as is true of e-mail. However, an international company is restricted to the transmission technologies that are supported by the telecommunication infrastructures of the countries where its subsidiaries are located. Information system managers must select the right network and protocol to manage connections and minimize error rates.

Information system managers must also consider the company's objectives when determining the network architecture. For example, if the company's international communication requirements only entail simple file sharing and if response time is not a critical factor, half-duplex transmission (one direction at a time) used with a value-added network is probably adequate. However, if the company uses multimedia applications (such as video conferencing and electronic meeting systems) in addition to normal file and database sharing, full-duplex transmission (both directions simultaneously) is more efficient. Furthermore, a private network or a dedicated leased line provides stability in transmission protocols when there are inadequate telecommunication infrastructures, as there often are in developing countries.

A network's main function is to allow users to share information. After a global network is in place, therefore, an international company must decide which types of information-sharing technology it will be using, such as electronic meeting systems or video conferencing, group support systems, FTP, data synchronization, and application sharing. (See the information box on the next page on video conferencing.)

VIDEO CONFERENCING SYSTEMS SUPPORT GLOBALIZATION

By 2018, the global market for video conferencing systems is expected to reach $4.5 billion. The driving force behind this growth is the increase in globalization and workforce mobility. Video conferencing is the next best alternative to conducting face-to-face business meetings, for a fraction of the cost, and it can be used in different time zones and locations around the world. Due to the economic downturn of the last few years, it has become increasingly popular.

Video conferencing systems provide a more effective environment for business meetings compared to e-mail and telephone conversations. Therefore, they are being increasingly used by global companies to maintain business relationships between headquarters and regional offices spread across the globe. In recent years, the capabilities of desktop video conferencing have significantly improved, with a much lower cost compared to the traditional video conferencing systems. With desktop video conferencing, participants can have multiple video windows open at one time. They also have interfaces to a conference installed on their workstations, so these systems are easier for employees to use. This cost reduction has made this technology more affordable for smaller companies.

Key players in the video conferencing market include Cisco Systems, Polycom, Avaya, and Sprint.[14]

While making these decisions, information system managers should keep in mind that standardized software and hardware are the ideal but not always feasible. For example, using the same hardware in another country might be desirable, but it is not as simple as shipping the system to the other country and plugging it in. Vendors might not offer technical support in that country, or the other country's electrical standards may differ. As for using the same software in other countries, that becomes more complicated because of differences in language, business methods, and **transborder data flow (TDF)**, which is subject to restrictions on how the data can be captured and transmitted. TDF comprises national laws and international agreements on privacy protection and data security. However, thanks to cooperation and coordination among countries, these problems are becoming more manageable.[15]

9-2b Requirements of Global Information Systems

What makes an information system global? A GIS must be capable of supporting complex global decisions. This complexity stems from the global environment in which **multinational corporations (MNCs)** operate. MNCs are organizations with assets and operations in at least one country other than their home country. They deliver products and services across national borders and are usually centrally managed from their headquarters. A global environment includes four kinds of factors: legal (intellectual property laws, patent and trademark laws, TDF regulations, and so forth), cultural (languages, ethical issues, and religious beliefs), economic (currency, tax structure, interest rates, monetary and fiscal policies), and political (government type and stability, policies toward MNCs, and so on).[16]

In international business planning, it is critical to understand the global risks of operating an MNC—specifically, the political, foreign exchange, and market risks. Political risks include the problems caused by an unstable government, which is an important

Transborder data flow (TDF) restricts what type of data can be captured and transmitted in foreign countries.

A **multinational corporation (MNC)** is an organization with assets and operations in at least one country other than its home country. An MNC delivers products and services across national borders and is usually centrally managed from its headquarters.

© iStockphoto.com / Loops7

consideration given the many political uprisings of recent years. An unstable government can result in currency rates fluctuating, power changing hands rapidly and unpredictably, and other issues that affect company operations. A company considering political risks might not want to set up offices in many parts of the Middle East or Africa, for example, despite the low costs available in those countries, because of their unstable political situations. In addition, managing global operations requires considering potential conflicts between the governments of the country where the company is based (the parent government) and the country where the subsidiary is located (the host government).

A GIS, like any information system, can be classified according to the different kinds of managerial support it provides: operational, tactical, and/or strategic. Strategic support involves broad and long-term goals; tactical support concentrates on medium-range activities that move the organization toward achieving long-term goals; operational support involves day-to-day activities. Based on this classification, a GIS should collect, process, and generate different types of information in formats that are suitable for each type of support.

The complexities of global decision making mean that a GIS has some functional requirements that differ from a domestic information system's requirements. In addition, the line between operational and tactical management has blurred.

The first four of the following requirements are classified as operational, the remaining ones as strategic:

- *Global data access*—Online access to information from locations around the world allows management to monitor global operations from the company headquarters. Ideally, global networks provide a real-time communication link with global subsidiaries by integrating voice, data, and video. Several MNCs, such as Hewlett-Packard, General Electric, Texas Instruments, and IBM, have corporate databases linked worldwide.

- *Consolidated global reporting*—This is a crucial tool for managing overseas subsidiaries. These reports should include accounting and financial data, manufacturing updates, inventory, and so forth, and they enable management to compare financial information in all the subsidiaries. Because of differences in accounting procedures and regulatory standards, these comparisons can be difficult; however,

consolidated global reporting can help lessen these difficulties.

- *Communication between headquarters and subsidiaries*—To facilitate decision-making and planning processes, a GIS should provide an effective means of communication between the MNC's headquarters and its subsidiaries.

- *Management of short-term foreign exchange risks*—A mix of free-floating (no government intervention), managed-floating, and fixed-exchange rates characterizes today's international monetary system. Currency rates can change daily, so management must minimize the impact of currency fluctuations in countries where the parent company and the subsidiaries are located. To manage foreign exchange risks, many companies have developed expert systems and decision support systems (discussed in Chapters 12 and 13).

- *Strategic planning support*—This is the core of any GIS, a focus on regionalizing resources more effectively and responding to rapid environmental changes, such as increased political and foreign exchange risks and global competition.

- *Management of global tax risks*—Designing tax-risk management systems requires detailed knowledge of international finance, international monetary systems, and international tax law.

9-2c Implementation of Global Information Systems

Implementing a GIS can be difficult because countries differ in culture, politics, social and economic infrastructures, and business methods. International policies can vary, too, which affects communication and standardization processes. Furthermore, some have argued that a truly global corporation does not exist, much less a GIS to support its operations. Several issues must be addressed first, including the following[17,18]:

- The organization must identify its business opportunities in the global marketplace.

- Decision makers must justify the organization's investment in a GIS, given the substantial commitment of resources that will have to be made, usually years in advance.

- The organization's personnel need to be screened for technical and business expertise, because implementing a GIS is more challenging than implementing a domestic information system.

A GIS, like any information system, can be classified according to the different kinds of managerial support it provides: operational, tactical, and/or strategic.

> Decision makers must justify the organization's investment in a GIS, given the substantial commitment of resources that will have to be made, usually years in advance.

- Migration to the GIS needs to be coordinated carefully to help personnel move from the old familiar system to the new one.

Using information systems on a global scale is more challenging than doing so on a local scale. The challenges, discussed in more detail later in this chapter, involve such factors as infrastructure, languages, time zones, and cultures. To design a successful GIS, management must first determine the kind of information that global companies need to share. In addition, management cannot assume the company's products or services will continue selling the same way because of possible changes in customers' needs and preferences and global competition.[19] Considering the entire organization's operational efficiency is critical in coordinating international business activities, so global companies need to change their production and marketing strategies in an effort to respond to the global market.

The organization's structure usually determines the architecture of its GIS, as you will see in the following sections, which discuss each of these structural types.

9-3a Multinational Structure

In a **multinational structure** (see Exhibit 9.1), production, sales, and marketing are decentralized and financial management remains the parent company's responsibility. Tyco Corporation is an example of a company with a multinational structure.[20] Tyco's focus is local—responding to customers' needs in a subsidiary's location. So, the company's subsidiaries operate autonomously but regularly report to the parent company. Another company that has a multinational structure is Nestle, which uses 140 different financial systems at its subsidiaries around the world. The company's

9-3 ORGANIZATIONAL STRUCTURES AND GLOBAL INFORMATION SYSTEMS

Four types of organizations do business across national borders:

- Multinational organizations
- Global organizations
- International organizations
- Transnational organizations

In a **multinational structure**, production, sales, and marketing are decentralized, and financial management remains the parent's responsibility.

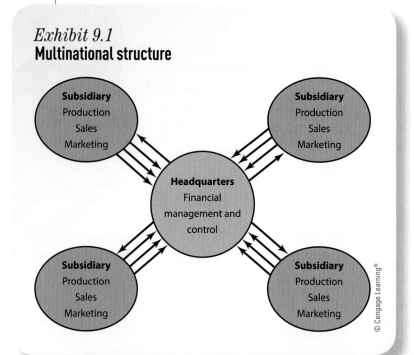

Exhibit 9.1
Multinational structure

© Cengage Learning®

multinational structure is an advantage because it reduces the need for communication between subsidiaries and headquarters, allowing subsidiaries to make many decisions on their own.[21]

Local hardware and software vendors influence which applications a multinational company chooses. Inevitably, each subsidiary operates on a different platform, and uniform connections are economically impractical.

9-3b Global Structure

An organization with a **global structure**, sometimes called a "franchiser," uses highly centralized information systems.[22] Subsidiaries have little autonomy and rely on headquarters for all process and control decisions as well as system design and implementation. Consequently, an extensive communication network is necessary to manage this type of organization, and a GIS fits well into this structure.

Unfortunately, the integration needed to manage production, marketing, and human resources is difficult to achieve with a global structure because of the heavy reliance on headquarters. To achieve organizational efficiency, duplicate information systems have to be developed.[23] Products are usually created, financed, and produced in the headquarters' country, and subsidiaries have the responsibility of selling, marketing, and tailoring the products to their countries' requirements and tastes. For example, McDonald's, in order to appeal to local tastes, changed the burgers it serves in India to a 100 percent vegetarian product consisting of potatoes, peas, carrots, and some Indian spices.

© Jonathan Larsen/Shutterstock.com

information from all over the world. General Motors also uses electronic meeting systems to coordinate its research and development efforts throughout the world.[25]

McDonald's, in order to appeal to local tastes, changed the burgers it serves in India to a 100 percent vegetarian product consisting of potatoes, peas, carrots, and some Indian spices.

In addition to McDonald's, Mrs. Field's Cookies and Kentucky Fried Chicken are companies that have a global structure.[24] Yet another example is General Motors, which uses a GIS to integrate inventory

A **global structure** (also known as franchiser) uses highly centralized information systems. Subsidiaries have little autonomy and rely on headquarters for all process and control decisions as well as system design and implementation.

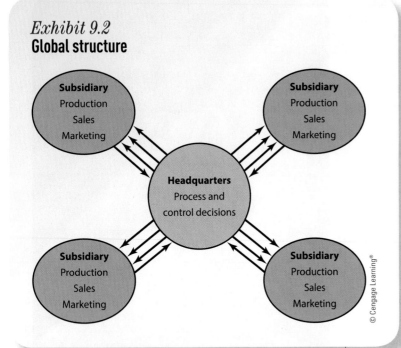

Exhibit 9.2
Global structure

Subsidiary
Production
Sales
Marketing

Subsidiary
Production
Sales
Marketing

Headquarters
Process and
control decisions

Subsidiary
Production
Sales
Marketing

Subsidiary
Production
Sales
Marketing

© Cengage Learning®

Exhibit 9.3 shows an international structure that uses two-way communication; for example, expertise information flows from headquarters to subsidiaries, and financial information flows from subsidiaries to headquarters. Heavy equipment manufacturers such as Caterpillar Corporation usually have this structure.

9-3d Transnational Structure

In an organization with a **transnational structure**, the parent company and all the subsidiaries work together in designing policies, procedures, and logistics for delivering products and services to the right market. This type of organization might have several regional divisions that share authority and responsibility, but in general it does not have its headquarters in a particular country. A transnational organization usually focuses on optimizing supply sources and using advantages available in subsidiary locations. Many companies do this when they look for manufacturing facilities in countries where labor is less expensive than it is in the parent country. For example, China, India, Vietnam, and other countries have cheaper labor costs than the United States does. Again, a GIS fits into this structure well by integrating global activities through cooperation and information sharing between headquarters and subsidiaries.

Exhibit 9.2 shows this structure, with a one-way flow of services, goods, information, and other resources.

9-3c International Structure

An organization with an **international structure** operates much like a multinational corporation, but subsidiaries depend on headquarters more for process and production decisions. Information systems personnel are regularly exchanged among locations to encourage joint development of applications for marketing, finance, and production. This exchange encourages a cooperative culture in geographically dispersed personnel, and using a GIS to support an international structure is more feasible because of this cooperative nature. Subsidiaries' GISs can be centralized or decentralized, depending on the extent to which they cooperate.

An organization with an **international structure** operates much like a multinational corporation, but subsidiaries depend on headquarters more for process and production decisions.

In an organization with a **transnational structure**, the parent and all subsidiaries work together in designing policies, procedures, and logistics for delivering products and services to the right market.

Exhibit 9.3
International structure

Subsidiary
Financial
information

Subsidiary
Financial
information

Headquarters
Expertise for
process and
production
decisions

Subsidiary
Financial
information

Subsidiary
Financial
information

© Cengage Learning®

The architecture of the GIS in a transnational structure requires a higher level of standardization and uniformity for global efficiency, and yet it must maintain local responsiveness. Universal data dictionaries and standard databases, for example, enhance the integration of GISs.

The level of cooperation and worldwide coordination needed for a transnational structure does not fully exist in today's global environment. However, with increasing cooperation between nations, this structure is becoming more feasible. Citigroup, Sony, and Ford, to name a few, have been striving to adopt a transnational structure. In fact, the globalizing trend is forcing many organizations to gravitate toward this structure because of competition with companies that already share innovations across borders and maintain responsiveness to local needs. These companies have increased efficiency in production costs because production can be spread across more locations.[26] Exhibit 9.4 shows this structure, with cooperation between headquarters and subsidiaries as well as between subsidiaries. Foreign exchange systems that allow traders and brokers from around the world to interact are an example of information systems that support this structure. The adjoining information box highlights the applications of GISs at Federal Express.

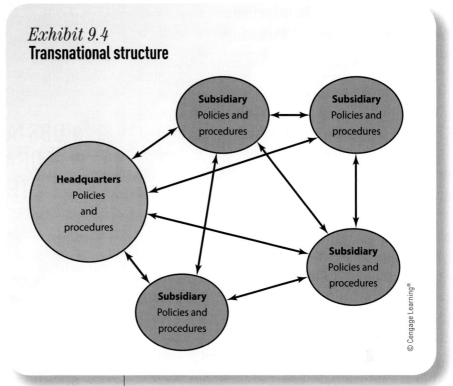

Exhibit 9.4
Transnational structure

© Cengage Learning®

GLOBAL INFORMATION SYSTEM AT FEDEX

To support its global operations in over 220 countries and territories around the world, FedEx uses a sophisticated GIS. In addition to the Internet and mobile networks, the FedEx system has several additional components. The following four components are among the main ones:

- *COSMOS (Customer Operations Service Master Online System)*—This component is a computerized package-tracking system that monitors every package from the time it is picked up from the customer until it is delivered to its destination.[27]
- *Command & Control: Delivery in Any Weather*—This is a satellite-to-ground-level operations system based in Memphis, Tennessee. It serves as a weather assistant, enabling FedEx to deliver packages in the most efficient and effective way regardless of the weather conditions. The central part of this component is a relational database that coordinates FedEx logistics throughout the world.[28]
- *FedEx Global Trade Manager*—Using www.fedex.com, customers can access this component to receive assistance for international shipping.[29]
- *FedEx INTrade, Traders Information Exchange System*—This component connects FedEx and the brokerage community. By doing so, it eliminates a number of tasks and functions that used to be done manually, including generating summary sheets, broker release notification reports, electronic commercial invoices, and so forth. The system makes all these documents available electronically.[30]

9-3e Global Information Systems Supporting Offshore Outsourcing

Offshore outsourcing is an alternative for developing information systems. With this approach, an organization chooses an outsourcing firm in another country that can provide needed services and products. Initially, offshore outsourcing was used mostly in manufacturing to find cheap labor, but now it is used for many information technology tasks, including the following:

- Medical diagnosis
- Tax preparation
- Programming
- Application development
- Web site development
- Help desk/user support
- Quality assurance/software testing

The widespread availability of the Internet, improved telecommunication systems, the reduced cost of communication, and increased bandwidth have made offshore outsourcing more attractive for all types of organizations. A GIS plays an important role in supporting offshore outsourcing by providing a global network that all participants can use for coordinating development activities, such as product design and global marketing campaigns. Table 9.1

> With **offshore outsourcing**, an organization chooses an outsourcing firm in another country that can provide needed services and products.

TABLE 9.1 TOP OFFSHORE LOCATIONS FOR OUTSOURCING IN 2012

Americas	Asia/Pacific	Europe, Middle East, and Africa (EMEA)
Argentina	Bangladesh	Belarus
Brazil	China	Bulgaria
Chile	India	Czech Republic
Colombia	Indonesia	Hungary
Costa Rica	Malaysia	Mauritius
Mexico	Philippines	Morocco
Peru	Sri Lanka	Poland
Uruguay	Thailand	Romania
	Vietnam	Russia
		Slovakia
		South Africa
		Turkey
		Ukraine

© Cengage Learning®

lists the top 30 offshore locations for outsourcing in 2012.[31] Some of the criteria used to rate countries include language proficiency, local government's support of offshore business, potential labor pool, existing infrastructure (roads, rail service, and airports), and quality of the educational system.

9-4 OBSTACLES TO USING GLOBAL INFORMATION SYSTEMS

A GIS helps an organization improve its global coordination, manage the factors that promote globalization, and maintain a competitive edge by supporting strategic planning. However, like any information system project, there are potential problems in implementing and maintaining a GIS. Companies planning to use GISs should analyze these problems and try to address them. Taking a proactive approach can increase the chance of success in using this technology. The following factors, some of which are discussed in more detail in the following sections, can hinder the success of a GIS[32,33]:

- Lack of standardization (including differences in time zones, taxes, language, and work habits)
- Cultural differences
- Diverse regulatory practices

© Holbox/Shutterstock.com

- Poor telecommunication infrastructures
- Lack of skilled analysts and programmers

A more subtle obstacle to GIS development is the organization's unwillingness to delegate control of information systems to host countries. To achieve true integration on an international scale, organizations must empower key personnel in other countries and rely on feedback and information-sharing technologies.

finding the right time to take the system offline for backup and maintenance.[34] A balance between international system development standards, allowing ease of integration, modularization, custom tailoring of systems, and applications for local responsiveness, is needed.

Sharing software is difficult and impractical when these factors are considered. Only 5–15 percent of a company's applications are truly global in nature. Most

> To achieve true integration on an international scale, organizations must empower key personnel in other countries and rely on feedback and information-sharing technologies.

9-4a Lack of Standardization

Lack of standardization can impede the development of a cohesive GIS that is capable of sharing information resources across borders. Electronic data interchange, e-mail, and telecommunication standards vary throughout the world, and trying to work with all the various standards is impractical. And although open-source systems are increasing in popularity and the technology to link diverse systems is available, few organizations can meet the costs of integrating different platforms.

Too much standardization can be a problem as well, decreasing an organization's flexibility in responding to local preferences—even time differences. For example, an organization should not insist that all its subsidiaries use the metric system. It should allow them to use the measuring systems they are familiar with, converting from one system to the other when necessary.

Time zones can also pose difficulties in managing a GIS. For example, information-systems personnel who are managing a centralized GIS under international standards and sharing information resources across time zones might have difficulties

> In some cultures, using technology is considered a boring, low-level task; in others, being technologically knowledgeable is seen as a sign of social importance.

applications are local in nature and cannot be integrated into a GIS infrastructure. Even if the software can be integrated globally, support and maintenance problems might result. If the network goes down, who is responsible for bringing the system back online? Moreover, employees calling the help desk might not speak the same language as the help desk personnel. Therefore, coordination and planning for variations in local needs are critical for using a GIS.[35]

9-4b Cultural Differences

Cultural differences in values, attitudes, and behaviors play an important role in using GISs. For example, in some cultures, using technology is considered a boring, low-level task; in others, being technologically knowledgeable is seen as a sign of social importance.

A travel-oriented Web site aimed at customers who make last-minute reservations at lower prices worked well in the United Kingdom but did not translate well to other countries, such as Germany, where advance planning is expected and last-minute reservations are not "rewarded" with lower prices.[36]

Organizations might also need to look at changing content or images on their Web sites; photos of women dressed in a certain way might be acceptable in the Western world but unacceptable in the Middle East, for instance. Cultural issues are best addressed with education and training.

9-4c Diverse Regulatory Practices

Diverse regulatory practices also impede the integration process. This obstacle does not necessarily apply to TDF (transborder data flow) regulations; it applies to policies on business practices and technological use. Many countries also restrict the type of hardware and software that can be imported or used, and the vendors that an organization normally deals with might not service certain countries.[37] For example, in August 2010, the United Arab Emirates, citing security concerns, announced that BlackBerry phones would not be allowed to access e-mail or the Web.[38]

Adopting open-source systems could eliminate part of this problem. However, as mentioned, few organizations are capable of adopting these systems.

Jurisdiction issues regarding the contents of a GIS can also be challenging. ISPs, content providers, servers, and organizations owning these entities might be scattered throughout the world and operating under different rules and regulations. For example, Yahoo! was sued in French courts because Nazi memorabilia were being sold on its auction site, which is an illegal activity in France. To date, French and U.S. courts have not agreed on the resolution or even on which court has jurisdiction.[39] Determining jurisdiction in cases involving cyberspace is still difficult.

The nature of intellectual property laws and how they are enforced in different countries also varies. Software piracy is a problem in all countries, but several have piracy rates higher than 90 percent. This problem has resulted in an estimated loss of $40 billion worldwide.[40] Other legal issues include privacy and cybercrime laws as well as censorship and government control, which vary widely from country to country.

9-4d Poor Telecommunication Infrastructures

As mentioned earlier, before adding a GIS, international companies must take into consideration the telecommunication infrastructures of the countries where subsidiaries are located. An organization might have the resources and skills to implement a worldwide integrated system but not be able to change an existing telecommunication infrastructure. Furthermore, the differences in telecommunication systems make consolidating them difficult. Implementing a GIS that encompasses 25 countries, for instance, is expensive and cumbersome when each country has different service offerings, price schedules, and policies.

In countries where Internet access is slower or more costly, Web pages should not have content with lots of graphics and animation that require more bandwidth. However, people in countries such as South Korea, where high-speed access is common, expect sophisticated Web sites with many graphic features.

Even when the telecommunication infrastructure in two countries is comparable, differences in standards can cause problems. For example, a company with branches in the United States and Egypt might face the problems of different Internet protocols, higher costs, slower speed, and less reliability in Egypt.

9-4e Lack of Skilled Analysts and Programmers

Having skilled analysts and consultants with the knowledge to implement a GIS is critical, particularly with the severe shortage of qualified information systems professionals in the United States and Western Europe. When forming integrated teams, companies must consider the nature of each culture and differences in skills in other countries.[41] For example, experts from Singapore and Korea have been regarded as the best consultants in Asia because of their work ethic and their broad skill base. Germans are recognized for their project management skills, and Japanese are known for their quality process controls and total quality management. Ideally, an organization would link the skills of people from different countries to form a "dream team." However, cultural and political differences can affect the cooperative environment needed for global integration. Training and certification programs, many of which are offered through the Internet, are one possible solution for narrowing this skills gap in developing nations.

The Industry Connection highlights the SAP Corporation as a leader in enterprise computing and global information systems.

Industry Connection: SAP Corporation[42]

SAP, founded by five former IBM employees in 1972, is one of the leading providers of business software. Its applications can be used to manage finances, assets, production operations, and human resources. The latest version of its enterprise resource planning software, SAP ERP 7.0, includes comprehensive Web-enabled products. SAP offers a Web interface for customers, called www.mySAP.com, and e-business applications, including customer relationship management (CRM) and supply chain management (SCM) systems. From the beginning, SAP products have been designed to be used with multiple languages and currencies, which is particularly useful for companies that need to support global operations. In addition, SAP includes upgrades for global aspects of information systems, such as making the transition from European currencies to euros. In addition to products for enterprise resource planning and SCM (both discussed in Chapter 11), SAP offers the following software:

- *SAP Supplier Relationship Management*—Helps manage a company's relationships with its suppliers by automating procurement processes, managing the supply chain, and creating a collaborative environment between the company and its suppliers.

- *SAP Product Lifecycle Management (PLM)*—Includes services for coordinating manufacturing processes, from developing product prototypes to producing the final product to ensuring compliance with industry standards and regulations.

- *SAP-HANA*—Is used for in-memory business analytics, streamlining of applications, and predictive analysis in real time. Because it does its calculations in memory, the software delivers extremely fast response times. It can be used on site or in the cloud.

STUDY TOOLS 9

LOCATED AT BACK OF THE TEXTBOOK
- ☐ Rip out Chapter Review Card

LOCATED AT WWW.CENGAGE.COM/LOGIN
- ☐ Review Key Terms Flash Cards (Print or Online)
- ☐ Download audio and visual summaries to review

- ☐ Complete both Practice Quizzes to prepare for tests
- ☐ Play "Beat the Clock" and "Quizbowl" to master concepts
- ☐ Complete "Crossword Puzzle" to review key terms
- ☐ Watch video on "Method" for a real company example

KEY TERMS

global information system (GIS) 185
global structure 191
international structure 192
multinational corporations (MNCs) 188

multinational structure 190
offshore outsourcing 194
transborder data flow (TDF) 188
transnational structure 192

REVIEWS AND DISCUSSIONS

1. What are two reasons that companies of all sizes want to become global?
2. What is the role of the Internet in promoting the globalization trend?
3. What are the three recommendations for making a Web site more appealing to global audiences?
4. What are two advantages of high coordination when used in global operations?
5. What are two main components of a global information system?
6. Which type of organizational structure is the most popular type and why?
7. What is the relationship between outsourcing and global information systems?
8. What are four obstacles in using global information systems?

PROJECTS

1. Like FedEx, UPS is a major player in global logistics. After reading the information presented in this chapter and other sources, write a one-page paper that describes three IT tools that UPS uses in its global logistics.
2. The availability and speed of Internet connections play a major role in the success of a GIS. After reading the information presented in this chapter and other sources, write a one-page paper that lists the 10 countries with the fastest Internet connections. Is the United Sates among the top 10?
3. Offshore outsourcing increases with the use of GISs. After reading the information presented in this chapter and other sources, write a one-page paper that describes four factors a company should consider when looking for a country to provide it with IT services.
4. Information-sharing technologies are among the major components of a successful GIS. After reading the information presented in this chapter and other sources, write a one-page paper that lists three such technologies; also mention the specific task and decision-making support that each technology provides.
5. Understanding cultural differences can play a major role in the success of a GIS. After reading the information presented in this chapter and other sources, write a two-page paper that identifies five such differences between the United States and Saudi Arabia. How would you resolve such differences?
6. After reading the information presented in this chapter and other sources, write a one-page paper describing four reasons that companies of all sizes should become global. How are small companies able to participate in the globalization trend?

ARE YOU READY TO MOVE ON?

1. Designing and implementing a global database is usually easier than designing a domestic one. True or False?
2. TDF comprises national laws and international agreements on privacy protection and data security. True or False?
3. Cultural differences are irrelevant when designing a global Web site. True or False?
4. Which continent has the highest number of Internet users?
   a. Asia
   b. North America
   c. Europe
   d. Africa
5. Which of the following is not an example of a GIS organizational structure?
   a. Multinational
   b. Global
   c. International
   d. Cross national
6. Which of the following is not an obstacle to using GISs?
   a. Lack of standardization
   b. Fast Internet access
   c. Cultural differences
   d. Diverse regulatory practices

Global Information Systems at Toyota Motor Company

The Toyota Motor Company is a global automobile manufacturer that operates on five different continents. In the United States, it runs five major assembly plants. To manage its operations efficiently and effectively around the globe, Toyota uses several types of information systems. It uses the Internet and global networks to communicate with its offices, plants, and dealerships around the globe.[43]

Toyota is one of the founders of the Toyota Production System, an early version of the Just In Time (JIT) inventory system.[44] This system allows Toyota to have on hand the exact number of components needed at any time in order to continue its operations, given that waste often occurs when components are inventoried and stored. To use a JIT inventory system, Toyota's GIS must be capable of managing real-time inventory, not only within its own manufacturing facilities but in all of its suppliers' facilities. Because of this, Toyota requires all of its suppliers to have a system capable of interfacing with the one Toyota uses for its own operations.

Toyota worked with Dell, Microsoft, and WorldCom to develop a "Dealer Daily" system that offers a centralized data

© iStockphoto.com/Tom England

center for the more than 1,100 Toyota and Lexus dealers in the United States. This system allows dealers to spend more time focusing on selling cars and less time on paperwork. For example the system is capable of providing a response to a financing application in 15 seconds.[45]

In 2011, using Oracle E-Business Suite 12.1, Toyota Motor Europe (TME) implemented Toyota's Vehicle Order Management (VOM) system. This system, which encompasses 13 countries, enables TME to improve its European operations by reducing delivery time to customers and managing inventory more efficiently. As a result, the system reduces operating costs.[46]

Answer the following questions:

1. What role do global networks play in the effective implementation of JIT?
2. What role does the Dealer Daily system play in Toyota's implementation of a GIS?
3. What is the function of Vehicle Order Management (VOM)?
4. How many countries will be impacted by VOM and in which part of the world?

Information Technologies Support Global Supply Chain

An efficient and effective supply chain plays a major role in the success of a multinational corporation (MNC). MNCs deal with a myriad of issues, including quality of raw materials, scarcity of materials, the locations where the materials' ingredients or components are grown or manufactured, and the counterfeiting of materials (especially in the pharmaceutical industry). MNCs also face environmental challenges, such as tsunamis in the Philippines, monsoons in Thailand, and volcanos in Indonesia. Piracy and theft of intellectual properties are yet another set of issues that MNCs must deal with.[47]

Because supply chains, in many cases, are outsourced and subcontracted, controlling each point of the supply chain becomes a challenge. Although IT cannot solve all of these problems, it can provide timely information that executives in MNCs can use to quickly respond to any of these issues. MNCs are using data warehouses and analytics to analyze, pinpoint, and quickly deliver key information related to parts of the supply chain. RFIDs, GPSs, and

© 3d Brained/Shutterstock.com

QR codes have significantly improved the efficiency of these supply chains.

Ports America, for example, uses OCR cameras to scan each container, RFID tags to match the contents with the trucks, and GPS sensors to identify equipment locations and cargo movements. Computer networks and the Internet have created unparalleled connections throughout the world. Major logistics companies such as UPS and FedEx are using state-of-the-art information technologies to efficiently deliver products and services around the globe.[48]

Answer the following questions:

1. What are some of the challenges that MNCs face as far as intellectual properties are concerned?
2. What are some of the environmental challenges MNCs face?
3. How can information technology assist with the counterfeit problem?
4. How can information technology assist with logistics?

PART 3

10 | Building Successful Information Systems

LEARNING OUTCOMES

After studying this chapter, you should be able to:

10-1 Describe the systems development life cycle (SDLC) as a method for developing information systems.

10-2 Explain the tasks involved in the planning phase.

10-3 Explain the tasks involved in the requirements-gathering and analysis phase.

10-4 Explain the tasks involved in the design phase.

10-5 Explain the tasks involved in the implementation phase.

10-6 Explain the tasks involved in the maintenance phase.

10-7 Describe new trends in systems analysis and design, including service-oriented architecture, rapid application development, extreme programming, and agile methodology.

After you finish this chapter, go to **PAGE 220** for the **STUDY TOOLS**

© iStockphoto.com/iGlobalStock

This chapter explains the systems development life cycle (SDLC), a model for developing a system or project. The cycle is usually divided into five phases, and you learn about the tasks involved in each phase. For example, in the first phase—planning—a feasibility study is typically conducted, and the SDLC task force is formed. You also learn about two alternatives to the SDLC model: self-sourcing and outsourcing. Finally, you review new trends in systems analysis and design, such as service-oriented architecture, rapid application development, extreme programming, and agile methodology.

10-1 SYSTEMS DEVELOPMENT LIFE CYCLE: AN OVERVIEW

In the information systems field, system failure can happen for several reasons, including missed deadlines, users' needs that are not met, dissatisfied customers, lack of support from top management, and exceeding the budget. Using a system development method can help prevent these failures. Designing a successful information system requires integrating people, software, and hardware. To achieve this integration, designers often follow the **systems development life cycle (SDLC)**, also known as the "waterfall model." It is a series of well-defined phases performed in sequence that serves as a framework for developing a system or project. Exhibit 10.1 shows the phases of the SDLC, which are explained throughout this chapter. In this model, each phase's output (results) becomes the input for the next phase. When following this model, keep in mind that the main goal of an information system is delivering useful information in a timely manner to the right decision maker.

Systems planning today is about evaluating all potential systems that need to be implemented. A preliminary analysis of requirements for each is done, and a feasibility study is conducted for each system. Then the

> **Systems development life cycle (SDLC)**, also known as the "waterfall model," is a series of well-defined phases performed in sequence that serves as a framework for developing a system or project.

Exhibit 10.1
Phases of the SDLC

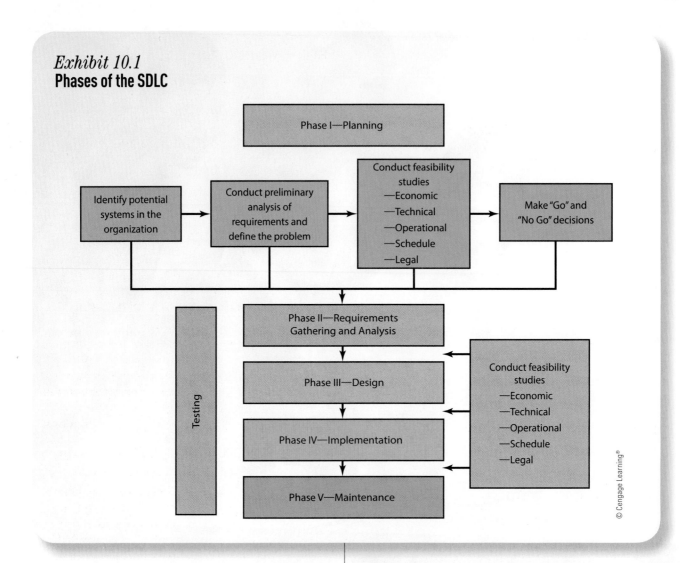

organization decides which ones are a "go" and proceeds to the next phase.

Information system projects are often an extension of existing systems or involve replacing an old technology with a new one. However, sometimes an information system needs to be designed from scratch, and the SDLC model is particularly suitable in these situations. For existing information systems, some phases might not be applicable, although the SDLC model can still be used. In addition, when designing information systems, projecting the organization's growth rate is important; otherwise, the system could become inefficient shortly after it is designed.

> During the **planning phase**, which is one of the most crucial phases of the SDLC model, the systems designer must define the problem the organization faces, taking care not to define symptoms rather than the underlying problem.

10-2 PHASE 1: PLANNING

During the **planning phase**, which is one of the most crucial phases of the SDLC model, the systems designer must define the problem the organization faces, taking care not to define symptoms rather than the underlying problem. The problem can be identified internally or externally. An example of an internally identified problem would be management voicing concern about the organization's lack of a competitive edge in the marketplace. An example of an externally identified problem would be suppliers noting inefficiency in the inventory control procedure.

After identifying the problem, an analyst or team of analysts assesses the current and future needs of the organization or a specific group of users by answering the following questions:

- Why is this information system being developed?
- Who are the system's current and future users?

- Is the system new, or is it an upgrade or extension of an existing system?
- Which functional areas (departments) will be using the system?

As part of this assessment, analysts must examine the organization's strategic goals, how the proposed system can support these goals, which factors are critical to the proposed system's success, and the criteria for evaluating the proposed system's performance. Establishing evaluation criteria ensures objectivity throughout the SDLC process.

In addition, analysts must get feedback from users on the problem and the need for an information system. During this phase, they need to make sure users understand the four Ws:

- *Why*—Why is the system being designed? Which decisions will be affected?
- *Who*—Who is going to use the system? Is it going to be used by one decision maker or a group of decision makers? This question is also about types of users. For example, will the Marketing Department be using the system? And will the Manufacturing Department be using the system as suppliers or as consumers of information?
- *When*—When will the system be operational? When in the business process (in what stages) will the system be used?
- *What*—What kind of capabilities will the system provide? How will these capabilities be used?

The end result of this phase should give users and top management a clear view of what the problem is and how the information system will solve the problem. As an example, here is a look at how ABC Furniture is planning for an information system to solve the problem of inaccurate inventory forecasts.

Currently, ABC Furniture buys wood from New England Wood (NEW).

- *Why*—ABC Furniture needs an information system to track inventory, generate a more accurate forecast of product demand, and track requirements for wood to be ordered from NEW. Clearly, a more accurate inventory will help reduce inventory costs, improve ABC Furniture's relationship with NEW and with distributors, ensure the company's products are available for retailers, and improve ABC's image in the marketplace.
- *Who*—The main users of the information system will be the procurement group responsible for placing orders with NEW, the manufacturing division responsible for tracking inventory and ensuring that demand for finished goods is met, the sales personnel who take orders from distributors, and possibly distributors who take orders from retailers.
- *When*—The system must become operational within the next 4 months because the company's main competitor is planning to open a new store in 6 months. Furthermore, the system must support the materials-ordering stage, the production-planning stage, and the shipping stage of the manufacturing process. It must also supply information for the marketing campaign that ABC Furniture is planning to run in 5 months and support ABC's expansion into a new region.
- *What*—On the inbound side, the system must track pending and received deliveries, quantities of raw materials, orders placed for raw materials, and raw material levels from all of ABC's suppliers, including NEW. On the operations side, the system must provide information on inventory levels of all products, raw materials, work in progress at each stage of manufacturing, quality of raw materials received, quality of finished goods inspected, and rejects. On the outbound side, the system must track placed orders, unfulfilled orders, and fulfilled orders for each finished product as well as the order history for each distributor and retailer demand.

10-2a Formation of the Task Force

To ensure an information system's success, users must have input in the planning, requirements-gathering and analysis, design, and implementation phases. For this reason, a task force is formed, consisting of representatives from different departments (including IT), systems analysts, technical advisors, and top management. This team collects user feedback and works toward getting users involved from the beginning.

The system designers and analysts should explain the goals and benefits of the new system so the task force knows what to look for in user input. Generally, an information system has two groups of users from whom the task force should gather feedback: internal and external. **Internal users** are employees who will use the system regularly, and they can offer important feedback on the system's strengths and weaknesses. **External users** are not employees but do use the system; they include customers, contractors, suppliers, and other business partners. Although external users are not normally part of the task force, their input is essential.

Using a task force for designing an information system is similar to using the joint application design approach. **Joint application design (JAD)** is a collective activity involving users, top management, and IT professionals. It centers on a structured workshop (called a JAD session) in which users and system professionals

requirements for the application are not too narrow and one-dimensional in focus.[1]

10-2b Feasibility Study

Feasibility is the measure of how beneficial or practical an information system will be to an organization; it should be measured continuously throughout the SDLC process (see the information box below). Upper management is often frustrated by information systems that are unrelated to the organization's strategic goals or have an inadequate payoff, by poor communication between system users and designers, and by designers' lack of consideration for users' preferences and work habits. A detailed feasibility study that focuses on these factors can help ease management's frustration with investing in information systems.[2]

> Feasibility is the measure of how beneficial or practical an information system will be to an organization; it should be measured continuously throughout the SDLC process.

come together to develop an application. It involves a detailed agenda, visual aids, a leader who moderates the session, and a scribe who records the specifications. It results in a final document containing definitions for data elements, workflows, screens, reports, and general system specifications. An advantage of the JAD approach is that it incorporates varying viewpoints from different functional areas of an organization to help ensure that collected

Internal users are employees who will use the system regularly, and they can offer important feedback on the system's strengths and weaknesses.

External users are not employees but do use the system; they include customers, contractors, suppliers, and other business partners. Although they are not normally part of the task force, their input is essential.

Joint application design (JAD) is a collective activity involving users, top management, and IT professionals. It centers on a structured workshop (called a JAD session) where users and system professionals come together to develop an application.

A FEASIBLE PROJECT BECOMES UNFEASIBLE

WestJet Airlines, a Canadian discount airline in Calgary, announced in July 2007 that it was stopping development on a new reservation system called AiRES, even though it had invested $30 million in the project. The problem was not with Travelport, the company developing the system. WestJet simply grew faster than anticipated, and the original specifications for the reservation system did not address this fast growth. Management wanted to add features, such as the capability to partner with international carriers, but the system had been planned to fit a small discount airline, not a large international airline. WestJet suspended the work of 150 internal IT specialists and about 50 outside consultants. This example shows the need for conducting feasibility studies through a project's life cycle. If WestJet had continued the project, the potential losses would have been more than $30 million.[3]

During the planning phase, analysts investigate a proposed solution's feasibility and determine how best to present the solution to management in order to obtain funding. The tool used for this purpose is a **feasibility study**, and it usually has five major dimensions, discussed in the following sections: economic, technical, operational, scheduling, and legal.

Economic Feasibility

Economic feasibility assesses a system's costs and benefits. Simply put, if implementing the system results in a net gain of $250,000 but the system costs $500,000, the system is not economically feasible. To conduct an economic feasibility study, the systems analyst team must identify all costs and benefits—tangible and intangible—of the proposed system. The team must also be aware of opportunity costs associated with the information system. Opportunity costs measure what you would miss by not having a system or feature. For example, if your competitor has a Web site and you do not, what is the cost of not having a site, even if you do not really need one? What market share are you likely to lose if you do not have a Web site?

To assess economic feasibility, the team tallies tangible development and operating costs for the system and compares them with expected financial benefits of the system. Development costs include the following:

- Hardware and software
- Software leases or licenses
- Computer time for programming, testing, and prototyping
- Maintenance costs for monitoring equipment and software
- Personnel costs—Salaries for consultants, systems analysts, network specialists, programmers, data entry clerks, computer operators, secretaries, and technicians
- Supplies and other equipment
- Training employees who will be using the system

Operating costs for running the system are typically estimated, although some vendors and suppliers can supply costs. These costs can be fixed or variable (depending on rate of use). After itemizing these costs, the team creates a budget. Many budgets do not allow enough for development costs, especially technical expertise (programmers, designers, and managers), and for this reason, many information system projects go over budget.

An information system's scope and complexity can change after the analysis or design phases, so the team should keep in mind that an information system project that is feasible at the outset could become unfeasible later. Integrating feasibility checkpoints into the SDLC process is a good idea to ensure the system's success. Projects can always be canceled or revised at a feasibility checkpoint, if needed.

To complete the economic feasibility study, the team must identify benefits of the information system, both tangible and intangible. Tangible benefits can be quantified in terms of monthly or annual savings, such as the new system allowing an organization to operate with three employees rather than five or the new system resulting in increased profits. The real challenge is assessing intangible costs and benefits accurately; attaching a realistic monetary value to these factors can be difficult.

Intangible benefits are difficult to quantify in terms of dollar amounts, but if they are not at least identified, many information system projects cannot be justified. Examples of intangible benefits include improved employee morale, better customer satisfaction, more efficient use of human resources, increased flexibility in business operations, and improved communication. For example, you could quantify customer service as maintaining current total sales and increasing them by 10 percent to improve net profit. Other measures have been developed to assess intangibles, such as quantifying employee morale with rates of on-time arrival to work or working overtime. Customer satisfaction, though intangible, can be measured by using satisfaction surveys, and the Internet has made this method easier.

After collecting information on costs and benefits, the team can do a cost-effectiveness analysis. This analysis is based on the concept that a dollar today is worth more than a dollar 1 year from now. If the system does not produce enough return on the investment, the money can be better

> A **feasibility study** analyzes a proposed solution's feasibility and determines how best to present the solution to management. It usually has five major dimensions: economic, technical, operational, scheduling, and legal.
>
> **Economic feasibility** assesses a system's costs and benefits.

spent elsewhere. The most common analysis methods are payback, net present value (NPV), return on investment (ROI), and internal rate of return (IRR). The final result of this task is the cost-benefit analysis (CBA) report, used to sell the system to top management. This report can vary in format but should include the following sections: executive summary, introduction, scope and purpose, analysis method, recommendations, justifications, implementation plans, summary, and appendix items, which can include supporting documentation. Some examples of useful supporting documentation are organizational charts, workflow plans, floor plans, statistical information, project sequence diagrams, and timelines or milestone charts.

Technical Feasibility

Technical feasibility is concerned with the technology that will be used in the system. The team needs to assess whether the technology to support the new system is available or feasible to implement. For example, a full-featured voice-activated monitoring system is not technically feasible at this point. However, given the pace of technological development, many of these problems will eventually have solutions. Lack of technical feasibility can also stem from an organization lacking the expertise, time, or personnel to implement the new system. This problem is also called "a lack of organizational readiness." In this case, the organization can take steps to address its shortcomings and then consider the new system. Extensive training is one solution to this problem.

Operational Feasibility

Operational feasibility is the measure of how well the proposed solution will work in the organization and how internal and external customers will react to it. The major question to answer is whether the information system is worth implementing. To assess operational feasibility, the team should address the following questions:

- Is the system doing what it is supposed to do? For example, will the information system for ABC reduce orders for raw materials by tracking inventory more accurately?
  - Will the information system be used?
  - Will there be resistance from users?
  - Will top management support the information system?
  - Will the proposed information system benefit the organization?
  - Will the proposed information system affect customers (both internal and external) in a positive way?

> The major question to answer when conducting operational feasibility is whether the information system is worth implementing.

Scheduling Feasibility

Scheduling feasibility is concerned with whether the new system can be completed on time. For example, an organization might need a wireless network immediately because of a disaster that destroyed the existing network. However, if the new system cannot be delivered in time, the loss of customers could force the organization out of business. In this case, the proposed system is not feasible from a schedule viewpoint. The problem of missing deadlines is common in the information systems field, but designers can often minimize this problem by using project management tools (discussed later in the chapter).

Legal Feasibility

Legal feasibility is concerned with legal issues; it typically addresses questions such as the following:

- Will the system violate any legal issues in the country where it will be used?
- Are there any political repercussions of using the system?

Technical feasibility is concerned with the technology to be used in the system. The team needs to assess whether the technology to support the new system is available or feasible to implement.

Operational feasibility is the measure of how well the proposed solution will work in the organization and how internal and external customers will react to it.

Scheduling feasibility is concerned with whether the new system can be completed on time.

Legal feasibility is concerned with legal issues, including political repercussions and meeting the requirements of the Information Privacy Act.

HealthCare.gov: Feasibility Issues

When the government Web site HealthCare.gov was launched on October 1, 2013 as a result of the Affordable Care Act, it quickly faced several technical problems. The system is designed to provide insurance information for millions of Americans who are looking to connect to sellers of insurance. It should be noted that many of the initial problems were fixed before the re-launching of the Web site on December 1, 2013 .[4,5,6,7]

Applying the information covered in this chapter, it's easy to see that the SDLC and/or feasibility studies were not carefully applied in the first stage of this project. The schedule, technical, and operational feasibilities were not met. Although we do not have all the figures, the initial budget for this project proved to be inadequate.

From the perspective of the SDLC, several things were not carefully considered. The project's duration was incorrectly estimated. It started three months late. Not enough time was allocated at the beginning to deal with unexpected problems. User and systems specifications were not clearly defined. The system was not set up to handle peak periods. In fact, it frequently went down whenever several hundred users logged onto the system.

System testing was not done in a comprehensive fashion. Although most independent modules were working properly, the system was not able to correctly interface with other modules. Clear and definable milestones were not established to monitor the system's progress throughout the life cycle. When the initial deadline was not met, the government assigned multiple developers to the project. This did not expedite the completion of the project; it may even have delayed it because of the need to complete most of the system components in a sequential fashion.[8] The next big challenge for a fully operational HealthCare.gov is protecting the security and privacy of all the participants and keeping hackers at bay .[9]

- Is there any conflict between the proposed system and legal requirements? For example, does the system take the Information Privacy Act into account?

The information box above discusses the launch of HealthCare.gov, in which several dimensions of feasibility were not carefully analyzed.

10-3 PHASE 2: REQUIREMENTS GATHERING AND ANALYSIS

In the **requirements-gathering and analysis phase**, analysts define the problem and generate alternatives for solving it. During this phase, the team attempts to understand the requirements for the system, analyzes these requirements to determine the main problem with the current system or processes, and looks for ways to solve problems by designing the new system.

The first step in this phase is gathering requirements. Several techniques are available for this step, including interviews, surveys, observations, and the JAD approach described earlier in the chapter. The intent is to find out the following:

- What users do
- How they do it
- What problems they face in performing their jobs
- How the new system would address these problems
- What users expect from the system

- What decisions are made
- What data is needed to make decisions
- Where data comes from
- How data should be presented
- What tools are needed to examine data for the decisions that users make

All this information can be recorded, and the team uses this information to determine what the new system should do (process analysis) and what data is needed for this process to be performed (data analysis).

The team uses the information collected during the requirements-gathering phase to understand the main problems, define the project's scope—including what it should and should not do—and create a document called the "system specifications." This document is then sent to all key users and task force members for approval. The creation of this document indicates the end of the analysis phase and the start of the design phase.

There are two major approaches to the analysis and design of information systems: the structured systems analysis and design (SSAD) approach and the object-oriented approach. (The object-oriented approach was introduced in Chapter 3 with the discussion of object-oriented databases.) The onset of the Web plus the release of Java, an

> In the **requirements-gathering and analysis phase**, analysts define the problem and generate alternatives for solving it.

TABLE 10.1 EXAMPLES OF TOOLS USED IN SSAD ANALYSIS MODELS

Modeling Tool	What Is Analyzed	What It Is Used For
Data flow diagram (DFD)	Process analysis and design	Helps break down a complex process into simpler, more manageable, and more understandable subprocesses; shows how data needed by each process flows between processes and what data is stored in the system; also helps define the system's scope
Flowchart	Process analysis	Illustrates the logical steps in a process but does not show data elements and associations; can supplement a DFD and help analysts understand and document how a process works
Context diagram	Process analysis and design	Shows a process at a more general level and is helpful for showing top management and the task force how a process works
Conceptual data model (such as an entity relationship model)	Data analysis	Helps analysts understand the data requirements a system must meet by defining data elements and showing the associations between them

© Cengage Learning®

object-oriented language, created the push for a different approach than SSAD. To understand the difference between the two approaches, first realize that any system has three parts: process, data, and user interface. Analyzing requirements in the analysis phase is done from the perspective of the process and data. The SSAD approach treats process and data independently and is a sequential approach that requires completing the analysis before beginning the design. The object-oriented approach combines process and data analysis, and the line between analysis and design is so thin that analysis and design seem to be a single phase instead of the two distinct phases shown in Exhibit 10.1.

These two approaches use different tools for creating analysis models. Table 10.1 shows some examples of tools used in the SSAD approach.

Exhibit 10.2 shows an example of a data flow diagram for ABC's inventory management system, and Exhibit 10.3 shows a context diagram.

Exhibit 10.2
Data flow diagram for ABC's inventory management system

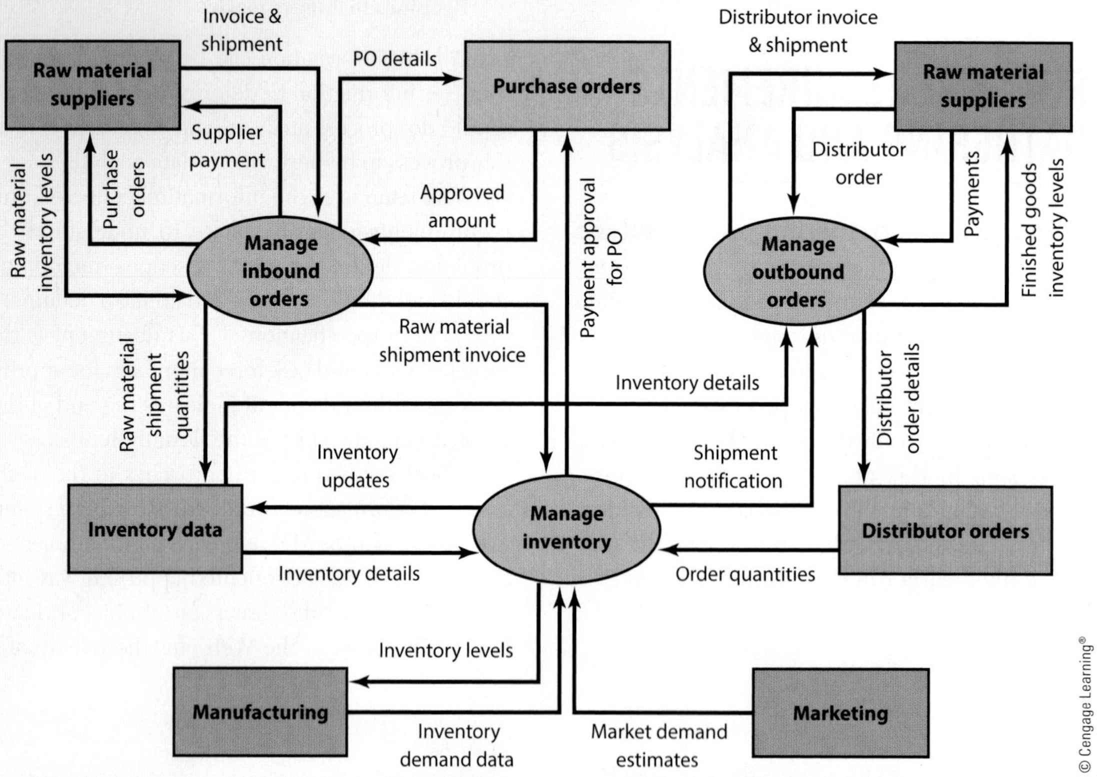

© Cengage Learning®

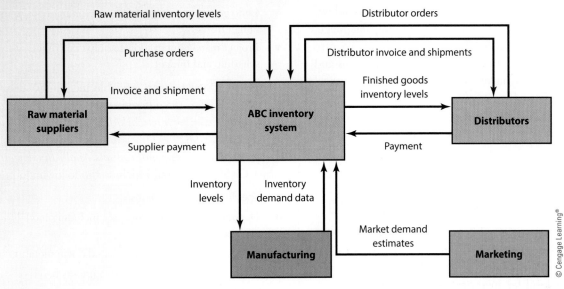

Exhibit 10.3
Context diagram for ABC's inventory management system

Raw material inventory levels

Distributor orders

Purchase orders

Distributor invoice and shipments

Invoice and shipment

Finished goods
inventory levels

**Raw material
suppliers**

**ABC inventory
system**

Distributors

Supplier payment

Payment

Inventory
levels

Inventory
demand data

Market demand
estimates

Manufacturing

Marketing

© Cengage Learning®

Notice in Exhibit 10.2 that processes are indicated with a circle. Anything that interacts with the system but is not part of it is considered an "external entity" and is shown as a blue rectangle. Data stores (databases, file systems, even file cabinets) are shown as grey rectangles.

In Exhibit 10.3, the DFD has been simplified into a context diagram, also called a "Level 0 diagram." Each process in this context diagram could be broken down into a separate diagram called "Level 1."

Both modeling tools show data flows between processes and external entities, and the DFD also shows data flows between processes and data stores. These data flows are general, so they do not show specific data elements. For example, "Purchase order" is shown in Exhibit 10.2 instead of all the pieces of data making up a purchase order, such as order number, order date, item number, and item quantity.

The models created during the analysis phase constitute the design specifications. After confirming these specifications with users, analysts start designing the system.

> **During the design phase, analysts choose the solution that is the most realistic and offers the highest payoff for the organization.**

10-4 PHASE 3: DESIGN

During the **design phase**, analysts choose the solution that is the most realistic and offers the highest payoff for the organization. Details of the proposed solution are outlined, and the output of this phase is a document with exact specifications for implementing the system, including files and databases, forms and reports, documentation, procedures, hardware and software, networking components, and general system specifications. For large projects in particular, CASE tools (discussed in the next section) are helpful in the analysis and design phases.

The design phase consists of three parts: conceptual design, logical design, and

During the **design phase**, analysts choose the solution that is the most realistic and offers the highest payoff for the organization. Details of the proposed solution are outlined, and the output of this phase is a document with exact specifications for implementing the system, including files and databases, forms and reports, documentation, procedures, hardware and software, networking components, and general system specifications.

physical design. The conceptual design is an overview of the system and does not include hardware or software choices. The logical design makes the conceptual design more specific by indicating hardware and software, such as specifying Linux servers, Windows clients, an object-oriented programming language, and a relational DBMS. These choices usually require changing the conceptual design to fit the platforms and programming languages chosen. Finally, the physical design is created for a specific platform, such as choosing Dell servers running Ubuntu Linux, Dell laptops running Windows 8 and Internet Explorer, Java for the programming language, and SQL Server 2014 for the relational DBMS.

10-4a Computer-Aided Systems Engineering

Systems analysts use **computer-aided systems engineering (CASE)** tools to automate parts of the application development process. These tools are particularly helpful for investigation and analysis in large-scale projects because they automate parts of the design phase. Analysts can use them to modify and update several design versions in an effort to choose the best version. CASE tools support the design phase by helping analysts do the following:

- Keep models consistent with each other
- Document models with explanations and annotations
- Ensure that models are created according to specific rules
- Create a single repository of all models related to a single system, which ensures consistency in analysis and design specifications
- Track and manage changes to the design
- Create multiple versions of the design

CASE tools are similar to computer-aided design (CAD) tools used by architects and engineers. Their capabilities vary, depending on the product, but generally include the following:

- Graphics tools, such as data flow diagrams, to illustrate a system's operation

- Dictionary tools designed to record the system's operation in detail
- Prototyping tools for designing input and output formats, forms, and screens
- Code generators to minimize or eliminate programming efforts
- Project management tools to help control the system's schedule and budget

Several CASE tools are available, including CA ERwin Process Modeler (*http://erwin.com*), Oracle Designer (*www.oracle.com/technetwork/developer-tools/designer/overview/index-082236.html*), and Visible System's Visible Analyst (*www.visible.com/Products/Analyst/vacorporate.htm*). CASE tools usually include the following output:

- Specifications documents
- Documentation of the analysis, including models and explanations
- Design specifications with related documentation
- Logical and physical design documents based on the conceptual design
- Code modules that can be incorporated into the system

10-4b Prototyping

Prototyping has been around for many years in physical science because building a small working model first is easier and less expensive than building the entire system. Prototypes can also be tested to detect potential problems and devise solutions.

Prototyping has gained popularity in designing information systems because needs can change quickly and lack of specifications for the system can be a problem. Typically, a small-scale version of the system is developed, but one that is large enough to illustrate the system's benefits and allow users to offer feedback. Prototyping is also the fastest way to put an information system into operation. Prototypes are usually used for the following purposes:

- *Gathering system requirements*—During the planning phase, designing a prototype and showing it to users is a good way to gather additional information and refine requirements for the proposed system.
- *Helping to determine system requirements*—If users are not sure about the type of information system they want, a prototype can serve as a valuable tool for demonstrating the system's functional capabilities, after which users can be asked for their reactions.
- *Determining a system's technical feasibility*—If a system is not technically feasible or appears to be unfeasible, a prototype can be used to show users that

a particular task can be done. This type of prototype is called a **proof-of-concept prototype**.

- *Selling the proposed system to users and management*—Prototypes are sometimes used to sell a proposed system to users and management by showing some of its features and demonstrating how beneficial it could be to the organization. This type of prototype is called a **selling prototype**.

Prototyping is usually done in four steps[10]:

1. Define the initial requirements.
2. Develop the prototype.
3. Review and evaluate the prototype.
4. Revise the prototype.

Defining the initial requirements involves agreement between users and designers that prototyping is the most suitable approach for solving a problem. After agreeing on the approach, users and designers work together to gather information about the prototype's components and how these components relate to one another. The team might decide on one of the following approaches for constructing the prototype: using an external vendor, using software packages or fourth-generation programming languages, or using high-level programming languages and developing the prototype from scratch.

Including users and top management in the construction phase is essential because some problems that crop up during construction can be solved only by users or top management. For example, top management typically must solve problems of financing a system, and lack of specifications is a problem better suited for users to solve. In addition, during this phase, users and top management can learn more about the problems the information system will solve, and the team of users and designers can learn a lot about decision making in the organization.

After completing the prototype, users begin using it and evaluating its performance. Depending on the outcome, one of the following decisions is made: revise the prototype, cancel the information system project, develop a new prototype, or build a complete system based on the prototype. Regardless of the decision, the prototype has provided useful information to the team of users and designers. At this point, the problem is better defined, and the system's operations are understood more clearly.

Prototyping Development Tools

Numerous tools can be used for constructing a system prototype. Widely used tools include spreadsheet packages, such as Microsoft Excel, and database management packages, such as Microsoft Access, whereas Visual Basic is commonly used to code the logic required for

processes. CASE tools and third- and fourth-generation programming languages can be used to quickly develop prototypes. Prototyping tools for user interface design include GUImagnets (*www.guimagnets.com*), Designer Vista (*http://designervista.com*), and GUI Design Studio (*www.carettasoftware.com/guidesignstudio*).

Advantages and Disadvantages of Prototyping

As mentioned, prototyping offers several advantages:

- It provides a method for investigating an environment in which the problem is poorly defined and information is difficult to gather.
- It reduces the need to train information system users because the users are involved in developing the system.
- It reduces costs because building a model is less expensive than building the complete system. If users and top management decide the system should not be developed, the organization has not lost all the money that would have been spent on building a complete system.
- It increases the system's chance of success by encouraging users' involvement.
- It is easier to modify a prototype than a complete system.
- It improves documentation because users and designers can walk through several versions of the system.
- It improves communication among users, top management, and information systems personnel because seeing a concrete model often prompts potential users of the system to ask questions, express opinions, point out shortcomings and strengths, and so forth.

Even with all these advantages, prototyping has some disadvantages:

- It might require more support and assistance from users and top management than they are willing to offer.
- The prototype might not reflect the final system's actual operation and, therefore, could be misleading.
- Developing a prototype might lead analysts and designers to forego comprehensive testing and documentation. If the prototype works, the team might be convinced that the final system will work, too, and this assumption can be misleading.

A **proof-of-concept prototype** shows users how a particular task that was not technically feasible can be done.

A **selling prototype** is used to sell a proposed system to users or management by showing some of its features.

10-5 PHASE 4: IMPLEMENTATION

During the **implementation phase**, the solution is transferred from paper to action, and the team configures the system and procures components for it. A variety of tasks takes place in the implementation phase, including the following:

- Acquiring new equipment
- Hiring new employees
- Training employees
- Planning and designing the system's physical layout
- Coding
- Testing
- Designing security measures and safeguards
- Creating a disaster recovery plan

When an information system is ready to be converted, designers have several options:

- *Parallel conversion*—In **parallel conversion**, the old and new systems run simultaneously for a short time to ensure the new system works correctly. However, this approach is costly and can be used only if an operational system is already in place.
- *Phased-in-phased-out conversion*—In **phased-in-phased-out conversion**, as each module of the new system is converted, the corresponding part of the old system is retired. This process continues until the entire system is operational. Although this approach is not suitable in all situations, it can be effective in accounting and finance.
- *Plunge (direct cutover) conversion*—In **plunge (direct cutover) conversion**, the old system is stopped and the new system is implemented. This approach is risky if there are problems with the new system, but the organization can save on costs by not running the old and new systems concurrently.
- *Pilot conversion*—In **pilot conversion**, the analyst introduces the system in only a limited area of the organization, such as a division or department. If the system works correctly, it is implemented in the rest of the organization in stages or all at once.

10-5a Project Management Tools and Techniques

The implementation of an information system can be a complex task. To manage this complexity and keep the implementation plan under budget and on schedule, systems analysts employ project management tools and techniques. These help systems analysts solve scheduling problems, plan and set goals, and highlight potential bottlenecks. Project management software such as Microsoft Project and Micro Planning International's Micro Planner enable the systems analyst to study the cost, time, and resource impact of schedule changes. Project management techniques are also used, including PERT (Program Evaluation Review Technique), CPM (Critical Path Method), and Gantt charts.

PERT and CPM techniques work by determining the "critical path" for the completion of a series of interrelated activities. This includes all those activities that are extremely crucial for the completion of the project, with zero slack time. If any of the activities is delayed, the entire project is delayed. Activities that are not on the critical path are more flexible and can be delayed without delaying the project.

To establish a PERT or CPM network, the analyst identifies all the activities needed for the completion of the project, identifies and establishes a prerequisite list (the activities that have to be accomplished first), and calculates the critical path duration.

Exhibit 10.4 shows several paths that lead from the beginning to the end of a project. The duration of each path is determined by the durations of the activities that make up that path. Here are some examples:

$$1 \rightarrow 2 \rightarrow 4 \rightarrow 6 \rightarrow 7 = 5 + 15 + 6 + 11 = 37$$
$$1 \rightarrow 4 \rightarrow 6 \rightarrow 7 = 13 + 6 + 11 = 30$$
$$1 \rightarrow 4 \rightarrow 5 \rightarrow 6 \rightarrow 7 = 13 + 9 + 10 + 11 = 43$$
$$1 \rightarrow 3 \rightarrow 4 \rightarrow 5 \rightarrow 6 \rightarrow 7 = 7 + 10 + 9 + 10 + 11 = 47$$
$$1 \rightarrow 3 \rightarrow 4 \rightarrow 6 \rightarrow 7 = 7 + 10 + 6 + 11 = 34$$
$$1 \rightarrow 2 \rightarrow 4 \rightarrow 5 \rightarrow 6 \rightarrow 7 = 5 + 15 + 9 + 10 + 11 = 50$$

During the **implementation phase**, the solution is transferred from paper to action, and the team configures the system and procures components for it.

In **parallel conversion**, the old and new systems run simultaneously for a short time to ensure the new system works correctly.

In **phased-in-phased-out conversion**, as each module of the new system is converted, the corresponding part of the old system is retired. This process continues until the entire system is operational.

In **plunge (direct cutover) conversion**, the old system is stopped and the new system is implemented.

In **pilot conversion**, the analyst introduces the system in only a limited area of the organization, such as a division or department. If the system works correctly, it is implemented in the rest of the organization in stages or all at once.

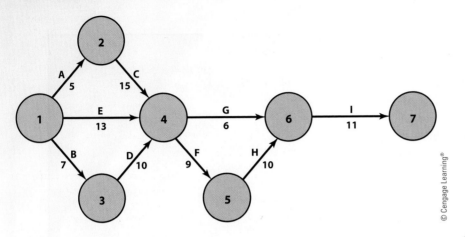

Exhibit 10.4
PERT network

© Cengage Learning®

completed as well. The activities on the other paths can be delayed for some time, and the project will still be completed on time. However, all the activities on the last path must be completed on time if the project is to be finished on time.

Using the critical path, the systems analyst can establish a Gantt chart. A Gantt chart lists the completion time (sometimes called the "milestone") on the x-axis and all the activities on the y-axis. This allows the systems analyst to monitor the progress of the project and detect any delay in the daily operation of the project. If a delay is spotted, the systems analyst must consider additional resources if the project is to be completed on schedule. Exhibit 10.5 illustrates a Gantt chart.

In these examples, the last path is the critical path because it takes the longest to be completed. While this path is being completed, the other paths will be

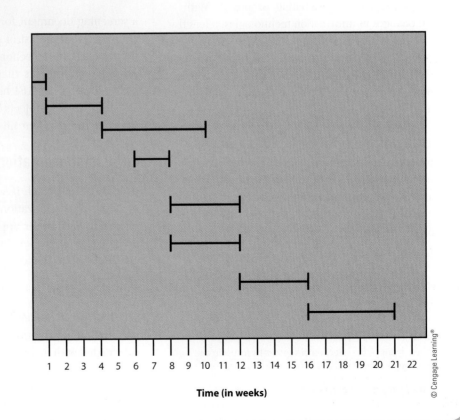

Exhibit 10.5
Gantt chart

Activity

Define the problem

Conduct the feasibility

Redefine the problem

Create and evaluate alternative solutions

Select new hardware/software

Install new hardware/software

Construct the prototype

Implement the new system

1 2 3 4 5 6 7 8 9 10 11 12 13 14 15 16 17 18 19 20 21 22

Time (in weeks)

© Cengage Learning®

10-5b Request for Proposal

A **request for proposal (RFP)** is a written document with detailed specifications that is used to request bids for equipment, supplies, or services from vendors. It is usually prepared during the implementation phase and contains detailed information about the functional, technical, and business requirements of the proposed information system. Drafting an RFP can take 6 to 12 months, but with software, the Internet, and other online technologies, time and costs can be reduced.

A crucial part of this process is comparing bids from single and multiple vendors. Using a single vendor to provide all the information system's components is convenient, but the vendor might not have expertise in all areas of the information system's operations.

The main advantage of an RFP is that all vendors get the same information and requirements, so bids can be evaluated more fairly. Furthermore, all vendors have the same deadline for submitting bids, so no vendor has the advantage of having more time to prepare an offer. RFPs are also useful in narrowing down a long list of prospective vendors.

A major disadvantage of an RFP is the time involved in writing and evaluating proposals. With the rapid changes in information technologies, a lengthy time frame makes RFPs less appealing. Many companies cannot wait 6 to 12 months to decide on a vendor for an information system. Exhibit 10.6 shows the main components of an RFP. You can find free templates for RFPs by searching at Business-in-a-Box (*www.biztree.com*), TEC (*http://technologyevaluation.com*), and Klariti (*www.klariti.com*).

Given the need to complete information system projects as quickly as possible, shortening the time needed to write and evaluate proposals is often necessary. One alternative to an RFP is a **request for information (RFI)**,

Exhibit 10.6
Main components of an RFP for ABC Furniture's inventory system

1. Introduction
   a. Background and organizational goals of ABC
2. System Requirements
   a. Problems to be solved
   b. Details of preliminary analyses
   c. Key insights gained
3. Additional Information
   a. Hardware available
   b. Software preferences
   c. Other existing systems and integration requirements
   d. Understanding of benefits to be gained
   e. Business understanding necessary (for the bidder to work with ABC)
   f. Technology and technical know-how necessary
4. Project Time Frame
5. Contact Information and Submission Procedures

© Cengage Learning®

a screening document for gathering vendor information and narrowing the list of potential vendors. An RFI can help manage the selection of vendors by focusing on the project requirements that are crucial to selecting vendors. However, an RFI has its limitations. It is not suitable for complex projects because it can be used only for selecting three or four finalists from a list of candidates.

10-5c Implementation Alternatives

The SDLC approach is sometimes called **insourcing**, meaning an organization's team develops the system internally. Two other approaches for developing information systems are self-sourcing and outsourcing, which are discussed in the following sections.

Self-Sourcing

The increasing demand for timely information has put pressure on information systems teams, who are already overloaded with maintaining and modifying existing systems. In many organizations, the task of keeping existing systems running takes up much of the available computing resources and personnel, leaving few resources for developing new systems. The resulting inability

A **request for proposal (RFP)** is a written document with detailed specifications that is used to request bids for equipment, supplies, or services from vendors.

A **request for information (RFI)** is a screening document for gathering vendor information and narrowing the list of potential vendors. It can help manage the selection of vendors by focusing on the project requirements that are crucial to selecting vendors.

Insourcing happens when an organization's team develops the system internally.

to respond to users' needs has increased employee dissatisfaction and caused a backlog in systems development in both well-managed and poorly managed organizations. In recent years, therefore, more end users have been developing their own information systems with little or no formal assistance from the information systems team. These users might not know how to write programming code, but they are typically skilled enough to use off-the-shelf software, such as spreadsheet and database packages, to produce custom-built applications.[11,12] This trend, called **self-sourcing** (or end-user development), has resulted from long backlogs in developing information systems, the availability of affordable hardware and software, and organizations' increasing dependence on timely information.

With the help of development tools, such as query languages, report generators, and fourth-generation programming languages, self-sourcing has become an important part of information system resources. It is also useful in creating one-of-a-kind applications and reports. Self-sourcing can help reduce the backlog in producing information systems and improve flexibility in responding to users' information needs. Backlogs, however, are just the tip of the iceberg. When the backlog list is long, end users often stop making new requests for many of the applications they need because they believe these requests would just make the list longer. The list of applications that are not requested is often longer than the backlog, and it is called the "invisible" backlog.

- Lack of documentation for the applications and systems that end users develop
- Inadequate security for the applications and systems that end users develop
- Applications developed by end users not up to information systems standards
- Lack of support from top management
- Lack of training for prospective users

Self-sourcing gives end users the power to build their own applications in a short time and create, access, and modify data. This power can be destructive, however, if the organization does not apply control and security measures. For example, end users' access to computing resources must be controlled to prevent interfering with the efficiency of the organization's information-processing functions.

To prevent the proliferation of information systems and applications that are not based on adequate systems-development principles, organizations should develop guidelines for end users and establish criteria for evaluating, approving or rejecting, and prioritizing projects. Criteria could include asking questions such as "Can any existing application generate the proposed report?" or "Can the requirements of multiple users be met by developing a single application?"

Classifying and cataloging existing applications are necessary to prevent end users from developing applications that basically handle the same functions as an existing application; this redundancy can be costly. In addition, data administration should be enforced

> With the help of development tools such as query languages, report generators, and fourth-generation programming languages, self-sourcing has become an important part of information system resources.

Although self-sourcing can solve many current problems, managers are concerned about end users' lack of adequate systems analysis and design background and loosening of system development standards. Other disadvantages of self-sourcing include the following:

- Possible misuse of computing resources
- Lack of access to crucial data

Self-sourcing is when end users develop their own information systems, with little or no formal assistance from the information systems team. These users might not know how to write programming code, but they are typically skilled enough to use off-the-shelf software, such as spreadsheet and database packages, to produce custom-built applications.

to ensure the integrity and reliability of information. Creating private data should be minimized, if not eliminated. Sometimes, for the sake of efficient data processing, redundant data can exist; however, it should be monitored closely. This task is becoming more difficult, however, because the number of end users using diverse data is growing. The best approach to control the proliferation of invalid and inconsistent data in corporate databases is controlling the flow of data, such as with rigorous data entry procedures that the database administrator establishes.

Outsourcing

With the **outsourcing** approach, an organization hires an external vendor or consultant who specializes in providing development services. This approach can save the cost of hiring additional staff while meeting the demand for more timely development of information systems projects. Companies offering outsourcing services include IBM Global Services, Accenture, Infosys Technologies, and Computer Sciences Corporation.

With the development of Web 2.0, another form of outsourcing has become popular: **crowdsourcing**. This refers to the process of outsourcing tasks that are traditionally performed by employees or contractors to a large group of people (a crowd) through an open call. Say your town's City Hall is developing a Web site in order to better serve the community. Using crowdsourcing, it would invite everybody to participate in the design process. Crowdsourcing has become popular with publishers, journalists, editors, and businesses that want to take advantage of the collaborative capabilities offered by Web 2.0. InnoCentive (*www2.innocentive.com*) is a company that is very active in crowdsourcing. It works with organizations to solve their problems, taking advantage of the power of diverse thinking inside and outside the organization.

An outsourcing company that employs the SDLC approach has the following options:

- *Onshore outsourcing*—The organization chooses an outsourcing company in the same country.

- *Nearshore outsourcing*—The organization chooses an outsourcing company in a neighboring country, such as when a U.S. organization chooses a company in Canada or Mexico.

- *Offshore outsourcing*—The organization chooses an outsourcing company in any part of the world (usually a country farther away than a neighboring country), as long as it can provide the needed services.

Although outsourcing has the advantages of being less expensive, delivering information systems more quickly, and giving an organization the flexibility to concentrate on its core functions as well as other projects, it does have some disadvantages. They include the following:

- *Loss of control*—Relying on the outsourcing company to control information system functions can result in the system not fully meeting the organization's information requirements.

- *Dependency*—If the organization becomes too dependent on the outsourcing company, changes in the outsourcing company's financial status or managerial structure can have a major impact on the organization's information system.

- *Vulnerability of strategic information*—Because third parties are involved in outsourcing, the risk of leaking confidential information to competitors increases.

The information box on the next page highlights some of the reasons that China will be a major IT outsourcing destination for many years to come.

© Dusit/Shutterstock.com

10-6 PHASE 5: MAINTENANCE

During the **maintenance phase**, the information system is operating, enhancements and modifications to the system have been developed and tested, and hardware and software components have been added or replaced. The maintenance team assesses how the system is working and takes steps to keep the system up and running. As part of this phase, the team collects performance data and gathers information on whether the system is meeting its objectives by talking with users, customers, and other people affected by the new system. If the system's objectives are not being met, the team must take corrective action. Creating a help desk to support users is another important task in this phase. With the ongoing nature of the SDLC approach, maintenance can lead to starting the cycle over at the planning phase if the team discovers the system is not working correctly.

10-7 NEW TRENDS IN SYSTEMS ANALYSIS AND DESIGN

The SDLC model might not be appropriate in the following situations:

- There is a lack of specifications—that is, the problem under investigation is not well defined.
- The input–output process cannot be identified completely.
- The problem is "ad hoc," meaning it is a one-time problem that is not likely to reoccur.

- Users' needs keep changing, which means the system undergoes several changes. The SDLC model might work in the short term, but in the long term, it is not suitable in this situation.

For these situations, other approaches, described in the following sections, are more suitable.

10-7a Service-Oriented Architecture

Service-oriented architecture (SOA) is a philosophy and a software and system development methodology that focuses on the development, use, and reuse of small, self-contained blocks of codes (called services) to meet the software needs of an organization. SOA attempts to solve software development issues by recognizing, accepting, and leveraging the existing services. Checking shipping status, customer credit, or the inventory status are a few examples of such services. More specifically, a service could be a database table, a set of related database tables, one or more data files in any format, or data obtained from another service.

The fundamental principle behind SOA is that the "blocks of codes" can be reused in a variety of different applications, allowing new business processes to be created

During the **maintenance phase**, the information system is operating, enhancements and modifications to the system have been developed and tested, and hardware and software components have been added or replaced.

Service-oriented architecture (SOA) is a philosophy and a software and system development methodology that focuses on the development, use, and reuse of small, self-contained blocks of codes (called services) to meet the software needs of an organization.

from a pool of existing services. These services should be organized so they can be accessed when needed via a network. SOA offers many potential benefits to organizations, including reduced application development time, greater flexibility, and an improved return on investment.

In any business organization, there are things that do not change very often, such as an order processing system. These often represent a major part of a business and are therefore called core business functions. At the same time, there are functions and activities that change on a regular basis, such as taxes and contents of a marketing campaign. SOA advocates that core business functions and the dynamic functions that change all the time should be decoupled. SOA allows an organization to pick and choose those services that respond most effectively to the customer's needs and market demands. Services or "blocks of codes" can be replaced, changed, or even combined.

Many organizations use SOA as a philosophy and methodology. For example, Starwood Hotels and Resorts Worldwide is replacing its legacy room-reservation system with an SOA-based system. By using SOA, they can offer as many as 150 service-based applications built on Web standards. T-Mobile is also employing SOA, both for internal integration and reuse and for external, revenue-generating services. This enables T-Mobile to work effectively with third-party content providers, such as Time Warner and the Bertelsmann Group, to deliver services to customers.[14]

10-7b Rapid Application Development

Rapid application development (RAD) concentrates on user involvement and continuous interaction between users and designers. It combines the planning and analysis phases into one phase and develops a prototype of the system. RAD uses an iterative process (also called "incremental development") that repeats the design, development, and testing steps as needed, based

on feedback from users. After the initial prototype, the software library is reviewed, reusable components are selected from the library and integrated with the prototype, and testing is conducted. After these steps, the remaining phases are similar to the SDLC approach. One shortcoming of RAD is a narrow focus, which might limit future development. In addition, because these applications are built quickly, the quality might be lower.

10-7c Extreme Programming

Extreme programming (XP) is a recent method for developing software applications and information system projects. Kent Beck, during his work on the Chrysler Comprehensive Compensation System, created this method as a way of establishing specific goals and meeting them in a timely manner. XP divides a project into smaller functions, and developers cannot go on to the next phase until the current phase is finished. Analysts write down features the proposed system should have—called the "story"—on index cards. The cards include the time and effort needed to develop these features, and then the organization decides which features should be implemented and in what order, based on current needs.[15] Each function of the overall project is developed in a step-by-step fashion. At the beginning, it is similar to a jigsaw puzzle; individually, the pieces make no sense, but when they are combined, a complete picture can be seen. The XP method delivers the system to users as early as possible and then makes changes that the user suggests. In the XP environment, programmers are usually organized into teams of two, sharing a workstation and working on the same code. This is called **pair programming** (also referred to as "sharing a keyboard"), each programmer performing the actions the other programmer is not currently performing. In this way, they can detect and correct programming mistakes as they go, which is faster than correcting them after the entire program has been written. There is also better communication between programmers during code development.

XP is a major departure from traditional software development, such as the SDLC model, which looks at the project as a whole. XP uses incremental steps to improve a system's quality, addressing major issues that have not been examined before. SDLC develops the entire system at once. Like RAD, XP uses a software library for reusable pieces that can be integrated into the new system. IBM, Chrysler, and Microsoft, among others, have used this method successfully. Its key features are:

- Simplicity
- Incremental process

- Responsiveness to changing requirements and changing technology
- Teamwork
- Continuous communication among key players
- Immediate feedback from users

The information box above highlights the use of XP at Sabre Holdings Corp. and other companies.

10-7d Agile Methodology

Agile methodology is similar to XP in focusing on an incremental development process and timely delivery of working software. However, there is less emphasis on team coding and more emphasis on limiting the project's scope. Agile methodology focuses on setting a minimum number of requirements and turning them into a working product. The Agile Alliance organization (*www.agilealliance.org*) has developed guidelines for this method, which emphasizes collaboration between programmers and business experts, preferably with face-to-face communication, and working in teams. Goals of this step-by-step approach include responding to changing needs instead of sticking to a set plan and developing working, high-quality software. The agile methodology also strives to deliver software quickly to better meet customers' needs.

The Agile Alliance has written a manifesto that includes the following principles[17]:

- Satisfy the customer through early and continuous delivery of valuable software.
- Welcome changing requirements, even late in development.
- Have business people and developers work together daily throughout the project.
- Build projects around motivated individuals. Give them the environment and support they need, and trust them to get the job done.
- Always attend to technical excellence. Good design enhances agility.
- At regular intervals, the team should reflect on how to become more effective, then tune and adjust its behavior accordingly.

The information box below discusses how agile methodology is used at HomeAway, Inc.

The Industry Connection highlights CA Technologies, which offers several systems development tools.

> **Agile methodology** is similar to XP in focusing on an incremental development process and timely delivery of working software. However, there is less emphasis on team coding and more emphasis on limiting the project's scope.

AGILE METHODOLOGY AT HOMEAWAY, INC.

Based in Austin, Texas, HomeAway, Inc. is the world's leading online marketplace for vacation rentals, its sites representing approximately 720,000 vacation rental-home listings throughout 168 countries and territories around the world.[18]

At HomeAway, the 250 software developers use agile as the methodology for software and systems development. In the past, individual developers and system designers had been using different systems development methodologies; but for the past 6 years, agile has been the company-wide systems development methodology. According to Jack Yang, HomeAway vice president of engineering, no one was forced to move into agile, but over time the systems designers found that it was the best methodology for keeping up with change, especially unexpected change that needs to be implemented quickly. If HomeAway's experience is any indication, agile methodology is continuing to become a mainstream tool for software and systems development. Companies and system developers choose it for its team-driven, collaborative, and modular aspects over the nonflexible, sequence-driven SDLC approach.[19]

STUDY TOOLS 10

LOCATED AT BACK OF THE TEXTBOOK
- [ ] Rip out Chapter Review Card

LOCATED AT WWW.CENGAGE.COM/LOGIN
- [ ] Review Key Terms Flash Cards (Print or Online)
- [ ] Download audio and visual summaries to review

- [ ] Complete both Practice Quizzes to prepare for tests
- [ ] Play "Beat the Clock" and "Quizbowl" to master concepts
- [ ] Complete "Crossword Puzzle" to review key terms
- [ ] Watch video on "Method" for a real company example

KEY TERMS

agile methodology 219
computer-aided systems engineering (CASE) 210
crowdsourcing 216
design phase 209
economic feasibility 205
external users 204
extreme programming (XP) 218
feasibility study 205
implementation phase 212
insourcing 214
internal users 204
joint application design (JAD) 204
legal feasibility 206
maintenance phase 217

operational feasibility 206
outsourcing 216
pair programming 218
parallel conversion 212
phased-in-phased-out conversion 212
pilot conversion 212
planning phase 202
plunge (direct cutover) conversion 212
proof-of-concept prototype 211
prototyping 210
rapid application development (RAD) 218
request for information (RFI) 214
request for proposal (RFP) 214
requirements-gathering and analysis phase 207

REVIEWS AND DISCUSSIONS

1. What are the five phases in the SDLC, which must be done in sequence?

2. Why must the planning phase of the SDLC be done first?

3. What are the five major dimensions of a feasibility study?

4. What are two examples of tools and techniques that are used in project management?

5. What are two disadvantages of RFP?

6. What are two benefits of prototyping?

7. What are three types of outsourcing? Which one is the most popular?

8. What are the differences between extreme programming and agile methodology?

PROJECTS

1. A startup Internet company has generated the following cash balance for the first six years of its IS projects: –$250,000, –$180,000, $225,000, $340,000, $410,000, and $425,000. Using the NPV function in Excel, calculate the net present value of this project at an 8.5 percent interest rate. What is the IRR of this project? If a bank is willing to give the company a loan at 15 percent to implement these projects, should the company accept the loan (assuming there are no other conditions)? Why or why not?

2. The CTO of the company mentioned in Project 1 has decided to acquire a CASE tool to help expedite and better document the Web site that is being designed for his company's e-commerce operations. The CTO has narrowed these choices to CA ERwin Process Modeler and Oracle Designer. Based on the information provided in this chapter and other sources, which tool would you recommend? What are the three major factors behind your recommendation?

3. The CTO of the company mentioned in Project 1 also wants to invest in project management software. The choices have been narrowed to Microsoft Project and Micro Planning International's Micro Planner.

Based on the information provided in this chapter and other sources, which software would you recommend? What are the three major factors behind your recommendation?

4. After reading the information presented in this chapter and other sources, write a one-page paper that identifies two companies (in addition to those mentioned in this book) that have been using XP as a systems development methodology. What specific advantages has XP offered these companies?

5. After reading the information presented in this chapter and other sources, write a one-page paper that identifies two companies (in addition to those mentioned in this book) that have been using crowdsourcing as an alternative to a traditional systems development approach. What are two advantages of using crowdsourcing for developing a system or application?

6. After reading the information presented in this chapter and other sources, write a one-page paper describing three countries that are prime candidates for offshore outsourcing for systems development projects. What are the three main criteria that should be considered in choosing a suitable candidate?

ARE YOU READY TO MOVE ON?

1. The first phase of the SDLC is requirements gathering and analysis. True or False?

2. Operational feasibility is an assessment of a system's costs and benefits. True or False?

3. Simplicity and incremental process are two key features of XP. True or False?

4. Which of the following is not among the dimensions of a feasibility study?

    a. Economic

    b. Technical

    c. Schedule

    d. Programming

5. Which of the following is not an example of a modeling tool?
   a. Agile
   b. DFD
   c. Flowchart
   d. Context diagram

6. Which of the following is not among the new trends in systems analysis and design?
   a. SOA
   b. SDLC
   c. XP
   d. RAD

CASE STUDY 10-1

Systems Development At Seb Latvia

Skandinaviska Enskilda Banken (SEB), one of northern Europe's largest banking groups, operates in several countries, including Germany, Poland, and Russia. In the past, many of SEB's decisions were made as a result of an approval process that involved circulating paper documents. In addition, important data was stored on decentralized systems that were accessible to only certain decision makers. Delays and bottlenecks resulted, with a consequent effect on the speed and effectiveness of customer service. To solve these problems, SEB Latvia chose the IBM Lotus Domino messaging and collaboration platform. After the system

© 6169758076/Shutterstock.com

was installed, Exigen (an IBM business partner) and the SEB Latvia's IT team started working with users to gather requirements and automate processes to increase efficiency. The Lotus system now provides centralized information access via a Web interface that all decision makers can use.[21]

Answer the following questions:
1. What processes did the Lotus platform automate and streamline?
2. What are some of the benefits of the new Web interface?
3. Which of the development methods discussed in this chapter would be most suitable at SEB Latvia?

CASE STUDY 10-2

Crowdsourcing Pays Off

Crowdsourcing is not only used for relatively simple tasks such as designing a Web site or designing a case for a smartphone. Today, it is increasingly being used in more complex designs as well. Precyse Technologies provides supply-chain products and solutions that assist organizations in tracking their inventories using RFID technology. It was having difficulty improving the performance and battery life of a particular RFID device; there were not adequate internal resources to tackle the problem. So, Precyse asked InnoCentive, a crowdsourcing provider, for assistance, believing that crowdsourcing would provide quick access to a worldwide talent pool and a higher return on

© ARENA Creative/Shutterstock.com

investment (ROI). Ultimately, the company received more than 300 ideas from global experts for solving its problem. It narrowed the pool down to three finalists, from which a winner was chosen. The whole process took approximately 4 months, and Precyse estimated that it saved the company $250,000.[22]

Answer the following questions:
1. What are some typical applications of crowdsourcing?
2. What are some advantages of crowdsourcing?
3. How could crowdsourcing reduce the systems design cost?

11 | Enterprise Systems

LEARNING OUTCOMES

After studying this chapter, you should be able to:

11-1 Explain how supply chain management is used.

11-2 Describe customer relationship management systems.

11-3 Describe knowledge management systems.

11-4 Describe enterprise resource planning systems.

After you finish this chapter, go to **PAGE 239** for the **STUDY TOOLS**

© Designdepot/Shutterstock.com

An **enterprise system** is an application that is used in all the functions of a business and supports decision making throughout the organization. For example, an enterprise resource planning system is used to coordinate operations, resources, and decision making among manufacturing, production, marketing, and human resources. As you have learned in previous chapters, intranets and Web portals are used by many organizations to improve communication among departments as well as to increase overall efficiency. Enterprise systems are another way to make important information readily available to decision makers throughout an organization.

In this chapter, you learn about the following enterprise systems: supply chain management (SCM), customer relationship management

(CRM), knowledge management systems, and enterprise resource planning (ERP). With each type of enterprise system, you review the system's goals, the information technologies used for it, and any relevant issues.

11-1 SUPPLY CHAIN MANAGEMENT

A **supply chain** is an integrated network consisting of an organization, its suppliers, transportation companies, and brokers used to deliver goods and services to customers. As Exhibit 11.1 shows, in a manufacturing firm's supply chain, raw materials flow from suppliers to manufacturers (where they are transformed into finished goods), then to distributors, then to retailers, and

An **enterprise system** is an application used in all the functions of a business that supports decision making throughout the organization.

A **supply chain** is an integrated network consisting of an organization, its suppliers, transportation companies, and brokers used to deliver goods and services to customers.

Exhibit 11.1
Supply chain configuration

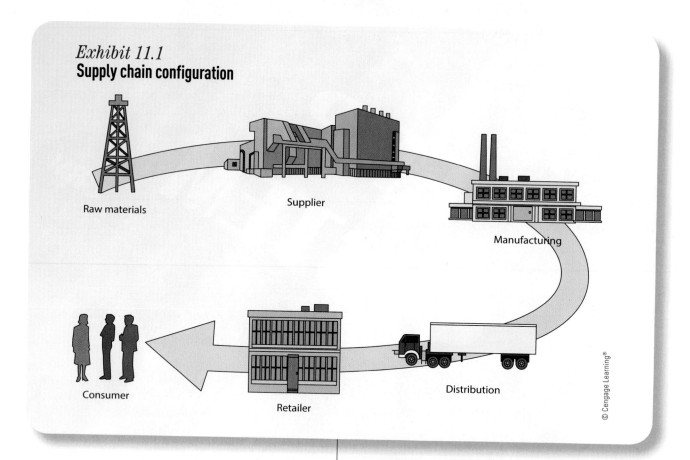

Raw materials

Supplier

Manufacturing

Distribution

Retailer

Consumer

© Cengage Learning®

finally to consumers. Supply chains exist in both service and manufacturing organizations, although the chain's complexity can vary widely in different organizations and industries. In manufacturing, the major links in the supply chain are suppliers, manufacturing facilities, distribution centers, retailers, and customers. In service organizations—such as those in real estate, the travel industry, temporary labor, and advertising—these links include suppliers (service providers), distribution centers (such as travel agencies), and customers.

Supply chain management (SCM) is the process of working with suppliers and other partners in the supply chain to improve procedures for delivering products and services. An SCM system coordinates the following functions:

- Procuring materials (in service organizations, can include resources and information)
- Transforming materials into intermediate and finished products or services
- Distributing finished products or services to customers

Supply chain management (SCM) is the process of working with suppliers and other partners in the supply chain to improve procedures for delivering products and services.

© Rade Kovac/Shutterstock.com

SUPPLY CHAIN MANAGEMENT AT COCA-COLA COMPANY

The Coca-Cola Company is the world's largest beverage company, its 500 sparkling and nonsparkling brands sold in more than 200 countries.[1] The company needed a more modern SCM system that would work well with its other systems. The new SCM system would also need to provide processes and information to assist with acquisitions of other companies, and it would need to integrate the information from manufacturing with its retail operations. To achieve these goals, Coca-Cola decided to implement a new supply chain management system at its 17 European plants, replacing several of the existing systems.

SAP ERP was adopted as the main system, with certain applications outsourced to CSC (*www.csc.com*). The new SCM implementation has succeeded by providing more automation and by streamlining various SCM processes. It also assists in bringing newly acquired companies online in a timely manner. More importantly, the new SCM system links supply-and-demand data into a single system. Overall, the new system provides key decision makers with timely and relevant information that improves the overall efficiency and effectiveness of the entire supply chain operation.[2]

In a manufacturing firm's SCM system, communication takes place among the following areas:

- *Product flow*—Managing the movement of goods all the way from suppliers to customers, including customer service and support
- *Information flow*—Overseeing order transmissions and delivery status updates throughout the order-processing cycle
- *Finances flow*—Handling credit terms, payment schedules, and consignment and title ownership arrangements

An SCM system must manage communication in all these areas as part of overseeing the manufacturing process and managing inventory and delivery. Four key decisions in SCM related to manufacturing are:

- *Location*—Where should manufacturing facilities be placed?
- *Inventory*—When should an order be placed? How much should be ordered?
- *Production*—What should be produced? How much should be produced?
- *Transportation*—Which transportation systems will reduce costs and expedite the delivery process?

For organizations that do not have in-house resources to develop an SCM system, several vendors offer comprehensive solutions: SAP (*www.sap.com/solution/lob/scm/software/overview/highlights.html*), Oracle (*www.oracle.com/us/solutions/scm/index.htm*), JDA Software (*www.jda.com*), Ariba (*www.ariba.com*), and Manhattan Associates (*www.manh.com*). In addition, hosting services are now available for SCM systems; this trend is called "software as a service"

(SaaS, discussed in Chapter 14). The information box above highlights the supply chain at Coca-Cola Company.

11-1a SCM Technologies

Information technologies and the Internet play a major role in implementing an SCM system. These tools are explained in the following sections.

Electronic Data Interchange

Electronic data interchange (EDI) enables business partners to send and receive information on business transactions. Many companies substitute EDI for printing, mailing, and faxing paper documents, such as purchase orders, invoices, and shipping notices. By using the Internet and established Web protocols for the electronic exchange of information, companies can improve the efficiency and effectiveness of the supply chain process. EDI expedites the delivery of accurate information in the following processes, among others:

- Transaction acknowledgments
- Financial reporting
- Invoice and payment processing
- Order status
- Purchasing
- Shipping and receiving
- Inventory management and sales forecasting

> Many companies substitute EDI for printing, mailing, and faxing paper documents, such as purchase orders, invoices, and shipping notices.

Electronic data interchange (EDI) enables business partners to send and receive information on business transactions.

In addition, using the Internet and Web protocols for EDI lowers the cost of transmitting documents. This method is called "Web-based EDI" or "Open EDI." It also has the advantage of being platform independent and easy to use.

Nevertheless, transmitting across the Internet does involve more security risks than traditional EDI, which uses proprietary protocols and networks. Using EDI does have some drawbacks. For instance, EDI uses proprietary standards. An EDI provider sets up an EDI network (as a VPN), and organizations enroll in the network. EDI is more beneficial when there are more companies in the EDI network, because when the number of partners is small, the cost per partner is higher. For this reason, large companies tend to insist on their suppliers and distributors becoming part of the same EDI network, which many small suppliers and distributors cannot afford. With the advent of XML, organizations can use the Internet and Open EDI to perform the same function that EDI performs, so traditional EDI has declined in popularity.

Internet-Enabled SCM

Internet-enabled SCM improves information sharing throughout the supply chain, which helps reduce costs for information transmission and improves customer service. For instance, many companies use point-of-sale (POS) systems that scan what is being sold and collect this data in real time. This information helps organizations decide what to reorder to replenish stock, and it is sent via the Internet to suppliers so they can synchronize production with actual sales. Internet-enabled SCM can improve the following SCM activities:

- *Purchasing/procurement*—Purchasing and paying for goods and services online, bargaining and renegotiating prices and term agreements, using global procurement strategies

- *Inventory management*—Providing real-time stock information, replenishing stock quickly and efficiently, tracking out-of-stock items

- *Transportation*—Allowing customers to use the Internet for shipping and delivery information

- *Order processing*—Checking order placement and order status, improving the speed and quality of order processing, handling returned goods and out-of-stock notifications to customers

An **e-marketplace** is a third-party exchange (B2B model) that provides a platform for buyers and sellers to interact with each other and trade more efficiently online.

- *Customer service*—Responding to customers' complaints, issuing notifications (such as product recalls), providing around-the-clock customer service

- *Production scheduling*—Coordinating just-in-time (JIT) inventory programs with vendors and suppliers, coordinating production schedules between companies and their vendors and suppliers, conducting customer demand analysis

E-Marketplaces

An **e-marketplace** is a third-party exchange (a B2B business model) that provides a platform for buyers and sellers to interact with each other and trade more efficiently online. E-marketplaces help businesses maintain a competitive edge in the supply chain in the following ways:

- Providing opportunities for sellers and buyers to establish new trading partnerships

- Providing a single platform for prices, availability, and stock levels that is accessible to all participants

- Solving time-constraint problems for international trade and making it possible to conduct business around the clock

© Indeed/Stockbyte/Jupiter Images

- Making it easy to compare prices and products from a single source instead of spending time contacting each seller
- Reducing marketing costs more than traditional sales channels can

E-distributors are common examples of e-marketplaces. An e-distributor is a marketplace owned and operated by a third party that provides an electronic catalog of products. For example, an e-distributor might offer a catalog containing a variety of hardware and software products so that a network administrator can order all the equipment and applications needed for an organization's network instead of purchasing components from several different vendors. Another common offering from e-distributors is maintenance, repair, and operations (MROs) services; a company can purchase an MRO package that might include services from different vendors, but the e-distributor coordinates them into one package for customers. This packaging is an example of a horizontal market, which concentrates on coordinating a business process or function involving multiple vendors. E-distributors offer fast delivery of a wide selection of products and services, usually at lower prices, and they help companies reduce the time and expense of searching for goods.

concept. By using the Internet, it brings traditional auctions to customers around the globe and makes it possible to sell far more goods and services than at a traditional auction. It is based on the brokerage business model discussed in Chapter 8, which brings buyers and sellers together in a virtual marketplace. Typically, the organization hosting the auction collects transaction fees for the service. Online auctions are particularly cost-effective for selling excessive inventory. Some companies use **reverse auctions**, which invite sellers to submit bids for products and services. In other words, there is

E-distributors offer fast delivery of a wide selection of products and services, usually at lower prices, and they help companies reduce the time and expense of searching for goods.

As you learned in Chapter 8, third-party exchanges bring together buyers and sellers in vertical as well as horizontal markets. Buyers can gather information on products and sellers, and sellers have access to more potential buyers. Examples of third-party exchanges include PowerSource Online (*www.powersourceonline.com*) and Farms.com (*www.farms.com*).

Online Auctions

Auctions help determine the price of goods and services when there is no set price in the marketplace. An **online auction** is a straightforward yet revolutionary business

one buyer and many sellers: a one-to-many relationship. The buyer can choose the seller that offers the service or product at the lowest price.

By using the Internet, an **online auction** brings traditional auctions to customers around the globe and makes it possible to sell far more goods and services than at a traditional auction.

A **reverse auction** invites sellers to submit bids for products and services. In other words, there is one buyer and many sellers: a one-to-many relationship. The buyer can choose the seller that offers the service or product at the lowest price.

Collaborative Planning, Forecasting, and Replenishment

Collaborative planning, forecasting, and replenishment (CPFR) is used to coordinate supply chain members through point-of-sale (POS) data sharing and joint planning (see Exhibit 11.2). In other words, any data collected with POS systems is shared with all members of the supply chain, which is useful in coordinating production and planning for inventory needs. The goal is to improve operational efficiency and manage inventory. With a structured process of sharing information among supply chain members, retailers can compare customer demands or sales forecasts with a manufacturer's order forecast, for example. If there is a discrepancy between forecasts, members can get together and decide on the correct quantity to order. One main obstacle to improving supply chain performance is companies not knowing enough about what customers want, which can lead to lost sales and unsold inventory for retailers and manufacturers. CPFR has the advantage of decreasing merchandising, inventory, and logistics costs for all supply chain members.

Coordinating the supply chain can be difficult. To understand these problems, recall the example of ABC Furniture Company, which was used in Chapter 10. ABC Furniture buys wood from New England Wood and buys hardware (nuts, bolts, and so forth) from Vermont Hardware. ABC Furniture also uses a distributor,

> **Collaborative planning, forecasting, and replenishment (CPFR)** is used to coordinate supply chain members through point-of-sale (POS) data sharing and joint planning.

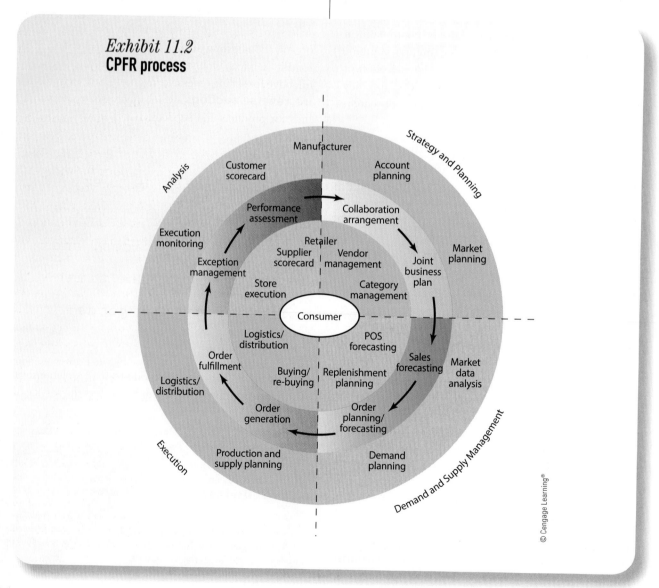

Exhibit 11.2
CPFR process

© Cengage Learning®

Furniture Distribution Company (FDC), to send the final products to retailers. Therefore, ABC Furniture's supply chain includes New England Wood and Vermont Hardware on what's called the "upstream" side of the supply chain and includes FDC, retailers, and customers on the "downstream" side of the supply chain.

The only partner in this supply chain that knows exactly how many pieces of furniture have been sold and how many are available in inventory is the retailer, and without agreements in place, most retailers do not share this information. The distributor, FDC, knows only how many pieces retailers order to replenish their stock, which does not indicate exact numbers sold. So, FDC places an order with ABC Furniture based on its forecasts of orders from retailers. ABC Furniture does not know exactly how much inventory FDC has on hand, only the quantity of orders FDC has placed. So, ABC Furniture orders supplies from New England Wood and Vermont Hardware based on the distributor's forecasts.

If a retailer gets more products than it can sell or stock, it returns the products to FDC, which absorbs part of the cost. FDC, in turn, sends some products back to ABC Furniture, which absorbs this cost. ABC Furniture now has too many products in its inventory and must slow down manufacturing. However, it cannot return lumber to New England Wood, so it ends up paying for carrying extra raw materials in inventory. As a result, ABC Furniture might suffer from having the highest costs in the supply chain, and retailers suffer the least.

CPFR ensures that inventory and sales data are shared across the supply chain so everyone knows the exact sales and inventory levels. The collaboration part of this process is the agreement between all supply chain partners that establishes how data is shared, how problems with overstock are solved, and how to ensure that costs for each partner are shared or minimized. The agreement also encourages retailers to share important data with the distributor and manufacturer, often by offering them better discounts. Retailers are also motivated to sell more to give themselves more leverage with ABC Furniture, which works for ABC Furniture because the more products it sells, the better its bottom line.

Even with an agreement in place, unforeseen problems can crop up, so planning for these "exceptions" is important. Handling unforeseen problems is called "exception management," and lessons learned during this process can be used in future planning.

11-2 CUSTOMER RELATIONSHIP MANAGEMENT

Customer relationship management (CRM) consists of the processes a company uses to track and organize its contacts with customers. The main goal of a CRM system is to improve services offered to customers and use customer contact information for targeted marketing. Businesses know that keeping and maintaining current customers is less expensive than attracting new customers, and an effective CRM system is useful in meeting this goal.

Marketing strategies in a CRM system focus on long-term relationships with customers instead of transactions. These strategies include identifying customer segments, improving products and services to meet customers' needs, improving customer retention, and identifying a company's most profitable (and loyal) customers. To get the most out of these strategies, a CRM system helps organizations make better use of data, information, and knowledge to understand their customers.[3] A CRM system captures information about customer interactions for sales personnel and customer service representatives so they can perform their jobs more effectively and efficiently. This information can include customers' preferences, background, income, gender, and education.

CRM is more than just tracking and organizing contacts with customers. It gives organizations more complete pictures of their customers. CRM systems include tools for conducting complex analyses on customer data, such as a data warehouse and data-mining tools, discussed in Chapter 3. With these systems, organizations can integrate demographic and other external data with customers' transaction data to better understand customer behavior. Based on this analysis, organizations can better target products to customers and manage customer issues, which increases customer satisfaction and retention. In addition, organizations can

> Marketing strategies in a CRM system focus on long-term relationships with customers instead of transactions.

Customer relationship management (CRM) consists of the processes a company uses to track and organize its contacts with customers. It improves services offered to customers and uses customer contact information for targeted marketing.

classify customers based on how valuable they are to the organization and manage them accordingly.

A grocery store offering loyalty cards with discounts to its customers is an example of how transaction data can be used in a CRM system. Knowing that a customer bought 4 gallons of milk the previous week does not give a grocery store much information, but with loyalty cards, the store can track all sorts of information on specific customers. When customers apply for loyalty cards, for example, they can be asked to give demographic information, such as name, age, marital status, and address. So, instead of knowing that "Customer 49 bought 4 gallons of milk last week," a store can learn that "James Smith, 35 years old, married and residing in zip code 11223, bought 4 gallons of milk last week." With this information, the store can assume James Smith has young children (or clearly is not lactose intolerant!). In addition, if James Smith purchases no cereal that same week, the store can assume he is buying cereal from another store (because with the purchase of that amount of milk and the assumption that he has young children, it is likely his children are eating cereal). Therefore, the store decides

© Diego Cervo/Shutterstock.com

to send coupons for discounts on cereal to James Smith. This is referred to as "cross-selling"—getting the customer to buy additional products. The store might also send James Smith coupons for a more expensive brand of milk, in the hope that his family will decide it prefers that brand. This practice is called "upselling."

Organizations can also pay external agencies for additional data about their potential customers. This data might be public or semiprivate, such as whether they own their homes, the value of their homes, and their estimated mortgage or rent payments. This gives organizations more information to analyze.

With a CRM system, an organization can do the following[4]:

- Provide services and products that meet customers' needs.
- Offer better customer service through multiple channels (traditional as well as the Internet).
- Increase cross-selling and upselling of products to increase revenue from existing customers.
- Help sales personnel close deals faster by offering data on customers' backgrounds.
- Retain existing customers, and attract new ones.

Several IT tools discussed throughout this book are used to improve customer service. For example, e-mail, the Internet, Web portals, and automated call centers have played a major role in CRM systems. E-commerce sites use e-mail to confirm items purchased, confirm shipping arrangements, and send notifications on new products and services. Web portals and extranets, such as FedEx.com, allow customers to perform tasks, such as checking the status of shipments and arranging a package pickup. Database systems, data warehouses, and data-mining tools are effective in tracking and analyzing customers' buying patterns, which helps businesses meet customers' needs. Yet this information could be used to generate predictive analytics that an organization can use for future planning in offering new products and services. The emergence of big data and the Internet of Everything may open up additional channels for reaching customers for increasing revenue and improving customer service. A CRM system includes the following activities:

- Sales automation
- Order processing
- Marketing automation
- Customer support
- Knowledge management
- Personalization technology

- *Salesforce automation*—Assists with such tasks as controlling inventory, processing orders, tracking customer interactions, and analyzing sales forecasts and performance. It also assists with collecting, storing, and managing sales contacts and leads.

- *eCRM or Web-based CRM*—Allows Web-based customer interaction and is used to automate e-mail, call logs, Web site analytics, and campaign management. Companies use campaign management to customize marketing campaigns, such as designing a marketing campaign tailored to customers in southern California or customers in the 18–35 age bracket.

- *Survey management*—Automates electronic surveys, polls, and questionnaires, which is useful for gathering information on customers' preferences.

- *Automated customer service*—Used to manage call centers and help desks and can sometimes answer customers' queries automatically.

These activities, performed by CRM software, are discussed in more detail in the following sections. The information box above highlights a real-life application of a CRM system.

11-2a CRM Applications

Typically, CRM applications are implemented with one of two approaches: on-premise CRM or Web-based CRM. Organizations with an established IT infrastructure often choose an on-premise CRM, which is implemented much like any other IT system. With Web-based CRM, the company accesses the application via a Web interface instead of running the application on its own computers and pays to use CRM software as a service (SaaS), which is similar to Web-hosting services. The SaaS vendor also handles technical issues. (SaaS is covered in more detail in Chapter 14.) Several software packages are available for setting up a CRM system, including Amdocs CRM (*www.amdocs.com/About/Pages/default.aspx*), Optima Technologies ExSellence (*www.optima-tech.com*), Infor CRM (*www.infor.com/solutions/crm*), SAP CRM, (*http://help.sap.com/CRM*), and OracleCRM (*www.oracle.com/us/solutions/crm/overview/index.html*). Although these packages vary in capabilities, they share the following features:

11-2b Personalization Technology

Personalization is the process of satisfying customers' needs, building customer relationships, and increasing profits by designing goods and services that meet customers' preferences better. It involves not only

> **Personalization** is the process of satisfying customers' needs, building customer relationships, and increasing profits by designing goods and services that meet customers' preferences better. It involves not only customers' requests, but also the interaction between customers and the company.

customers' requests but also the interaction between customers and the company. You are probably familiar with Web sites that tailor content based on your interests and preferences. Amazon, for example, suggests products you might enjoy, based on your past browsing and purchasing habits.

Customization, which is somewhat different from personalization, allows customers to modify the standard offering, such as selecting a different home page to be displayed each time you open your Web browser. As another example, after registering with Yahoo!, you can customize the start page by choosing your preferred layout, content, and colors. You can find many examples of customization in retail, too, such as Build-A-Bear Workshops, where children can design their own teddy bears, or Nike, which allows customers to create their own shoes by selecting styles and colors.[6]

Amazon is known for using personalization to recommend products to customers. You are probably familiar with the message "Customers who bought this item also bought" followed by a list of suggestions. Amazon's recommendation system is made up of a huge database containing customers' previous purchases and a recommendation algorithm. When a customer logs on to Amazon, the recommendation system first checks the customer's purchase history and that of similar customers. Using this information, a list of recommended products is displayed, based on the customer's shopping history and choices by other customers who have similar purchase histories. In addition, Amazon gives customers an opportunity to rate the recommendations. The more items the customer purchases and the more recommendations the customer rates, the better the recommendations are tailored to the customer.[7]

> Customization, which is somewhat different from personalization, allows customers to modify the standard offering, such as selecting a different home page to be displayed each time you open your Web browser.

Because personalization and customization help companies meet customers' preferences and needs, customers often experience a more efficient shopping process and, as a result, are less likely to switch to competitors to get similar products or services. However, using personalization requires gathering a lot of information about customers' preferences and shopping patterns, and some customers get impatient with answering long surveys about their preferences. In addition, collecting this information might affect customers' sense of privacy. For example, drugstore customers might be concerned that the drugstore has their prescription histories, that the information might be misused and even affect their insurance coverage. To ease these concerns, companies should include clear privacy policies on their Web sites stating how personal information is collected and used.

Many other companies use personalization technology to improve customer service. For example, if you buy a suit from Nordstrom.com, the site might suggest shoes or a tie that goes with the suit or a similar suit in the same category. If you buy a song from Apple iTunes, other songs purchased by listeners like you are suggested. Google also provides personalized services for Google account holders. Users can get personalized search results that are reordered based on their searching histories. Avni Shah, Google's product manager, has explained that if a user has "fly fishing" in his or her search history and then searches on "bass," more weight is given to search results that point to Web pages about fish rather than pages about musical instruments. Google also has a bookmark feature so users can save useful search results for later use. Unlike Yahoo!'s MyWeb feature, which saves the text of Web pages, this feature simply saves the link to the page.[8]

To implement a personalization system, several IT tools are needed, including the Internet, databases, data warehouse/data marts, data-mining tools, mobile

Customization allows customers to modify the standard offering, such as selecting a different home page to be displayed each time you open your Web browser.

networks, and collaborative filtering. **Collaborative filtering (CF)** is a search for specific information or patterns using input from multiple business partners and data sources. It identifies groups of people based on common interests and recommends products or services based on what members of the group purchased or did not purchase. It works well for a single product category, such as books, computers, and so forth. One drawback of CF is that it needs a large sample of users and content to work well. In addition, it is not useful for making recommendations across unrelated categories, such as predicting that customers who liked a certain CD would also like a particular computer.[9]

One application of collaborative filtering is making automatic predictions about customers' preferences and interests based on similar users. For example, if a user rates several movies and is then added to a database that contains other users' ratings, a CF system can predict the user's ratings for movies he or she has not evaluated. You may have seen this feature used on Netflix.com, where lists of other movies you might like are displayed. Recently, Netflix paid $1 million to the team that won a contest to come up with the best algorithm for improving the accuracy of the Netflix recommendation system. Other Web sites that use CF systems to improve customer services are Amazon, Barnes and Noble, and Half.com.

The information box above explains how Amazon uses personalization technologies to increase the efficiency and effectiveness of its marketplace.

11-3 KNOWLEDGE MANAGEMENT

Knowledge management (KM) is a technique used to improve CRM systems (and many other systems) by identifying, storing, and disseminating "know-how"—facts about how to perform tasks. Know-how can be explicit knowledge (formal, written procedures) or tacit knowledge (personal or informal knowledge). Knowledge is an asset that should be shared throughout an organization to generate business intelligence and maintain a competitive advantage in the marketplace. Knowledge management, therefore, draws on concepts of organizational learning, organizational culture, and best practices to convert tacit knowledge into explicit knowledge, create a knowledge-sharing culture in an organization, and eliminate obstacles to sharing knowledge. In this respect, knowledge management shares many of

Collaborative filtering (CF) is a search for specific information or patterns, using input from multiple business partners and data sources. It identifies groups of people based on common interests and recommends products or services based on what members of the group purchased or did not purchase.

Knowledge management (KM) draws on concepts of organizational learning, organizational culture, and best practices to convert tacit knowledge into explicit knowledge, create a knowledge-sharing culture in an organization, and eliminate obstacles to sharing knowledge.

the goals of information management but is broader in scope because information management tends to focus on just explicit knowledge.

Knowledge is more than information and data. It is also contextual. Explicit knowledge, such as how to close a sale, can be captured in data repositories and shared. Expert salespeople can document how they close sales successfully, and this documentation can be used to train new salespeople or those who are struggling with closing sales. Tacit knowledge, however, cannot be captured

their value to the organization would be diminished. To motivate them to share knowledge, rewards must be offered. A knowledge management system can track how often an employee participates in knowledge-sharing interactions with other employees and track any resulting improvements in performance. This information can be used to reward employees for sharing tacit knowledge. Reward systems can be set up for sharing explicit knowledge, too—by tracking how often an employee contributes to a company's internal wiki, for example.

> A knowledge management system can track how often an employee participates in knowledge-sharing interactions with other employees and track any resulting improvements in performance.

as easily. Knowledge that someone has gained through experience might vary depending on the situation in which it was used—the context. Typically, the best way to gather this information is interactively, such as asking the employee specific questions about how he or she would handle an issue. Because interaction is a key part of managing tacit knowledge, a knowledge management system must encourage open communication and the exchange of ideas, typically via e-mails, instant messaging, internal company wikis, videoconferencing, and tools such as WebEx or GoToMeeting, which create virtual instructional environments.

By storing knowledge captured from experts, a knowledge repository can be created for employees to refer to when needed. The most common example is creating a knowledge base of typical customer complaints and solutions. Dell Computer uses this type of knowledge base, so when customers call about problems their computers are having, the steps for solving the problem are documented and readily accessible, which shortens response times.

Knowledge bases can also be used when new products are being designed. A company can store past experiences with similar designs, mistakes made in testing, and so forth to help speed up the delivery timetable and avoid making the same mistakes. This use of knowledge bases is particularly helpful in designing software products and services.

Employees might be reluctant to share their expertise because, once everybody knows what they know,

A simple knowledge management system might consist of using groupware (discussed in Chapter 12), such as IBM Lotus Notes or Microsoft Sharepoint Server, to create, manage, and distribute documents in an organization. These documents include the kind of information discussed previously, such as outlines of procedures for customer service representatives or reports of past design efforts. Other tools and technologies might include DBMSs, data-mining tools, and decision support systems (discussed in Chapter 12). Knowledge management plays a key role in the success of a CRM system because it helps businesses use their knowledge assets to improve customer service and productivity, reduce costs, and generate more revenue. A knowledge management system should help an organization do one or more of the following[11]:

- Promote innovation by encouraging the free exchange of ideas.
- Improve customer service by reducing response time.
- Increase revenue by reducing the delivery time for products and services.
- Improve employee retention rates by rewarding employees for their knowledge.

Because of the importance of knowledge, knowledge management, and knowledge management systems, some organizations have created an executive position called chief knowledge officer (CKO). This individual is responsible for overseeing knowledge management within an organization. He/she makes sure that key

KNOWLEDGE MANAGEMENT IN ACTION

Goodwin Procter was a big law firm, but it had a big problem. With eight offices and 850 attorneys and staff members, it was billing clients to the tune of $611 million a year, but the firm was spending too much time putting together documents and pulling information from different sources. It needed a system to manage 2 terabytes of data, 60,000 cases, and nearly 10 million documents. Using Microsoft SharePoint, it developed a knowledge management system called "Matter Page System"—a hub through which attorneys could access business data and client information. With this platform, attorneys at all eight offices are able to share their work with one another. Before having this system, they would spend hours answering a client's question because multiple files and file systems had to be searched. The new system enables them to share past experiences when working on a case.[12]

knowledge resources are properly collected, stored, and disseminated among the key decision-makers and also makes sure that the organization profits from knowledge resources, including its employees, its processes, and its intellectual property. Finally, the CKO tries to maximize the return on investment (ROI) related to KM, KMS (knowledge management systems), and processes.

The information box above highlights the applications of knowledge management at the Goodwin Procter law firm.

> **Enterprise resource planning (ERP)** is an integrated system that collects and processes data and manages and coordinates resources, information, and functions throughout an organization.

11-4 ENTERPRISE RESOURCE PLANNING

Enterprise resource planning (ERP) is an integrated system that collects and processes data and manages and coordinates resources, information, and functions throughout an organization. A typical ERP system has many components, including hardware, software, procedures, and input from all functional areas. To integrate information for the entire organization, most ERP systems use a unified database to store data for various functions (see Exhibit 11.3). Table 11.1 summarizes the functions of these components.

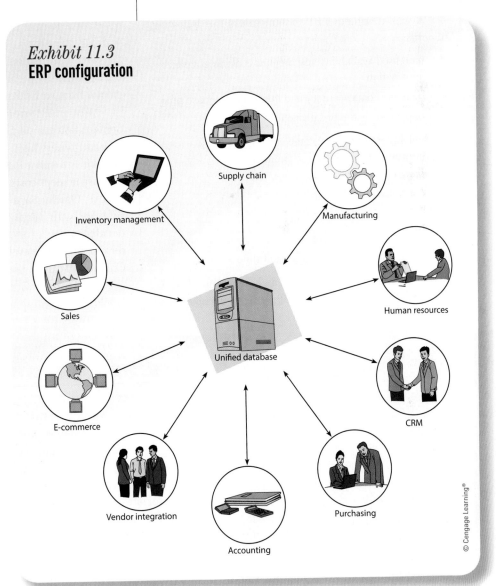

Exhibit 11.3
ERP configuration

TABLE 11.1 ERP COMPONENTS

Component	Function
Unified database	Collects and analyzes relevant internal and external data and information needed by other functions
Inventory management	Provides inventory status and inventory forecasts
Supply chain	Provides information on supply chain members, including suppliers, manufacturing, distribution, and customers
Manufacturing	Provides information on production costs and pricing
Human resources	Provides information on assessing job candidates, scheduling and assigning employees, and predicting future personnel needs
CRM	Provides information on customers and their needs and preferences
Purchasing	Provides information related to the purchasing function, including e-procurement
Accounting	Tracks financial information, such as budget allocations and debits and credits
Vendor integration	Integrates information for vendors, such as offering automated downloads of data on product pricing, specifications, and availability
E-commerce	Provides B2C information related to order status and B2B information related to suppliers and business partners
Sales	Provides information on sales and marketing

© Cengage Learning®

A well-designed ERP system offers the following benefits:

- Increased availability and timeliness of integrated information
- Increased data accuracy and improved response time
- Improved customer satisfaction
- Improved employee satisfaction
- Improved planning and scheduling
- Improved supplier relationship
- Improved reliability of information
- Reduction in inventory costs
- Reduction in labor costs
- Reduction in order-to-fulfillment time

Along with all its advantages, an ERP system also has drawbacks, such as high cost, difficulties in installation, a need for extensive training, and compatibility problems with legacy systems.

The information box below summarizes some of the benefits in operational efficiency that Naghi Group gained from an ERP system.

Most ERP systems are available as modules, so an organization can purchase only the components it needs and add others later, if needed. Having modular components is a major factor in the success of ERP systems because it keeps costs down. More than 40 vendors, such as SAP, Oracle, Sage Group, and Microsoft, offer ERP software with varying capabilities. If an organization decides to use a full-featured ERP system, the systems development life cycle (SDLC) method introduced in Chapter 10 can be used.

The Industry Connection highlights Salesforce.com as a leader in enterprise systems.

ERP STREAMLINES OPERATIONS AT NAGHI GROUP

Naghi Group, based in Jeddah, Saudi Arabia, operates several companies that together offer a wide range of products and services throughout the Middle East. Unfortunately, the legacy software that the firm was using was not able to communicate and integrate with the distribution and manufacturing software it was using. This lack of integration cost Naghi Group many hours each month as it tried to reconcile the data from various sources and generate critical financial-management reports. It needed an ERP system to integrate its major functional areas, including finance, sales, and supply chain management. It also needed to monitor inventory status and purchasing activity. The solution was ERP software—specifically, VAI's (Vormittag Associates, Inc.'s) S2K Distribution Suite. Now, the firm's managers are able to view financial data, keep track of inventory status, and analyze its customers' purchasing activities in real time. The ERP system has streamlined operations and improved customer service; it has also contributed to more timely business decisions.[13]

Salesforce.com, a leader in CRM services, offers enterprise applications that can be customized to meet companies' needs. Its products and services include the following:

CRM applications: Products such as Cloud Platform for CRM and Cloud Infrastructure for CRM are used for salesforce automation, sales management, and contact management.

Sales analytics: Sales Cloud enables management to discover which salespeople are closing the most deals and how long tasks take. Customizable dashboards offer instant access to real-time information, allow monitoring of critical factors on sales, marketing, service, and other departments, and produce consolidated analyses from a variety of data sources.

Chatter: A social networking and collaborative application that works with Sales Cloud. All users of Salesforce can access Chatter for no additional cost. Similar to Facebook Pages, Chatter enables groups to collaborate on projects, share information and documents, and control privacy so information is only shown to appropriate team members.

Service and support: Service Cloud offers a customer portal, call center, and knowledge base. With information from these features, users can analyze who is asking for support and how long responses take, examine employee performance, and determine which reps handle most of the customer inquiries.

Marketing automation: Includes Google Adwords, campaign management, marketing analytics, and marketing dashboards. Users can track multichannel campaigns, from generating sales leads to closing sales.

Force.com Builder: Allows developers to create add-on applications that can be integrated into Salesforce.com applications and hosted by Salesforce.com.

STUDY TOOLS 11

LOCATED AT BACK OF THE TEXTBOOK
☐ Rip out Chapter Review Card

LOCATED AT WWW.CENGAGE.COM/LOGIN
☐ Review Key Terms Flash Cards (Print or Online)
☐ Download audio and visual summaries to review

☐ Complete both Practice Quizzes to prepare for tests
☐ Play "Beat the Clock" and "Quizbowl" to master concepts
☐ Complete "Crossword Puzzle" to review key terms
☐ Watch video on "Method" for a real company example

KEY TERMS

collaborative filtering (CF) 235
collaborative planning, forecasting, and
    replenishment (CPFR) 230
customer relationship management (CRM) 231
customization 234
e-marketplace 228
electronic data interchange (EDI) 227
enterprise resource planning (ERP) 237

enterprise system 224
knowledge management (KM) 235
online auction 229
personalization 233
reverse auction 229
supply chain 225
supply chain management (SCM) 226

REVIEWS AND DISCUSSIONS

1. What are two examples of technologies used in an SCM?
2. What are the four key decisions in an SCM related to a manufacturing firm?
3. What is the difference between traditional auctions and reverse auctions?
4. Who benefits the most from a collaborative planning, forecasting, and replenishment (CPFR) system?
5. What are two key applications of a CRM system?
6. What is the difference between explicit and tacit knowledge? Which one is more challenging to gather and analyze?
7. What are two advantages of an ERP system?
8. What are three components of an ERP system?

PROJECTS

1. SenseAware by FedEx provides key information for improving the logistics part of a supply chain network. After reading the information presented in this chapter and other sources, write a one-page paper that explains the key features of this system. Which technologies are being used?
2. After reading the information presented in this chapter and other sources, write a one-page paper that identifies two companies (in addition to those mentioned in this book) that have implemented a supply chain network. What has been achieved by these systems? How can SCM help the bottom line?
3. After reading the information presented in this chapter and other sources, write a one-page paper that describes the job responsibilities of a chief knowledge officer (CKO) and a chief technology officer (CTO). What are the differences between these two jobs?
4. After reading the information presented in this chapter and other sources, write a two-page paper that describes the differences between a domestic supply chain and an international supply chain. Why is the international supply chain more challenging to manage?
5. In addition to Google, Netflix, and Amazon, identify two companies that use personalization technologies. How can personalization help the bottom line? What are some of the challenges for implementing these systems?
6. Most experts believe that an ERP system, with all the advantages it offers, also has drawbacks, such as high cost, difficulties in installation, a need for extensive training, and compatibility problems with legacy systems. After reading the information presented in this chapter and other sources, write a two-page paper that recommends some ways to minimize these drawbacks.

ARE YOU READY TO MOVE ON?

1. An e-marketplace is a third-party exchange that provides a platform for buyers and sellers to interact with each other and trade more efficiently online. True or False?
2. Personalization allows customers to modify the standard set-up, such as by selecting a different home page to be displayed each time they open their Web browsers. True or False?
3. To integrate all the information for an organization, most ERP systems use a unified database to store data for various functions. True or False?
4. Which of the following is not an SCM technology?
   a. EDI
   b. E-marketplace
   c. Online auction
   d. EDP
5. Which of the following activities is not included in a CRM system?
   a. CPRF automation
   b. Sales automation
   c. Order processing
   d. Marketing automation
6. Which of the following does a well-designed ERP system not offer?
   a. Increased availability and timeliness of integrated information
   b. Improved planning and scheduling
   c. Reduced implementation cost
   d. Increased data accuracy and improved response time

ERP at Johns Hopkins Institutions

Maryland's largest private employer, Johns Hopkins Institutions (JHI), has more than 45,000 full-time staff members at Johns Hopkins Hospital and Health System, Johns Hopkins University, and several other hospitals and institutions. To improve data quality and reporting, JHI decided to use a centralized ERP system instead of several different business applications that were not fully integrated. Getting an overall view of JHI's operations and performance was difficult with these different applications.

JHI faced technical challenges in adopting an ERP system. To overcome these challenges, it chose BearingPoint's SAP applications with IBM

© Richard Thornton/Shutterstock.com

Power 570 servers as the hardware platform. JHI also decided to use server clusters to make the system more fault tolerant, and it installed a storage area network (SAN) for high performance. The ERP system now provides a centralized method of gathering information for the entire organization.[15]

Answer the following questions:

1. What was the main reason for using an ERP system at JHI?
2. How did the ERP system that JHI adopted help address technical challenges?
3. What did the ERP system achieve at JHI?

CRM At Starbucks

With over 17,000 stores spread out over 50 countries, Starbucks uses a CRM system called My Starbucks Idea, which is powered by Salesforce.com.[16] The system, which includes an interactive forum, gives the Starbucks community an online presence, allowing the company to collect feedback from its customers. It also integrates the company's physical stores with social media sites such as Facebook and Twitter. On the system's Web site (*www.mystarbucksidea.com*), customers offer ideas, make comments, and cast votes on

© iStockphoto.com/mattjeacock

issues arranged into 13 categories, from the actual drinks that are served to what the store's atmosphere feels like. So far, the system has collected over 92,000 ideas, nearly 100 of which Starbucks has implemented.[17]

Answer the following questions:

1. What is My Starbucks Idea?
2. What has the Starbucks CRM achieved? How does it help the bottom line?
3. Which software platform is behind the Starbucks CRM system?

12 | Management Support Systems

LEARNING OUTCOMES

After studying this chapter, you should be able to:

12-1 Describe the phases of the decision-making process in a typical organization and the types of decisions that are made.

12-2 Describe a decision support system.

12-3 Explain an executive information system's importance in decision making.

12-4 Describe group support systems, including groupware and electronic meeting systems.

12-5 Summarize the uses for a geographic information system.

12-6 Describe the guidelines for designing a management support system.

After you finish this chapter, go to **PAGE 259** for the **STUDY TOOLS**

© Andrey_Popov/Shutterstock.com

This chapter begins by summarizing the phases of the decision-making process and the types of decisions that are made. Next, you learn about a decision support system (DSS), its components and its capabilities, and see how it can benefit an organization. In addition, you learn about other management support systems used in decision making: executive information systems (EISs), group support systems (GSSs), and geographic information systems (GISs). This chapter concludes with an overview of guidelines for designing a management support system.

12-1 TYPES OF DECISIONS IN AN ORGANIZATION

In a typical organization, decisions fall into one of these categories:

- *Structured decisions*—**Structured decisions**, or programmable tasks, can be automated because a well-defined standard operating procedure exists for these types of decisions. Record keeping, payroll, and simple inventory problems are examples of structured tasks. Information technologies are a major support tool for making structured decisions.

- *Semistructured decisions*—**Semistructured decisions** are not quite as well defined by standard operating procedures, but they include a structured aspect that benefits from information retrieval, analytical models, and information systems technology.

For example, preparing budgets has a structured aspect in calculating percentages of available funds for each department. Semistructured decisions are often used in sales forecasting, budget preparation, capital acquisition analysis, and computer configuration.

Structured decisions, or programmable tasks, can be automated because a well-defined standard operating procedure exists for these types of decisions.

Semistructured decisions include a structured aspect that benefits from information retrieval, analytical models, and information systems technology.

Exhibit 12.1
Organizational levels and types of decisions

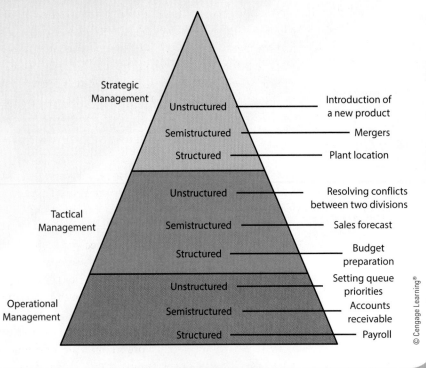

Strategic Management	Unstructured — Introduction of a new product
	Semistructured — Mergers
	Structured — Plant location
Tactical Management	Unstructured — Resolving conflicts between two divisions
	Semistructured — Sales forecast
	Structured — Budget preparation
Operational Management	Unstructured — Setting queue priorities
	Semistructured — Accounts receivable
	Structured — Payroll

© Cengage Learning®

these decisions. Areas involving unstructured decisions include research and development, hiring and firing, and introducing a new product.

Semistructured and unstructured decisions are challenging because they involve multiple criteria, and often users have to choose between conflicting objectives. For example, a manager might want to give raises to employees to boost morale and increase employee retention but has been asked to reduce the total cost of production. These two objectives conflict, at least in the short run. Artificial intelligence applications (discussed in Chapter 13) might be helpful in the future for handling qualitative decisions. Exhibit 12.1 shows organizational levels (operational, tactical, and strategic) and types of decisions.

> Semistructured and unstructured decisions are challenging because they involve multiple criteria, and often users have to choose between conflicting objectives.

- *Unstructured decisions*—**Unstructured decisions** are typically one-time decisions, with no standard operating procedure pertaining to them. The decision maker's intuition plays the most important role, as information technology offers little support for

Unstructured decisions are typically one-time decisions, with no standard operating procedure pertaining to them.

Management support systems (MSSs) are the different types of information systems that have been developed to support certain aspects and types of decisions. Each type of MSS is designed with unique goals and objectives.

Different types of information systems have been developed to support certain aspects and types of decisions. Collectively, these systems are called **management support systems (MSSs)**, and each type is designed with its own goals and objectives, as discussed in this chapter.

12-1a Phases of the Decision-Making Process

Herbert Simon, winner of the 1978 Nobel Prize in economics, defines three phases in the decision-making process: intelligence, design, and choice.[1] A fourth phase, implementation, can be added. The following sections explain these phases.

Intelligence Phase

In the **intelligence phase**, a decision maker (a marketing manager, for example) examines the organization's environment for conditions that need decisions. Data is collected from a variety of sources (internal and external) and processed. From this information, the decision maker can discover ways to approach the problem. This phase has three parts: First, you determine what the reality is—identify what is really going on in order to help define the problem. Second, you get a better understanding of the problem by collecting data and information about it. Third, you gather data and information needed to define alternatives for solving the problem.

As an example, say an organization has noticed a decrease in total sales over the past 6 months. To pinpoint the cause of the problem, the organization can collect data from customers, the marketplace, and the competition. After the data has been processed, analysis can suggest possible remedies. Information technologies, particularly database management systems, can help in this analysis. In addition, many third-party vendors, such as Nielsen and Dow Jones, specialize in collecting data about the marketplace, the competition, and the general status of the economy. The information they collect can support the intelligence phase of decision making.

Design Phase

In the **design phase**, the objective is to define criteria for the decision, generate alternatives for meeting the criteria, and define associations between the criteria and the alternatives. Criteria are goals and objectives that decision makers establish in order to achieve certain performance levels. For example, the criterion in the previous example of decreased sales might simply be to increase sales. To make this criterion more specific, you can state it as "Increase sales by 3 percent each month for the next 3 months." Next, the following alternatives could be generated:

- Assign more salespeople to the target market.
- Retrain and motivate current salespeople.
- Reassign current salespeople.
- Revamp the product to adjust to consumers' changing tastes and needs.
- Develop a new advertising campaign.
- Reallocate existing advertising to other media.

Defining associations between alternatives and criteria involves understanding how each alternative affects the criteria. For example, how would increasing the sales force increase sales? By how much does the sales force need to be increased to achieve a 3 percent increase in sales? Generally, information technology does not support this phase of decision making very much, but group support systems and electronic meeting systems, discussed later in this chapter, can be useful. Expert systems (covered in Chapter 13) are helpful in generating alternatives, too.

Choice Phase

The **choice phase** is usually straightforward. From the practical alternatives, the best and most effective course of action is chosen. It starts with analyzing each alternative and its relationship to the criteria to determine whether it is feasible. For instance, for each salesperson added, how are sales expected to increase? Will this result be economically beneficial? After a thorough analysis, the choice phase ends with decision makers recommending the best alternative. For the problem of decreased sales, the organization decided to use the first alternative, assigning more salespeople to the target market. A decision support system (DSS) can be particularly useful in this phase. DSSs are discussed later in the chapter, but these systems help sort through possible solutions to choose the best one for the organization. Typically, they include tools for calculating cost–benefit ratios, among others. For example, say an organization is deciding which of three transportation systems to use for shipping its products to retail outlets. A DSS can assess cost factors and determine which transportation system minimizes costs and maximizes profits. Generally, information technologies are more useful in the intelligence and choice phases than in the design phase.

In the **intelligence phase**, a decision maker examines the organization's environment for conditions that need decisions. Data is collected from a variety of sources (internal and external) and processed. From this information, the decision maker can discover ways to approach the problem.

In the **design phase**, the objective is to define criteria for the decision, generate alternatives for meeting the criteria, and define associations between the criteria and the alternatives.

During the **choice phase**, the best and most effective course of action is chosen.

© bloomua/Shutterstock.com

Implementation Phase

In the **implementation phase**, the organization devises a plan for carrying out the alternative selected in the choice phase and obtains the resources to implement the plan. In others words, ideas are converted into actions. Information technologies, particularly DSSs, can also be useful in this phase. A DSS can do a follow-up assessment on how well a solution is performing. In the previous example of selecting a transportation system, a DSS might reveal that the system the organization chose is not performing as well as expected and suggest an alternative.

In the **implementation phase**, the organization devises a plan for carrying out the alternative selected in the choice phase and obtains the resources to implement the plan.

A **decision support system (DSS)** is an interactive information system consisting of hardware, software, data, and models (mathematical and statistical) designed to assist decision makers in an organization. Its three major components are a database, a model base, and a user interface.

12-2 DECISION SUPPORT SYSTEMS

For the purposes of this book, a **decision support system (DSS)** is an interactive information system consisting of hardware, software, data, and models (mathematical and statistical) designed to assist decision makers in an organization. The emphasis is on semistructured and unstructured tasks. A DSS should meet the following requirements:

- Be interactive.
- Incorporate the human element as well as hardware and software.
- Use both internal and external data.
- Include mathematical and statistical models.
- Support decision makers at all organizational levels.
- Emphasize semistructured and unstructured tasks.

12-2a Components of a Decision Support System

A DSS, shown in Exhibit 12.2, includes three major components: a database, a model base, and a user interface. In addition, a fourth component, the DSS engine,

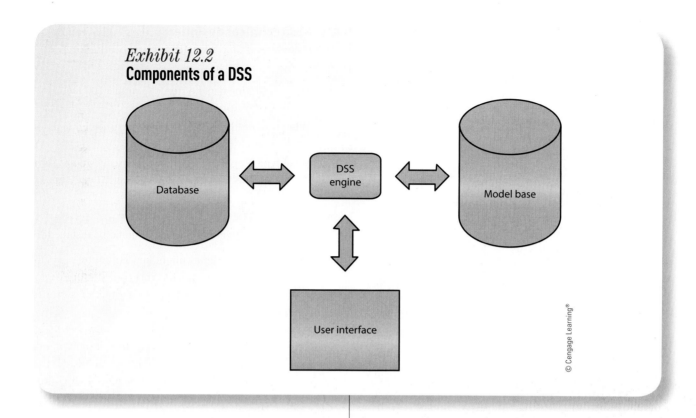

Exhibit 12.2
Components of a DSS

Database ⟷ DSS engine ⟷ Model base

DSS engine ⟷ User interface

© Cengage Learning®

manages and coordinates these major components. The database component includes both internal and external data, and a database management system (DBMS) is used for creating, modifying, and maintaining the database. This component enables a DSS to perform data analysis operations.

The **model base** component includes mathematical and statistical models that, along with the database, enable a DSS to analyze information. A model base management system (MBMS) performs tasks similar to a DBMS in accessing, maintaining, and updating models in the model base. For example, an MBMS might include tools for conducting what-if analysis so a forecasting model can generate reports showing how forecasts vary, depending on certain factors.

Finally, the user interface component is how users access the DSS, such as when querying the database or model base, for help in making decisions. From the end user's point of view, the interface is the most important part of a DSS and must be as flexible and user friendly as possible. Because most DSS users are senior executives with little computer training, user friendliness is essential in these systems.[2]

> **Because most DSS users are senior executives with little computer training, user friendliness is essential in these systems.**

12-2b DSS Capabilities

DSSs include the following types of features to support decision making:

- *What-if analysis*—This shows the effect of a change in one variable, answering questions such as "If labor costs increase by 4 percent, how is the final cost of a product affected?" and "If the advertising budget increases by 2 percent, what is the effect on total sales?"

- *Goal seeking*—This is the reverse of what-if analysis. It asks what has to be done to achieve a particular goal—for example, how much to charge for a product in order to generate $200,000 profit, or how much to advertise a product to increase total sales to $50,000,000.

- *Sensitivity analysis*—This enables you to apply different variables, such as determining the maximum price you could pay for raw materials and still make a profit, or determining how much the interest rate has to go down for you to be able to afford a $100,000 house with a monthly payment of $700.

> The **model base** component includes mathematical and statistical models that, along with the database, enable a DSS to analyze information.

- *Exception reporting analysis*—This monitors the performance of variables that are outside a defined range, such as pinpointing the region that generated the highest total sales or the production center that went over budget.

A typical DSS has many more capabilities, such as graphical analysis, forecasting, simulation, statistical analysis, and modeling analysis.

12-2c Roles in the DSS Environment

To design, implement, and use a DSS, several roles are involved. These include the user, managerial designer, technical designer, and model builder.[3]

Users comprise the most important category because they are the ones using the DSS; therefore, the system's success depends on how well it meets their needs. Users can include department or organizational units in addition to people.

A **managerial designer** defines the management issues in designing and using a DSS. These issues do not involve the technological aspects of the system; they are related to management's goals and needs. This person specifies data requirements, what models are needed, how these models might be used, and how users want to view the results (graphics, text, and so forth). This role addresses questions such as the following:

- What type of data should be collected, and from what sources?
- How recent should the collected data be?
- How should the data be organized?
- How should the data be updated?
- What should the balance between aggregated (lump sum) and disaggregated (itemized) data be?

The **technical designer** focuses on how the DSS is implemented and usually addresses the following questions:

- How should the data be stored (centralized, decentralized, or distributed)?
- What type of file structure should be used (sequential, random, or indexed sequential)?
- What type of user access should be used? Menu driven, such as QBE? Or command line, such as SQL?
- What type of response time is required?
- What types of security measures should be installed?

The technical designer might be a computer specialist or a consultant from outside the company and may use a commercial DSS package or write the system's code from scratch.

A **model builder** is the liaison between users and designers. For example, during the design phase, the model builder might explain users' needs to the managerial designer or technical designer. Later, during the implementation phase, this person might explain the output of a regression analysis to users, describing the assumptions underlying the model, its limitations, and its strengths. The model builder is responsible for supplying information on what the model does, what data inputs it accepts, how the model's output should be interpreted, and what assumptions go into creating and using the model. Typically, requirements for what the model should do come from the managerial designer, implementation of the model is carried out by the technical designer, and specifications for the model come from the model builder. The model builder can also suggest new or different applications of a DSS.

12-2d Costs and Benefits of Decision Support Systems

Some DSSs can be developed from resources already available in the organization, which can reduce costs, but many require new hardware and software. Before making this investment, organizations should weigh the costs and benefits of using a DSS. Costs and benefits can be difficult to assess, however, because these systems are focused on effectiveness rather than efficiency. In addition, a DSS facilitates improvements but does not necessarily cause them. For example, how do you assign a monetary value to facilitating communication or expediting problem solving?

Peter G. Keen, a former MIT professor, conducted an interesting study on how organizations use DSSs and concluded that the decision to build a DSS seems to be

A **managerial designer** defines the management issues in designing and using a DSS. These issues do not involve the technological aspects of the system; they are related to management's goals and needs.

The **technical designer** focuses on how the DSS is implemented and usually addresses questions about data storage, file structure, user access, response time, and security measures.

A **model builder** is the liaison between users and designers. He or she is responsible for supplying information on what the model does, what data inputs it accepts, how the model's output should be interpreted, and what assumptions go into creating and using the model.

> You can quantify the benefit of saving time, for instance, by measuring the 2 hours a manager wasted looking for information that a DSS could have made available immediately.

based on value rather than cost. He outlined the benefits of a DSS as follows[4]:

- Increase in the number of alternatives examined
- Fast response to unexpected situations
- Ability to make one-of-a-kind decisions
- New insights and learning
- Improved communication
- Improved control over operations, such as controlling the cost of production
- Cost savings from being able to make better decisions and analyze several scenarios (what-ifs) in a short period
- Better decisions
- More effective teamwork
- Time savings
- Better use of data resources

As this study indicates, most of the benefits are intangible and difficult to assess. However, they can be quantified to a degree, although the quantification might vary, depending on the person doing the calculations. You can quantify the benefit of saving time, for instance, by measuring the 2 hours a manager wasted looking for information that a DSS could have made available immediately. Of course, you would probably also notice that a

manager who did not have to waste this much time is less frustrated and more productive, but quantifying these results is harder, or at least requires more work, such as conducting interviews or surveys.

The benefit of improving communication and interactions between management and employees is perhaps the most difficult to quantify, but it is one of the most important.[5] DSSs can, and are, improving how decision makers view themselves, their jobs, and the way they spend time. Therefore, improving communication and expediting learning are among the main objectives of a DSS.[6,7]

A DSS is said to have achieved its goals if employees find it useful in doing their jobs. For example, a portfolio manager who uses a financial DSS to analyze different scenarios would certainly find the ease of analyzing a variety of variables, such as the interest rate and economic forecasts, to be useful. The manager can try different values for these variables quickly and easily to determine which variable has the greatest effect and decide which portfolio will be the most profitable. In addition, some DSSs result in saving on clerical costs, and others improve the decision-making process. The information box below describes the DSS applications at Family Dollar, a retail merchandise chain.

DECISION SUPPORT SYSTEMS AT FAMILY DOLLAR

Based in Charlotte, North Carolina, Family Dollar is a retail chain with over 7,000 stores in 45 states. Its stores are designed in a way that allows customers to quickly get in and out. About five years ago, the company noticed decelerating same-store sales growth and declining operating margins. The challenge was to tailor store offerings to customer preferences in each part of the country. The company needed a DSS that would increase sales by making sure each store carried merchandise that the customers of that particular store wanted and make sure that customers would come back. Family Dollar employed an SAS Integrated Merchandise Planning platform for this implementation. The system provides optimization, planning, and analytics across the enterprise and at the local-store level, and it has helped the company reduce inventories and increase inventory turnover. Using the system, company executives can now drill all the way down to determine which stores sell more of Brand A versus Brand B; they can even determine which items of a particular brand sell the most . Jody Crozier, Divisional Vice President of Merchandise Solutions, says the system "allows us to analyze the data in a way that is quickly actionable."[8]

12-3 EXECUTIVE INFORMATION SYSTEMS

Executive information systems (EISs), a branch of DSSs, are interactive information systems that give executives easy access to internal and external data and typically include "drill-down" features (explained in Chapter 3) and a digital dashboard for examining and analyzing information. (Although some experts consider executive support systems and executive management systems variations of EISs, this book considers them to fall under the term *EIS*.)

Ease of use plays an important role in the success of an EIS. Because most EIS users are not computer experts, simplicity of the system is crucial, and EIS designers should focus on simplicity when developing a user interface. Typically, GUIs are used, but adding features such as multimedia, virtual reality, and voice input and output can increase ease of use.

Another important factor in an effective EIS is access to both internal and external data so executives can spot trends, make forecasts, and conduct different types of analyses. For an EIS to be useful, it should also collect data related to an organization's "critical success factors"—issues that make or break a business. In banks, interest rates are considered a critical success factor; for car manufacturers, location of dealerships might be a critical success factor. An EIS should be designed to provide information related to an organization's critical success factors.

Most EISs include a **digital dashboard**, which integrates information from multiple sources

and presents it in a unified, understandable format, often as charts and graphs. Digital dashboards and scorecards offer up-to-the minute snapshots of information and assist decision makers in identifying trends and potential problems. Many digital dashboards are Web based, such as the one included in Microsoft SharePoint. Exhibit 12.3 shows an example of a digital dashboard.

The following are some important characteristics of an EIS[9,10]:

- Tailored to meet management's information needs
- Can extract, compress, filter, and track critical data
- Provides online status access, trend analysis, and exception reporting
- Offers information in graphical, tabular, and text formats
- Includes statistical analysis techniques for summarizing and structuring data
- Retrieves data in a wide range of platforms and data formats
- Contains customized application-development tools
- Supports electronic communication, such as e-mail and video conferencing

12-3a Reasons for Using EISs

An EIS can put a wealth of analytical and decision-making tools at managers' fingertips and includes graphical representations of data that helps executives make

> **Executive information systems (EISs)**, branches of DSSs, are interactive information systems that give executives easy access to internal and external data and typically include "drill-down" features and a digital dashboard for examining and analyzing information.
>
> A **digital dashboard** integrates information from multiple sources and presents it in a unified, understandable format, often as charts and graphs. It offers up-to-the minute snapshots of information and assists decision makers in identifying trends and potential problems.

Exhibit 12.3
Digital dashboard

Microsoft product screenshots used with permission from Microsoft Corporation.

critical decisions. In addition, executives can use EISs to share information with others more quickly and easily. Managers can use these tools to improve the efficiency and effectiveness of decision making in the following ways:

- Increase managers' productivity by providing fast and easy access to relevant information.

- Convert information into other formats, such as bar charts or graphs, to help managers analyze different business scenarios and see the effect of certain decisions on the organization.

- Spot trends and report exceptions, such as gathering data on profitability and production costs at a manufacturing plant to determine whether closing the plant is more beneficial than keeping it open.

12-3b Avoiding Failure in Design and Use of EISs

As with other management support systems, effective design and implementation of an EIS requires top-management support, user involvement, and the right technologies. The following factors can lead to a failed EIS[11,12]:

- The corporate culture is not ready, there is organizational resistance to the project, or the project is viewed as unimportant.

- Management loses interest or is not committed to the project.

- Objectives and information requirements cannot be defined clearly, or the system does not meet its objectives.

- The system's objectives are not linked to factors critical to the organization's success.

- The project's costs cannot be justified.

- Developing applications takes too much time, or the system is too complicated.

- Vendor support has been discontinued.

- Some of today's senior executives missed the computer revolution and might feel uncomfortable using computers. Ongoing

education and increasing computer awareness should solve this problem.

- Executives' busy schedules and frequent travel make long training sessions difficult, do not allow much uninterrupted time for system use, and often prevent daily use of an EIS. The result is that senior executives are unlikely to use systems that need considerable training and regular use to learn. A user-friendly interface can encourage executives to use an EIS more often, however.

- Some EISs do not contain the information that senior executives need because there is a lack of understanding about what executives' work involves. Designers must determine what types of information executives need before designing a system.

12-3c EIS Packages and Tools

EISs are generally designed with two or three components: an administrative module for managing data access, a builder module for developers to configure data mapping and screen sequencing, and a runtime module for using the system. Sometimes, administrative and builder modules are combined into one module. Some EIS packages provide a data storage system, and some simply package data and route it to a database, usually on a LAN. Most EIS packages come with a standard graphical user interface (GUI). Exhibit 12.4 shows

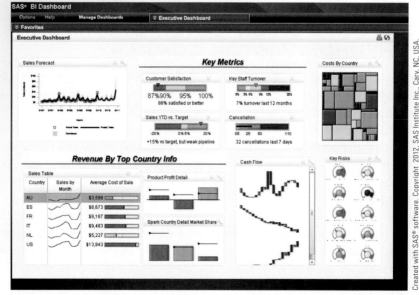

Exhibit 12.4
Business Intelligence interface

a screen from one EIS package, Business Intelligence (SAS Institute).

Generally, managers perform six tasks for which an EIS is useful: tracking performance, flagging exceptions, ranking, comparing, spotting trends, and investigating/exploring. Most EIS packages provide tools for these tasks, such as displaying summaries of data in report or chart format and sequencing screens to produce slide shows. Exception or variance reporting is another useful technique that managers use to flag data that is unusual or out of normal boundaries. Both unusual and periodic events can be defined to trigger visual cues or activate intelligent agents to perform a specific task. Intelligent agents, covered in Chapter 13, are "smart" programs that carry out repetitive tasks and can be programmed to make decisions based on specified conditions. Widely used EIS packages include SAS Business Intelligence (*www.sas.com/technologies/bi*), Datawatch (*www.datawatch.com*), and Cognos PowerPlay (*www-01.ibm.com/software/data/cognos/products/series7/powerplay*).

The information box below describes the EIS applications at Hyundai Motor Company.

Group support systems (GSSs) assist decision makers working in groups. These systems use computer and communication technologies to formulate, process, and implement a decision-making task and can be considered a kind of intervention technology that helps overcome the limitations of group interactions.

12-4 GROUP SUPPORT SYSTEMS

In today's business environment, decision makers often work in groups, so you hear the terms *group computing* or *collaborative computing* used often. All major software vendors are competing to enter this market or increase their market share in this fast-growing field. In this collaborative environment, there has been an increase in **group support systems (GSSs)**, which are intended to assist decision makers working in groups. DSSs are usually designed to be used by a

EXECUTIVE INFORMATION SYSTEMS AT HYUNDAI MOTOR COMPANY

Hyundai Motor Company (HMC) is one of the world's largest automotive manufacturers. It employs more than 47,000 employees and has several subsidiaries around the world as well as manufacturing plants in Seoul, Tokyo, Peking, Detroit, and Frankfurt.

HMC's executive information system, designed with SAS software, supports and facilitates executive decision-making throughout the corporation. Called the Enterprise Information and Management System (EIMS), the system has three main components: a warning system, a goal-oriented system, and a DSS. The warning system alerts executives about the current sales and production volume figures and compares them with past performance and reports any deviations. It brings to the executives' attention those areas that did not perform well. The goal-oriented system keeps executives informed regarding progress that the corporation is making towards long-term goals established by the company. The DSS component provides executives with timely and accurate information related to decision-making activities. In order to deliver the needed information, the system gathers data from human resources, domestic and international sales, and production. The EIMS has enabled executives to make decisions in a proactive versus a reactive manner.[13]

particular decision maker; a GSS is designed to be used by more than one decision maker. These systems use computer and communication technologies to formulate, process, and implement a decision-making task and can be considered a kind of intervention technology that helps overcome the limitations of group interactions. The intervention aspect of a GSS reduces communication barriers and introduces order and efficiency into situations that are inherently unsystematic and inefficient, such as group meetings and brainstorming sessions. A GSS, with the help of a human facilitator, enhances decision making by providing a clear focus for group discussion, minimizing politicking, and focusing attention on critical issues. The success of a GSS depends on the following:

- Matching the GSS's level and sophistication to the group's size and the scope of the task
- Providing supportive management (especially at the CEO level) that is willing to "champion" using a GSS in the organization

DSS, a GSS should include communication features so decision makers in many different locations can still work together to participate in the decision-making process.

12-4a Groupware

The goal of **groupware** is to assist groups in communicating, collaborating, and coordinating their activities. It is intended more for teamwork than for decision support. For the purposes of this book, groupware is a collection of applications that supports decision makers by providing access to a shared environment and information. A shared environment can consist of an e-mail message, a memo, a single file, or even an entire database.

Groupware is software that helps a group of decision makers work with the same application, regardless of their locations. Groupware tools include e-mail, chat applications, video conferencing, and database sharing.

The intervention aspect of a GSS reduces communication barriers and introduces order and efficiency into situations that are inherently unsystematic and inefficient, such as group meetings and brainstorming sessions.

Related technologies for group support, such as electronic meeting systems (EMSs), groupware, computer-mediated communication (CMC), computer-supported cooperative work (CSCW), and e-collaboration, are not considered full-function GSSs because they do not have decision-making tools, but they are less expensive and include communication and problem-solving mechanisms for effective team management.[14,15]

GSSs are useful for committees, review panels, board meetings, task forces, and decision-making sessions that require input from several decision makers. They can be used to find a new plant location, introduce a new product or advertising campaign, participate in an international bid, brainstorm alternatives, and other tasks. In addition to all the capabilities of a

IBM Lotus Notes, Microsoft SharePoint (see the information box on Microsoft Office SharePoint Server on the next page), and Novell GroupWise are common examples of groupware, the capabilities of which include the following:

- Audio and video conferencing
- Automated appointment books
- Brainstorming
- Database access

Groupware assists groups in communicating, collaborating, and coordinating their activities. It is a collection of applications that supports decision makers by providing access to a shared environment and information.

MICROSOFT OFFICE SHAREPOINT SERVER: A NEW TYPE OF GROUPWARE

Microsoft Office SharePoint Server 2013, part of the Microsoft Office suite, is used to improve collaboration, provide content management features, carry out business processes, and provide access to information that is essential to organizational goals (*http://sharepoint.microsoft.com/en-us/Pages/default.aspx*). You can create SharePoint sites that support content publishing, content management, records management, and business intelligence needs. You can also conduct searches for people, documents, and data as well as access and analyze large amounts of business data. SharePoint Server 2013 provides a single, integrated location where employees can collaborate with team members, find organizational resources, search for experts and information, manage content and workflow, and use the information they have found to make better decisions.[16]

- E-mail
- Online chat
- Scheduling
- To-do lists
- Workflow automation

LANs, WANs, and MANs, discussed in Chapter 6, are the network foundations for groupware. Although e-mail is not the same thing as groupware, it provides the main functions of groupware: transmitting text messages across a network. The information box below introduces an application of groupware in the health-care industry.

The Internet has become an important part of groupware. The most important advantage of a Web-based GSS is being able to use open network standards, meaning the GSS can be used on any operating system or type of workstation. The most notable disadvantages are speed limitations (because the Internet is often slower than a company's proprietary network) and security issues. Some examples of Web-based GSS tools are Microsoft Office SharePoint Server and IBM Lotus Domino. Another type of software used for e-collaboration is an electronic meeting system, such as Microsoft Live Meeting, Metastorm, and IBM FileNet.

GROUPWARE AND HEALTH IT

Hospital personnel have become accustomed to electronic health records (EHRs) over the last 25 years, but the technology, which presumes a client-server, physician-centric model, has never been very popular. Today, there is an alternative: IT systems that integrate and coordinate the data provided by every member of the healthcare team, from the doctor to the patient. Under the general heading of "clinical groupware," there are now software products available—RMDNetworks and Shared Health are two examples—that are Web based, networkable, and interactive, thereby improving health care through better communication and coordination. Clinical groupware is used by groups, not individuals, and the patient is considered a vital member of that group. The overall aim is to provide a unified view of the patient by collecting and analyzing data and information from a variety of sources. Clinical groupware has the following additional benefits[17]:

- Inexpensive to acquire and use
- Offers evidence-based guidance
- Routinely collects quality and performance measures
- Provides a collaborative workflow platform

12-4b Electronic Meeting Systems

Electronic meeting systems enable decision makers in different locations to participate in a group decision-making process. There are various types, but they all have the following features:

- *Real-time computer conferencing*—This allows a group of people to interact via their workstations and share files, such as documents and images. The conference often includes an audio link, but there is no video capability.

- *Video teleconferencing*—The closest thing to a face-to-face meeting, this requires special equipment and sometimes trained operators. Video cameras are used to transmit live pictures and sounds, which makes video teleconferencing more effective than phone conferencing but also more expensive. The main drawback is that participants cannot share text and graphics.

- *Desktop conferencing*—This combines the advantages of video teleconferencing and real-time computer conferencing. With desktop conferencing, participants can have multiple video windows open at one time. They also have interfaces to a conference installed on their workstations, so these systems are easier for employees to use. The information box below highlights new generations of electronic meeting systems.

12-4c Advantages and Disadvantages of GSSs

Advantages of GSSs include the following:

- Because decision makers do not have to travel as much (which includes paying for planes, hotels, and meals), costs as well as stress levels are reduced.

- Because decision makers are not traveling long distances, they have more time to talk with one another and solve problems.

- Shyness is not as much of an issue in GSS sessions as it is in face-to-face meetings.

- Increasing collaboration improves the effectiveness of decision makers.

Disadvantages of GSSs include the following:

- *Lack of the human touch*—Gestures, handshakes, eye contact, and other nonverbal cues can be lost, which can hinder the effectiveness of meetings. New developments in virtual reality technologies (discussed in Chapter 14) could solve this problem, however.

> **Electronic meeting systems** enable decision makers in different locations to participate in a group decision-making process.

NEW GENERATIONS OF ELECTRONIC MEETING SYSTEMS

Increasingly, consumers and businesspersons are using computers and mobile devices for one-to-one video calls. Group video calls are also on the rise. To use this technology, there are several options available, and they are either free or very inexpensive. Here are some popular examples:

- Microsoft's Skype—Offers individual video calls for free; group video calling requires a plan that costs about $10 a month. The maximum number of participants is 10.[18]
- Apple's FaceTime—Available free of charge on Apple's iPhone, iMac, iPad, and iPod Touch.[19]
- Tango Video Calls—Allows video chats among different devices, including Windows phones, Android and iOS phones, tablets, and PCs.[20]
- Google's Hangouts—Allows video chat for up to nine people with Hangouts on Google+.[21]
- ooVoo from ooVoo LLC—Allows up to 12 participants in group video chats that are free. The service is available on the Internet using PCs, Macs, iPhones, and other smartphones and tablets.[22]
- Zoom.us from Zoom Video Communications—Allows high-definition group video chat for up to 15 participants. It works over wired and Wi-Fi Internet connections or cellular 3G and 4G networks.[23,24]

- *Unnecessary meetings*—Because arranging a GSS session is easy, there is a tendency to schedule more meetings than are necessary, which wastes time and energy.
- *Security problems*—GSS sessions have the same security problems as other data communication systems, so there is the possibility of private organizational information falling into the hands of unauthorized people. Tight security measures for accessing GSS sessions and transferring data are essential.

12-5 GEOGRAPHIC INFORMATION SYSTEMS

Executives often need to answer questions such as the following:

- Where should a new store be located to attract the most customers?
- Where should a new airport be located to keep the environmental impact to a minimum?
- What route should delivery trucks use to reduce driving time?
- How should law enforcement resources be allocated?

A well-designed **geographic information system (GIS)** can answer these questions and more. This system captures, stores, processes, and displays geographic information or information in a geographic context, such as showing the location of all city streetlights on a map. A GIS uses spatial and nonspatial data and specialized techniques for storing coordinates of complex geographic objects, including networks of lines (roads, rivers, streets) and reporting zones (zip codes, cities, counties, states). Most GISs can superimpose the results of an analysis on a map, too. Typically, a GIS uses three geographic objects:

- *Points*—The intersections of lines on a map, such as the location of an airport or a restaurant

> A GIS can provide all kinds of information that enables you to zero in on specific customers.

- *Lines*—Usually a series of points on a map (a street or a river, for example)
- *Areas*—Usually a section of a map, such as a particular zip code or a large tourist attraction

Digitized maps and spatially oriented databases are two major components of a GIS. For example, say you want to open a new store in southwest Portland, Oregon, and would like to find out how many people live within walking distance of the planned location. With a GIS, you can start with the map of the United States and zoom in repeatedly until you get to the street map level. You can mark the planned store location on the map and draw a circle around it to represent a reasonable walking distance. Next, you can request a summary of U.S. census data on everyone living inside the circle who meet certain conditions, such as a particular income level, age, marital status, and so forth. A GIS can provide all kinds of information that enables you to zero in on specific customers. A GIS can perform the following tasks:

- Associate spatial attributes, such as a manufacturing plant's square footage, with points, lines, and polygons on maps.
- Integrate maps and database data with queries, such as finding zip codes with a high population of senior citizens with relatively high income.

A GIS can support some sophisticated data management operations, such as the following:

- Show the customers who live within a 5-mile radius of the Super Grocery at the corner of 34th and Lexington. A database cannot answer this question because it cannot determine the latitude/longitude coordinates of the store, compute distances using the specified location as the center, identify all zip codes within this circle, and pull out the customers living in these zip codes.
- Show the customers whose driving route from work to home and back takes them through the intersection of 34th and Lexington. A database cannot address this query, either. In this case, the GIS maps customers' home and work locations and determines all possible routes. It can then narrow down the customer list by picking only those whose shortest route takes them through the specified intersection.

A GIS with analytical capabilities evaluates the impact of decisions by interpreting spatial data.

Modeling tools and statistical functions are used for forecasting purposes, including trend analysis and simulations. Many GISs offer multiple windows so you can view a mapped area and related nonspatial data simultaneously, and points, lines, and polygons can be color coded to represent nonspatial attributes. A zoom feature is common for viewing geographic areas in varying levels of detail, and map overlays can be useful for viewing such things as gas lines, public schools, or fast-food restaurants in a specified region. A buffering feature creates pin maps that highlight locations meeting certain criteria, such as finding a new store location based on population density. Exhibit 12.5 shows an example of output from Environmental Systems Research Institute (ESRI), a major vendor of GIS software. To see examples of different types of maps, visit *www.esri.com*.

A common example of a GIS—and one you have probably used often—is getting driving directions from Google Maps. It is an interactive GIS that identifies routes from start to destination, overlays routes on a map, shows locations of nearby landmarks, and estimates distances and driving times. It is also considered a DSS because you can change routes by dragging different points and have what-if analysis performed on alternative routes (such as taking back roads instead of the highway), including estimates of driving time to help you decide which route is best. Google Maps has a user-friendly interface that helps you visualize the route, and after you make a decision, you can print driving directions and a map.

12-5a GIS Applications

GISs integrate and analyze spatial data from a variety of sources. Although they are used mainly in government and utility companies, more businesses are using them,

particularly in marketing, manufacturing, insurance, and real estate. No matter into what category a GIS falls, most applications require a GIS to handle converting data to information, integrating data with maps, and conducting different types of analysis. GIS applications can be classified in the following categories, among several others:

- *Education planning*—Analyzing demographic data toward changing school district boundaries or deciding where to build new schools.

- *Urban planning*—Tracking changes in ridership on mass-transit systems and analyzing demographic data to determine new voting districts, among many other uses.

- *Government*—Making the best use of personnel and equipment while dealing with tight budgets, dispatching personnel and equipment to crime and fire locations, and maintaining crime statistics.

- *Insurance*—Combining data on community boundaries, street addresses, and zip codes with search capabilities to find information (some from federal and state agencies) on potential hazards, such as natural disasters, auto-rating variables, and crime rate indexes.[25, 26]

- *Marketing*—Pinpointing areas with the greatest concentration of potential customers, displaying sales statistics in a geographic context, evaluating demographic and lifestyle data to identify new markets, targeting new products to specific populations, analyzing market share and population growth in relation to new store locations, and evaluating a company's market position based on geographic location.[27, 28] For example, Pepsico uses a GIS to find the best locations for new Pizza Hut and Taco Bell outlets.

- *Real estate*—Finding properties that meet buyers' preferences and price ranges, using a combination of census data, multiple listing files, mortgage information, and buyer profiles. GISs also help establish selling prices for homes by surveying an entire city to identify comparable neighborhoods and average sales prices. GISs can be used for appraisal purposes to determine relationships between national, regional, and local economic trends and the demand for local real estate.[29]

  - *Transportation and logistics*—Managing vehicle fleets, coding delivery addresses, creating street networks for predicting driving times, and developing maps for scheduling routing and deliveries.[30]

  The information box on the next page discusses how GISs can be used to monitor and reduce the spread of disease.

Exhibit 12.5
Locations of child care centers: an example of a pin map

GISS FOR FIGHTING DISEASE

Public health officials and government agencies around the globe use GISs, demographic information, remote-sensing data, and even Google Maps to help fight diseases such as avian flu, malaria, H1N1 (swine flu), SARS, and more. With a GIS, health officials can map the spread of epidemics and identify high-risk population areas before the epidemic reaches them, which can help decrease the death rate and reduce the spread of infectious diseases by tracking their origins. GISs can also be useful in the following tasks[31,32]:

- Locating contaminated water sources for waterborne diseases
- Plotting confirmed and suspected cases of a disease on a map
- Identifying existing medical infrastructures
- Determining how far people have to travel to reach a healthcare center and whether public transportation is available
- Monitoring virus mutations and their locations during an outbreak and assisting in early identification of infected people or animals
- Determining the geographic distribution of diseases
- Planning and targeting interventions
- Routing healthcare workers, equipment, and supplies to remote locations
- Locating the nearest healthcare facility

12-6 GUIDELINES FOR DESIGNING A MANAGEMENT SUPPORT SYSTEM

Before designing any management support system, the system's objectives should be defined clearly, and then the system development methods discussed in Chapter 10 can be followed. Because MSSs have a somewhat different purpose than other information systems, the important factors in designing one are summarized in the following list:

- *Support from the top*—Without a full commitment from top management, the system's chances of success are low.

- *Objectives and benefits clearly defined*—Costs are always in dollars, but benefits are qualitative. The design team should spend time identifying all costs and benefits in order to present a convincing case to top management. When benefits are intangible, such as improving customer service, the design team should associate the benefit with a measurable factor, such as increased sales.

- *Identifying executives' information needs*—Examine the decision-making process that executives use to find out what kinds of decisions they are making—structured, semistructured, or unstructured—and what kind of information they need to make these decisions.

- *Keeping lines of communication open*—This is important to ensure that key decision makers are involved in designing the MSS.

- *System's complexity hidden, interface kept simple*—Avoid using technical jargon when explaining the system to executives because they might lose interest if they think the system is too technical. Executives are not interested in the choice of platform or software, for example. Their main concern is getting the information they need in the simplest way possible. In addition, the system must be easy for executives to learn, with little or no training. To most executives, the interface is the system, so its ease of use is a crucial factor in the system's success.

- *Keeping the "look and feel" consistent*—Designers should use standard layouts, formats, and colors in windows, menus, and dialog boxes for consistency and ease of use. That way, a user who has learned the database portion of the system, for example, should be able to switch to the report-generating portion with little trouble because the interface is familiar. You can see this in Microsoft Office, which uses similar features, such as formatting toolbars, in all its applications. Users accustomed to Word, for example, can learn how to use Excel quickly because the interface is familiar.

- *Designing a flexible system*—Almost all aspects of an MSS, including the user interface, change over time because of rapid developments in technology and the dynamic business environment. A flexible system can incorporate changes quickly.

- *Making sure response time is fast*—MSS designers must monitor the system's response time at regular intervals, as executives rarely tolerate slow response times. In addition, when a system function takes more than a few seconds, make sure a message is displayed stating that the system is processing the request. Using a progress bar can help reduce frustration, too.

The Industry Connection highlights the software and services available from SAS Corporation, one of the leaders in decision support systems.

Industry Connection: SAS, Inc.[33]

SAS was founded in the early 1970s to analyze agricultural research data. Eventually, it developed into a vendor of software for conducting comprehensive analysis and generating business intelligence information for a variety of decision-making needs. Products and services offered by SAS include the following:

SAS Business Intelligence: Analyzes past data to predict the future, including features for reporting, queries, analysis, OLAP, and integrated analytics.

Data Integration: Gives organizations a flexible, reliable way to respond quickly to data integration requirements at a reduced cost.

SAS Analytics: Provides an integrated environment for modeling analyses, including data mining, text mining, forecasting, optimization, and simulation; provides several techniques and processes for collecting, classifying, analyzing, and interpreting data to reveal patterns and relationships that can help with decision making.

STUDY TOOLS 12

LOCATED AT BACK OF THE TEXTBOOK
- ☐ Rip out Chapter Review Card

LOCATED AT WWW.CENGAGE.COM/LOGIN
- ☐ Review Key Terms Flash Cards (Print or Online)
- ☐ Download audio and visual summaries to review

- ☐ Complete both Practice Quizzes to prepare for tests
- ☐ Play "Beat the Clock" and "Quizbowl" to master concepts
- ☐ Complete "Crossword Puzzle" to review key terms
- ☐ Watch video on "Method" for a real company example

KEY TERMS

choice phase 245
decision support system (DSS) 246
design phase 245

digital dashboard 250
electronic meeting systems 255
executive information systems (EISs) 250

REVIEWS AND DISCUSSIONS

1. What are three types of decisions in a typical business organization?

2. What are the four phases of the decision-making process?

3. What are the three main components of a DSS?

4. What are two tangible and two intangible benefits of a DSS?

5. What are three examples of tasks or applications that executives use EISs for?

6. What are two examples of commercial EMSs available on the market?

7. What are three business applications of GISs?

8. What are three recommendations for designing a successful MSS?

PROJECTS

1. After reading the information presented in this chapter and other sources, write a two-page paper that compares and contrasts Microsoft Live Meeting and IBM FileNet as two leading e-collaboration software applications. Which one would you recommend to a company with over 200 employees scattered through four offices in California? What are the bases of your recommendation?

2. After reading the information presented in this chapter and other sources, write a two-page paper that compares and contrasts Microsoft Office SharePoint Server and IBM Domino. Which one would you recommend to the company mentioned in Project 1? What are the bases of your recommendation?

3. After reading the information presented in this chapter and other sources, write a one-page paper that describes five key applications of GISs in the insurance industry. How can these applications help the bottom line?

4. Esri (*www.esri.com/products*) is one of the major providers of GIS software. What are two examples of Location Analytics software? What types of businesses will benefit the most from this type of GIS? What are some of the features of Esri Maps for Office?

5. After reading the information presented in this chapter and other sources, write a one-page paper that identifies five applications of MSSs in the retail industry. How can these applications help the bottom line?

6. SAS Business Intelligence (*www.sas.com/technologies/bi*) and Datawatch (*www.datawatch.com*) are among the popular packages used to design an EIS. After reading the information presented in this chapter and other sources, write a two-page paper that identifies four key features of each package. What types of businesses will benefit the most from these packages?

ARE YOU READY TO MOVE ON?

1. Structured decisions, or programmable tasks, can be automated because a well-defined standard operating procedure exists for these types of decisions. True or False?

2. A DSS includes three major components: a database, a model base, and a user interface. True or False?

3. A geographic information system (GIS) can only manipulate spatial data. True or False?

4. Which of the following is not among the phases of decision making?
   a. Intelligence
   b. Collaboration
   c. Choice
   d. Design

5. In general, which of the following tasks are EISs not used for?

   a. Flagging exceptions

   b. Tracking performance

   c. Spotting trends

   d. Eliminating decision making

6. Which of the following capabilities is not typically included in groupware?

   a. Audio and video conferencing

   b. Brainstorming

   c. Optimization modeling

   d. Database access

CASE STUDY 12-1

UPS Deploys Routing Optimization with a Big Payoff

The UPS ORION (On-Road Integrated Optimization and Navigation) system reduces delivery miles and fuel and sets the stage for enhanced customer service. The system not only reduces the overall delivery cost, it is also a great move toward sustainability. At the end of 2013, according to UPS, 10,000 delivery routes were optimized by ORION and 1.5 million gallons of fuel were saved. As a result, the emission of carbon dioxide was reduced by 14,000 metric tons. ORION combines more than 250 million address data points, online map data, and an optimization algorithm to find the best routes for pickups and deliveries.

The data for the system has been collected since 2008 by using GPS systems and telematics technologies installed in delivery trucks. A prototype of the ORION system was deployed in 11 sites between 2008 and 2011.

In order to challenge the system and further improve the algorithm, UPS invited its most-experienced drivers to try to "beat the system." Tim Ahn, a UPS driver says, "I get options that I would have actually never thought of before. It's a new way of thinking to make me more efficient." UPS is planning to complete the deployment of this technology on all of its 55,000 U.S. routes by the end of 2017, and global deployments will take place after that.[34]

Answer the following questions:

1. What does ORION stand for?
2. What are three advantages of using ORION at UPS?
3. What are three main components of ORION?
4. By which time will ORION be deployed in all of the UPS routes in the United States?

CASE STUDY 12-2

Mobile Medical Collaboration Tools Gaining in Popularity

Spending on electronic health record (EHR) systems and mobile health applications is on the rise. According to a study by the market research firm RNCOS, more than 72 percent of U.S. physicians now use smartphones for their work, and over 500 million people will be using mobile health apps by 2015.[35,36] There are more than 17,000 medical apps to choose from—over 6,000 available through the iTunes store alone. Mobile health apps allow doctors to monitor patients in real time, regardless of where they both are. The patients create and transmit their own

data using portable ECG machines, blood pressure monitors, and weight scales. These tools are relatively easy to use and quite portable. In fact, ECGs weigh just 3.5 ounces. As they continue to improve the efficiency and effectiveness of medical delivery, mobile apps are creating a true collaborative environment between doctors and patients.

Answer the following questions:

1. What percentage of U.S. doctors use smartphones for their work?
2. What can patients do with medical apps?
3. How many mobile medical apps are currently available?

13 | Intelligent Information Systems

LEARNING OUTCOMES

After studying this chapter, you should be able to:

13-1 Define artificial intelligence, and explain how AI technologies support decision making.

13-2 Describe an expert system, its applications, and its components.

13-3 Describe case-based reasoning.

13-4 Summarize the types of intelligent agents and how they are used.

13-5 Describe fuzzy logic and its uses.

13-6 Explain artificial neural networks.

13-7 Describe how genetic algorithms are used.

13-8 Explain natural-language processing and its advantages and disadvantages.

13-9 Summarize the advantages of integrating AI technologies into decision support systems.

13-10 Explain contextual computing.

After you finish this chapter, go to **PAGE 278** for the **STUDY TOOLS**

This chapter covers the use of intelligent information systems, beginning with artificial intelligence (AI). It discusses the ways AI technologies are used in decision making, with an overview of robotics as one of the earliest AI applications. Next, the chapter discusses expert systems—their components and the ways these systems are used. Case-based reasoning and intelligent agents are also discussed as applications of AI. Next, the chapter discusses fuzzy logic, artificial neural networks, genetic algorithms, and natural-language processing systems as well as the advantages of integrating AI technologies into decision support systems. Finally, it provides an overview of contextual computing.

13-1 WHAT IS ARTIFICIAL INTELLIGENCE?

Artificial intelligence (AI) consists of related technologies that try to simulate and reproduce human thought behavior, including thinking, speaking, feeling, and reasoning. AI technologies apply computers to areas that require knowledge, perception, reasoning, understanding, and cognitive abilities.[1] To achieve these capabilities, computers must be able to do the following:

- Understand common sense (see the information box on the next page).
- Understand facts and manipulate qualitative data.
- Deal with exceptions and discontinuity.
- Understand relationships between facts.

- Interact with humans in their own language.
- Be able to deal with new situations based on previous learning.

Information systems are concerned with capturing, storing, retrieving, and working with data, but AI technologies are concerned with generating and displaying

> **Artificial intelligence (AI)** consists of related technologies that try to simulate and reproduce human thought behavior, including thinking, speaking, feeling, and reasoning. AI technologies apply computers to areas that require knowledge, perception, reasoning, understanding, and cognitive abilities.

COMPUTERS UNDERSTANDING COMMON SENSE

Humans take common sense for granted. For example, we know that fish are found in the ocean, trees and bushes are found in the woods, and people eat food. And we know that when you drop a ball, it will fall. To computers, understanding such simple things is a challenge, but this is changing.

At Carnegie Mellon University, a computer called Never Ending Image Learner runs 24 hours a day and goes through the Web finding and analyzing images in order to build a visual database that can recognize associations among these images. There are billions of images of various types on the Web; Facebook alone contains more than 200 billion images.[2]

Since it was begun in July 2013, Never Ending Image Learner has analyzed more than three million images and identified 1,500 types of objects in half a million images and 1,200 types of scenes. As a result, the computer has learned 2,500 associations by connecting the dots. For example, it knows that buildings are vertical instead of lying on their sides and that most automobiles have four wheels. As the number of analyzed images grows, the computer will learn more and will increase the number of associations. The interesting aspect of this project is that researchers are enabling a computer to teach itself common sense without any human intervention.[3]

knowledge and facts. In the information systems field, as you have learned, programmers and systems analysts design systems that help decision makers by providing timely, relevant, accurate, and integrated information. In the AI field, knowledge engineers try to discover "rules of thumb" that enable computers to perform tasks usually handled by humans. Rules used in the AI field come from a diverse group of experts in areas such as mathematics, psychology, economics, anthropology, medicine, engineering, and physics. AI encompasses several related technologies discussed in this chapter, including robotics, expert systems, fuzzy logic systems, intelligent agents, artificial neural networks, genetic algorithms, and natural-language processing.

intelligence and human intelligence. The following section discusses possibilities for using AI technologies in decision-making processes.

13-1a AI Technologies Supporting Decision Making

As you know, information technologies are used to support many phases of decision making. The most recent developments in AI technologies promise new areas of decision-making support. Table 13.1 lists some applications of AI-related technologies in various organizations.[4,5,6]

Information systems are concerned with capturing, storing, retrieving, and working with data, but AI technologies are concerned with generating and displaying knowledge and facts.

Although these applications and technologies may not offer true human intelligence, they are certainly more intelligent than traditional information systems. ~~er~~ the years, the capabilities of these systems have ~~...~~ d in an attempt to close the gap between artificial

Decision makers use information technologies in the following types of decision-making analyses[7]:

- *What-is*—This analysis is commonly used in transaction-processing systems and management information systems. For example, if you enter a

TABLE 13.1 APPLICATIONS OF AI TECHNOLOGIES

Field	Organization	Applications
Energy	Arco and Tenneco Oil Company	Neural networks used to help pinpoint oil and gas deposits
Government	Internal Revenue Service	Software used to read tax returns and spot fraud
Human services	Merced County, California	Expert systems used to decide if applicants should receive welfare benefits
Marketing	Spiegel	Neural networks used to determine most likely buyers from a long list
Telecommunications	BT Group	Heuristic search used for a scheduling application that provides work schedules for more than 20,000 engineers
Transportation	American Airlines	Expert systems used to schedule the routine maintenance of airplanes
Inventory/forecasting	Hyundai Motors	Neural networks and expert systems used to reduce delivery time by 20 percent and increase inventory turnover from 3 to 3.4
Inventory/forecasting	SCI Systems	Neural networks and expert systems used to reduce on-hand inventory by 15 percent, resulting in $180 million in annual savings
Inventory/forecasting	Reynolds Aluminum	Neural networks and expert systems used to reduce forecasting errors by 2 percent, resulting in an inventory reduction of 1 million pounds
Inventory/forecasting	Unilever	Neural networks and expert systems used to reduce forecasting errors from 40 percent to 25 percent, resulting in a multimillion-dollar savings

© Cengage Learning®

customer account number, the system displays the customer's current balance. However, these systems lack the capability to report real-time information or predict what could happen in the future. For example, reports generated by accounting information systems that show performance over the preceding fiscal quarter consist of past events, so decision makers cannot do much with this information.

- *What-if*—This analysis is used in decision support systems. Decision makers use it to monitor the effect of a change in one or more variables. It is available in spreadsheet programs, such as Microsoft Excel.

In addition to these types of analysis, decision makers often need to answer the following questions about information: Why? What does it mean? What should be done? When should it be done? AI technologies have the potential to help decision makers address these questions.

13-1b Robotics

Robots are one of the most successful applications of AI. You are probably familiar with robots used in factories or ones you have seen on the news. They are far from being intelligent, but progress has been steady. They perform well at simple, repetitive tasks and can be used to free workers from tedious or hazardous jobs. Robots are currently used mainly on assembly lines in Japan and the United States as part of computer-integrated manufacturing, but they are also used in the military, aerospace, and medical industries as well as for performing such services as delivering mail to employees.

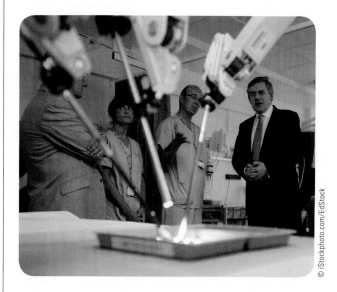

© iStockphoto.com/EdStock

The cost of industrial robots ranges from $100,000 to $250,000 or more. Typically, their mobility is limited. For example, they might have only a fixed arm that moves objects from one point to another. Some robots have limited vision that is useful for locating and picking up objects, as long as the objects are isolated from other objects. A robot's operation is controlled by a computer program that includes commands such as when and how far to reach, which direction to go or turn, when

Robots are one of the most successful applications of AI. They perform well at simple, repetitive tasks and can be used to free workers from tedious or hazardous jobs.

to grasp an object, and how much pressure to apply. Programming languages for controlling robots include Variable Assembly Language (VAL), Functional Robotics (FROB), and A Manufacturing Language (AML). These languages are usually proprietary, meaning they are specific to individual robot manufacturers.

One of the most advanced and most popular robots is Honda's Advanced Step in Innovative Mobility (ASIMO) (*http://world.honda.com/ASIMO*). Honda's intelligence technologies enable ASIMO to coordinate with other robots. It recognizes moving objects, sound, gestures, multiple environments, faces, and postures. ASIMO is also able to choose between stepping back and yielding the right-of-way or continuing to move forward based on the predicted movement of oncoming people. It is able to automatically recharge its battery when the charge falls below a certain level.

Personal robots have attracted a lot of attention recently. These robots have limited mobility, limited vision, and some speech capabilities. Currently, they are used mostly as prototypes to test such services as helping the elderly, bringing breakfast to the table, cooking, opening doors, and carrying trays and drinks. Examples include TWENDY-ONE, Motoman, ApriAttenda Robot, and PR2. One of the most successful and advanced personal robots on the market today, PR2 performs many ordinary tasks around the home and office.

Robots offer the following advantages over humans in the workplace:

- They do not fall in love with coworkers, get insulted, or call in sick.
- They are consistent.

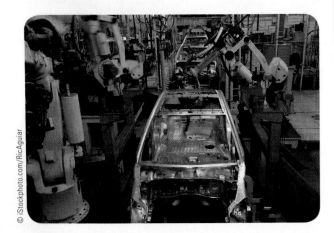
© iStockphoto.com/RicAguiar

- They can be used in environments that are hazardous to humans, such as working with radioactive materials.
- They do not spy for competitors, ask for a raise, or lobby for longer breaks.

Developments in AI-related fields, such as expert systems and natural-language processing, will affect the future development of the robotics industry. For example, natural-language processing will make it easier to communicate with robots in human languages.

13-2 EXPERT SYSTEMS

Expert systems have been one of the most successful AI-related technologies and have been around since the 1960s. They mimic human expertise in a particular field to solve a problem in a well-defined area. For the purposes of this book, an expert system consists of programs that mimic human thought behavior in a specific area that human experts have solved successfully. The first expert system, called DENDRAL, was developed in the mid-1960s at Stanford University to determine the chemical structure of molecules. For expert systems to be successful, they must be applied to an activity that human experts have already handled, such as tasks in medicine, geology, education, and oil exploration. PortBlue (PortBlue Corporation) is an example of an expert system that can be applied to various financial applications, including examination of complex financial structures, foreign exchange risk management, and more. COGITO (by Italian-based Expert System) is used for monitoring what consumers are saying in blogs, comment sections, message boards, and Web-based articles. It is also used in search engines to better understand users' queries.[8]

Decision support systems generate information by using data, models, and well-defined algorithms, but expert systems work with heuristic data. Heuristics consist of common sense, rules of thumb, educated guesses, and instinctive judgments, and using heuristic data encourages applying knowledge based on experience to solve or describe a problem. In other words, heuristic data is not formal knowledge, but it helps in finding a solution to a problem without following a rigorous algorithm.

13-2a Components of an Expert System

A typical expert system includes the components described in the following list, which are shown in Exhibit 13.1:

Exhibit 13.1
An expert system configuration

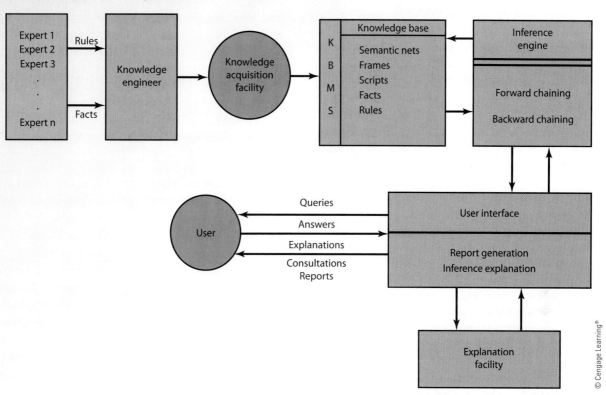

© Cengage Learning®

- *Knowledge acquisition facility*—A **knowledge acquisition facility** is a software package with manual or automated methods for acquiring and incorporating new rules and facts so the expert system is capable of growth. This component works with the KBMS (described later in this list) to ensure that the knowledge base is as up to date as possible.

- *Knowledge base*—A **knowledge base** is similar to a database, but in addition to storing facts and figures it keeps track of rules and explanations associated with facts. For example, a financial expert system's knowledge base might keep track of all figures constituting current assets, including cash, deposits, and accounts receivable. It might also keep track of the fact that current assets can be converted to cash within 1 year. An expert system in an academic environment might include facts about all graduate students, such as GMAT scores and GPAs, as well as a rule specifying that classified graduate students must have a GMAT of 650 or better and a GPA of 3.4 or better. To be considered part of a true expert system, the knowledge base component must include the following types of knowledge:

  - *Factual knowledge*—Facts related to a specific discipline, subject, or problem. For example,

facts related to kidney problems might include kidney size, blood levels of certain enzymes, and duration and location of pain.

- *Heuristic knowledge*—Rules related to a problem or discipline. For example, the general rules indicating that a patient has a kidney problem could include severe pain in the lower left or lower right of the back and high levels of creatinine and blood urea nitrogen.

- *Meta-knowledge*—Meta-knowledge is knowledge about knowledge. It enables an expert system to learn from experience and examine and extract relevant facts to determine the path to a solution. It also guides future planning or execution phases of an expert system. For example,

A **knowledge acquisition facility** is a software package with manual or automated methods for acquiring and incorporating new rules and facts so the expert system is capable of growth.

A **knowledge base** is similar to a database, but in addition to storing facts and figures it keeps track of rules and explanations associated with facts.

knowing how an expert system makes decisions is considered meta-knowledge. Although this type of knowledge is not currently available in expert systems, integrating neural networks into expert systems is one possibility for acquiring meta-knowledge.

- *Knowledge base management system*—A **knowledge base management system (KBMS)**, similar to a DBMS, is used to keep the knowledge base updated, with changes to facts, figures, and rules.

- *User interface*—This is the same as the user interface component of a decision support system. It provides user-friendly access to the expert system. Although GUIs have improved this component, one goal of AI technology is to provide a natural language (discussed later in the chapter) for the user interface.

- *Explanation facility*—An **explanation facility** performs tasks similar to what a human expert does by explaining to end users how recommendations are derived. For example, in a loan evaluation expert system, the explanation facility states why an applicant was approved or rejected. In a medical expert system, it explains why the system concluded that a patient has a kidney stone, for instance. This component is important because it helps give users confidence in the system's results.

- *Inference engine*—An **inference engine** is similar to the model base component of a decision support system (discussed in Chapter 12). By using different techniques, such as forward and backward chaining (discussed in the following paragraphs), an inference engine manipulates a series of rules. Some inference engines work from a matrix of facts that includes several rows of conditions and rules, similar to a decision table. In this case, rules are evaluated one at a time, then advice is provided. Some inference engines also learn from doing.

In **forward chaining**, a series of "if-then-else" condition pairs is performed. The "if" condition is evaluated first, then the corresponding "then-else" action is carried out. For example, "if" the temperature is less than 80°F and the grass is 3 inches long, "then" cut the grass or "else" wait. In a medical diagnostic expert system, the system could evaluate a problem as follows:

- If the patient's temperature is over 101°F and
- If the patient has a headache
- Then it's very likely (a 95 percent chance) that the patient has the flu, or else search for other diseases.

In **backward chaining**, the expert system starts with the goal—the "then" part—and backtracks to find the right solution. In other words, to achieve this goal, what conditions must be met? To understand the differences between these two techniques, consider the following example. In an expert system that provides financial investment advice for investors, the system might use forward chaining and ask 50 questions to determine which of five investment categories—oil-gas, bonds, common stocks, public utilities, and transportation—is more suitable for an investor.[9] In addition, each investor is in a specific tax bracket, and each investment solution provides a different tax shelter. In forward chaining, the system evaluates all the "if-then-else" conditions before making the final recommendation. In backward chaining, the system might start with the public utilities category, specified by the investor, and go through all the "if" conditions to see whether this investor qualifies for this investment category. The backward chaining technique can be faster in some situations because it does not have to consider irrelevant rules, but the solution the system recommends might not be the best one.

Other techniques are used for representing knowledge in the expert system's knowledge base, such as semantic (associative) networks that represent information as links and nodes, frames that store conditions or objects in hierarchical order, and scripts that describe a sequence of events. For a child's birthday party, for example, events might include buying a cake, inviting friends, lighting the candles, and serving the cake. A script for generating a purchase order might include events such as identifying the quantity to order, identifying the supplier and gathering data, generating the purchase order and sending it to the supplier, updating accounts payable, and informing the receiving department that an order has been placed.

A **knowledge base management system (KBMS)**, similar to a DBMS, is used to keep the knowledge base updated, with changes to facts, figures, and rules.

An **explanation facility** performs tasks similar to what a human expert does by explaining to end users how recommendations are derived.

An **inference engine** is similar to the model base component of a decision support system. By using different techniques, such as forward and backward chaining, it manipulates a series of rules.

In **forward chaining**, a series of "if-then-else" condition pairs is performed.

In **backward chaining**, the expert system starts with the goal—the "then" part—and backtracks to find the right solution.

13-2b Uses of Expert Systems

Many companies are engaged in research and development of expert systems, and these systems are now used in areas such as the following:

- *Airline industry*—American Airlines developed an expert system to manage frequent flier transactions.

- *Forensics lab work*—Expert systems are used to review DNA samples from crime scenes and generate results quickly and accurately, helping reduce the backlog in labs and get data entered in national crime databases faster.[10]

- *Banking and finance*—JPMorgan Chase developed a foreign currency trade expert system to assess historical trends, new events, and buying and selling factors.

- *Education*—Arizona State University developed an expert system to teach math and evaluate students' math skills.

- *Food industry*—Campbell's Soup Company developed an expert system to capture expertise that a highly specialized, long-time employee had about plant operations and sterilizing techniques.

- *Personnel management*—IBM developed an expert system to assist in training technicians; it has reduced training time.

- *Security*—Canada Trust Bank (now part of TD Canada Trust) developed an expert system to track credit card holders' purchasing trends and report deviations, such as unusual activity on a card.

- *U.S. government*—Expert systems have been developed to monitor nuclear power plants and assist departments such as the IRS, INS, U.S. Postal Service, Department of Transportation, Department of Energy, and Department of Defense in decision-making processes.

- *Agriculture*—The National Institute of Agricultural Extension Management has designed an expert system to diagnose pests and diseases in rice crops and suggest preventive measures.[11]

The information box above highlights a real-life application of an expert system, this one in burglary and crime detection. The system reduced the time and expenses involved in police operations.

13-2c Criteria for Using Expert Systems

An expert system should be used if one or more of the following conditions exists:

- A lot of human expertise is needed but a single expert cannot tackle the problem on his or her own. (An expert system can integrate the experience and knowledge of several experts more easily.)

Expert Systems in Baltimore County Police Department

In Baltimore County, an expert system was developed so detectives could analyze information about burglary sites and identify possible suspects. Data entered in the system included information about known burglars, records of 300 solved burglary cases, and records of 3,000 unsolved cases. Then, 18 detectives were interviewed to gather their knowledge about local burglaries. Detectives could enter statements about burglaries, such as neighborhood characteristics, the type of property stolen, and the type of entry used; they could also get information on possible suspects. The system is now used in other police departments in the United States.[12]

- The knowledge that is needed can be represented as rules or heuristics; a well-defined algorithm is not available.

- The decision or task has already been handled successfully by human experts, allowing the expert system to mimic human expertise.

- The decision or task requires consistency and standardization. (Because computers are more accurate at following standard procedures, an expert system can be preferable to humans in this situation.)

- The subject domain is limited. (Expert systems work better if the problem under investigation is narrow.)

- The decision or task involves many rules (typically between 100 and 10,000) and complex logic.

- There is a scarcity of experts in the organization, or key experts are retiring. (An expert system can be used to capture the knowledge and expertise of a long-time employee who is retiring.)

13-2d Criteria for Not Using Expert Systems

The following situations are unsuitable to expert systems:

- There are very few rules (less than 10) involved. Human experts are more effective at solving these problems.

- There are too many rules (usually more than 10,000) involved. Processing slows to unacceptable levels.

- There are well-structured numerical problems (such as payroll processing) involved, which means that standard transaction-processing methods can handle the situation more quickly and economically.

- A broad range of topics is involved, but there are not many rules. Expert systems work better when there are deep and narrow problem areas.
- There is a lot of disagreement among experts.
- The problems require human experts—for example, a combination of the five senses, such as taste and smell. Selecting a perfume is a problem better solved by human experts.

13-2e Advantages of Expert Systems

An expert system can have the following advantages over humans:

- It never becomes distracted, forgetful, or tired. Therefore, it is particularly suitable for monotonous tasks that human workers might object to.
- It duplicates and preserves the expertise of scarce experts and can incorporate the expertise of many experts.
- It preserves the expertise of employees who are retiring or leaving an organization.
- It creates consistency in decision making and improves the decision-making skills of nonexperts.

13-3 CASE-BASED REASONING

Expert systems solve a problem by going through a series of if-then-else rules, but **case-based reasoning (CBR)** is a problem-solving technique that matches a new case (problem) with a previously solved case and its solution, both stored in a database. Each case in the database is stored with a description and keywords that identify it. If there is no exact match between the new case and cases stored in the database, the system can query the user for clarification or more information. After finding a match, the CBR system offers a solution; if no match is found, even after supplying more information, the human expert must solve the problem. The new case and its solution are then added to the database.

Hewlett-Packard uses CBR to assist users of its printers; this system performs the role of a help desk operator. Users' complaints and difficulties over the past several years have been stored in a database as cases and solutions. This information is used in dealing with new users, who likely have the same problems as users in the past. In the long term, these systems can improve customer service and save money by reducing the number of help desk employees.

As another example, some banks use a CBR system to qualify customers for loans, using parameters from past customers stored in a database. These parameters include gross income, number of dependents, total assets, net worth, and amount requested for the loan. The database also stores the final decision on each application (acceptance or rejection). When a new customer applies for a loan, the CBR system can compare the application with past applications and provide a response. The new application and its outcome then become part of the database for future use.

13-4 INTELLIGENT AGENTS

Intelligent agents, also known as bots (short for robots), are software capable of reasoning and following rule-based processes; they are becoming more popular, especially in e-commerce. They are also called virtual agents (VAs) or intelligent virtual agents (IVAs). A sophisticated intelligent agent has the following characteristics[13]:

- *Adaptability*—Able to learn from previous knowledge and go beyond information given previously. In other words, the system can make adjustments.
- *Autonomy*—Able to operate with minimum input. The system can respond to environmental stimuli, make a decision without users telling it to do so, and take preemptive action, if needed.
- *Collaborative behavior*—Able to work and cooperate with other agents to achieve a common objective.
- *Humanlike interface*—Able to interact with users in a more natural language.
- *Mobility*—Able to migrate from one platform to another with a minimum of human intervention.
- *Reactivity*—Able to select problems or situations that need attention and act on them. An agent with this capability typically responds to environmental stimuli.

Most intelligent agents today fall short of these capabilities, but improvement is expected in the near

Case-based reasoning (CBR) is a problem-solving technique that matches a new case (problem) with a previously solved case and its solution, both stored in a database. After searching for a match, the CBR system offers a solution; if no match is found, even after supplying more information, the human expert must solve the problem.

Intelligent agents are software capable of reasoning and following rule-based processes; they are becoming more popular, especially in e-commerce.

future. One important application of intelligent agents that is already available is Web marketing. Intelligent agents can collect information about customers, such as items purchased, demographic information, and expressed and implied preferences. E-commerce sites then use this information to better market their products and services to customers. Other agents, called product-brokering agents, can alert customers to new products and services. Amazon has used these agents successfully.

reminders, or personal surfing assistants. Pricewatch (*www.pricewatch.com*) is a commercial shopping agent that finds the lowest price for many items and displays all competitive prices. Another example is BestBookBuys (*www.bestbookbuys.com*), which asks you to identify a book by its title, author, or ISBN, then finds all the online booksellers that carry this book and organizes them into a list from least expensive to most expensive. Another comparison-shopping agent is available at *www.mysimon.com*.

> Intelligent agents can collect information about customers, such as items purchased, demographic information, and expressed and implied preferences.

Intelligent agents are also used for smart or interactive catalogs, called "virtual catalogs." A virtual catalog displays product descriptions based on customers' previous experiences and preferences. Currently, Stanford University's Center for Information Technology (CIT) is working with IBM, Hewlett-Packard, and National Semiconductor to develop smart catalogs. The CIT project's goals include dynamic creation and updates for catalogs, the ability to search them by content instead of navigating via links, and cross-referencing so users can find items meeting their needs in multiple catalogs.

Intelligent agents that are currently available fall into these categories:

- Shopping and information agents
- Personal agents
- Data-mining agents
- Monitoring and surveillance agents

These are discussed in the following sections.

13-4a Shopping and Information Agents

Shopping and information agents help users navigate through the vast resources available on the Web and provide better results in finding information. These agents can navigate the Web much faster than humans and gather more consistent, detailed information. They can serve as search engines, site

Usenet and newsgroup agents have sorting and filtering features for finding information. For example, DogPile (*www.dogpile.com*) searches the Web by using several search engines, including Google, Yahoo!, and Yandex, to find information for users. DogPile can remove duplicate results and analyze the results to sort them with the most relevant results at the top.

Shopping and information agents help users navigate through the vast resources available on the Web and provide better results in finding information. These agents can navigate the Web much faster than humans and gather more consistent, detailed information. They can serve as search engines, site reminders, or personal surfing assistants.

13-4b Personal Agents

Personal agents perform specific tasks for a user, such as remembering information for filling out Web forms or completing e-mail addresses after the first few characters are typed. An e-mail personal agent can usually perform the following tasks:

- Generate auto-response messages.
- Forward incoming messages.
- Create e-mail replies based on the content of incoming messages.

13-4c Data-Mining Agents

Data-mining agents work with a data warehouse, detecting trends and discovering new information and relationships among data items that were not readily apparent. Volkswagen Group uses a data-mining agent that acts as an early-warning system about market conditions. For

© iStockphoto.com/jbk_photography

example, the data-mining agent might detect a problem that could cause economic conditions to worsen, resulting in delayed payments from customers. Having this information early enables decision makers to come up with a solution that minimizes the negative effects of the problem.

13-4d Monitoring and Surveillance Agents

Monitoring and surveillance agents usually track and report on computer equipment and network systems to predict when a system crash or failure might occur. NASA's Jet Propulsion Laboratory has an agent

---

Personal agents perform specific tasks for a user, such as remembering information for filling out Web forms or completing e-mail addresses after the first few characters are typed.

Data-mining agents work with a data warehouse, detecting trends and discovering new information and relationships among data items that were not readily apparent.

Monitoring and surveillance agents usually track and report on computer equipment and network systems to predict when a system crash or failure might occur.

INTELLIGENT AGENTS IN ACTION

Intelligent agents cost approximately 2 percent of what a live human assistant costs.
   Here are some examples of intelligent agents in real-life practice:

- SFR, which is a division of Vodafone, a mobile communications company, uses a virtual agent to facilitate 750,000 conversations a month. The intelligent agent answers customers' questions about their accounts and about the company's services and offerings.[15]
- The French division of eBay uses an intelligent agent named Louise to assist in over 200,000 customer conversations a day throughout six countries. According to eBay, Louise performs its assigned tasks with an 88 percent problem-solving rate. Currently, it handles 30 percent of the client contact for eBay.fr.[16]
- Apple uses an intelligent agent named Siri on its iPhone, iPad, and iPod Touch devices. Siri allows the user's voice to send messages, make calls, set reminders, and much more.[17]
- IBM Watson is being marketed as an intelligent agent. For a fee, it can be used at call centers, used in the medical field, in insurance companies, and much more.[18,19]
- IntelliResponse offers an intelligent agent that is used by banks, airlines, and telecommunications companies to answer customers' questions and provide other relevant information about products and services directly from the company's Web site.[20]

that monitors inventory, planning, and the scheduling of equipment to keep costs down.[14]

The information box on the previous page highlights real-life applications of several commercial intelligent agents.

13-5 FUZZY LOGIC

Have you ever been given a questionnaire that asks ambiguous questions but expects yes or no responses? Although you might be tempted to use words such as *usually*, *sometimes*, *only if*, and the like, you know the software used to analyze responses simply cannot deal with anything but clear-cut yes and no responses. However, with the development of fuzzy logic, a wide variety of responses is possible in questionnaires and other survey tools. **Fuzzy logic** allows a smooth, gradual transition between human and computer vocabularies and deals with variations in linguistic terms by using a degree of membership. A degree of membership shows how relevant an item or object is to a set. A higher number indicates it is more relevant, and a lower number shows it is less. For example, when heating water, as the temperature changes from 50°C to 75°C, you might say the water is warm. What about when the water's temperature reaches 85°C? You can describe it as warmer, but at what point do you describe the water as hot? Describing varying degrees of warmness and assigning them membership in certain categories of warmness involves a lot of vagueness.

Fuzzy logic is designed to help computers simulate vagueness and uncertainty in common situations. Lotfi A. Zadeh developed the fuzzy logic theory in the mid-1960s by using a mathematical method called "fuzzy sets" for handling imprecise or subjective information.[21] Fuzzy logic allows computers to reason in a fashion similar to humans and makes it possible to use approximations and vague data yet produce clear and definable answers.

Fuzzy logic works based on the degree of membership in a set (a collection of objects). For example, 4 feet, 5 feet, and 6 feet could constitute a set of heights for a population. Fuzzy sets have values between 0 and 1, indicating the degree to which an element has membership in the set. At 0, the element has no membership; at 1, it has full membership.

In a conventional set (sometimes called a "crisp" set), membership is defined in a black-or-white fashion; there's no room for gray. For instance, if 90 percent or higher

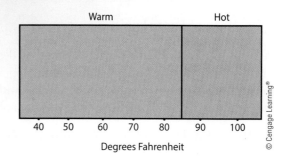

Exhibit 13.2
Example of a conventional set

means a "Pass" grade in a course, getting 89.99 percent does not give you membership in the "Pass" area of this crisp set. Therefore, despite getting 89.99 percent, you have failed the course. In this example, there is a very small difference between the two grades (0.01), but it means the difference between passing and failing. In other words, a small difference has a huge impact. This does not happen in a fuzzy logic environment. To help you understand the membership function better, Exhibit 13.2 shows an example of a conventional set. In this example, 84.9°F is considered warm and 85.1°F is considered hot. This small change in temperature can cause a large response in the system.

Exhibit 13.3 shows the same set but with fuzzy logic conventions. For example, 80°F has a membership of 30 percent in the fuzzy set "Warm" and 40 percent in

> **Fuzzy logic** allows a smooth, gradual transition between human and computer vocabularies and deals with variations in linguistic terms by using a degree of membership.

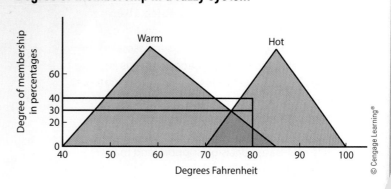

Exhibit 13.3
Degree of membership in a fuzzy system

the fuzzy set "Hot." All temperatures from 40° to 100°F make up the membership set.

13-5a Uses of Fuzzy Logic

Fuzzy logic has been used in search engines, chip design, database management systems, software development, and other areas.[22] You might be more familiar with its uses in appliances, as shown in the following examples:

- Dryers that convert information on load size, fabric type, and flow of hot air into drying times and conditions

- Refrigerators that set defrosting and cooling times based on usage patterns

- Shower systems that suppress variations in water temperature

- TVs that adjust screen color and texture for each frame and stabilize the volume based on the viewer's location in the room

- Video camcorders that eliminate shakiness in images (common with handheld video cameras) and adjust focus and lighting automatically[23, 24]

The information box below highlights how the Department of Defense uses fuzzy logic.

> Fuzzy logic has been used in search engines, chip design, database management systems, software development, and other areas.

FUZZY LOGIC IN ACTION

In addition to its increasing use in household appliances, fuzzy logic is being applied to multicriteria decision-making processes in which there is no exact data available. Today, most new cars let you know when it is time to change the oil based not on the traditional notion of doing it every 3,000 miles but on engine speed and the particular distance traveled. Likewise, the U.S. Department of Defense uses fuzzy logic to solve complex problems in such areas as assessing the cost and benefit of individual weapon systems and evaluating alternative weapons development strategies (e.g., spending more money on ground weapons than on air defense). Using qualitative as well as quantitative data, a panel of subject-matter experts combines ~nguistics, rule-based decision support, and maps.[25]

13-6 ARTIFICIAL NEURAL NETWORKS

Artificial neural networks (ANNs) are networks that learn and are capable of performing tasks that are difficult with conventional computers, such as playing chess, recognizing patterns in faces and objects, and filtering spam e-mail. Like expert systems, ANNs are used for poorly structured problems—when data is fuzzy and uncertainty is involved. Unlike an expert system, an ANN cannot supply an explanation for the solution it finds because an ANN uses patterns instead of the if-then-else rules that expert systems use.

An ANN creates a model based on input and output. For example, in a loan application problem, input data consists of income, assets, number of dependents, job history, and residential status. The output data is acceptance or rejection of the loan application. After processing many loan applications, an ANN can establish a pattern that determines whether an application should be approved or rejected.

Exhibit 13.4
Artificial neural network configuration

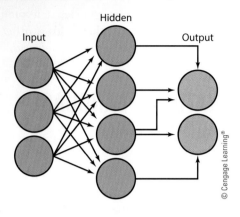

As shown in Exhibit 13.4, an ANN has an output layer, an input layer, and a middle (hidden) layer where learning takes place. If you are using an ANN for approving loans in a bank, the middle layer is trained by using past data (from old loan applications, in which the decisions are known) that includes both accepted and rejected applications. Based on the pattern of data entered in the input layer—applicant's information, loan amount, credit rating, and so on—and the results in the output layer (the accept or reject decision), nodes in the middle layer are assigned different weights. These weights determine how the nodes react to a new set of input data and mimic decisions based on what they have learned. Every ANN has to be trained, and when organizational policies change, the network needs to be retrained so it can mimic the new policies.

ANNs are used for many tasks, including the following:

- Bankruptcy prediction
- Credit rating
- Investment analysis
- Oil and gas exploration
- Target marketing

The information box below discusses real-life applications of neural networks.

NEURAL NETWORKS IN MICROSOFT AND THE CHICAGO POLICE DEPARTMENT

Microsoft is using neural network software to maximize the returns on direct mail. Each year, Microsoft sends out approximately 40 million pieces of direct mail to 8.5 million registered customers. The goal of these mailings is to encourage customers to upgrade their software or buy other Microsoft products. The first mailing goes out to all the customers in the company's database. The second mailing goes out only to those customers who are most likely to respond, and neural network software is used to cull the latter from the former. According to Microsoft spokesman Jim Minervino, the neural network software BrainMaker has increased the rate of response from 4.9 percent to 8.2 percent. This has resulted in a significant savings for the company—the same revenue at 35 percent less cost.[26]

The Chicago Police Department has used neural network software to predict which police officers are likely to engage in misconduct. Here, BrainMaker has compared the conduct of current officers with the conduct of those who have previously been terminated for disciplinary reasons, and this has produced a list of officers that might be "at risk."[27]

Several other real-life applications of neural networks are posted at *www.calsci.com/BrainIndex.html*.

13-7 GENETIC ALGORITHMS

Although they are not as widely accepted, **genetic algorithms (GAs)** have become more recognized as a form of artificial intelligence. They are used mostly to find solutions to optimization and search problems. John Holland developed genetic algorithms in the 1940s at MIT, and the term applies to adaptive procedures in a computer system that are based on Darwin's ideas about natural selection and survival of the fittest.[28]

Genetic algorithms are used for optimization problems that deal with many input variables, such as jet engine design, portfolio development, and network design. They find the combination of inputs that generates the most desirable outputs, such as the stock portfolio with the highest return or the network configuration with the lowest cost. Genetic algorithms can examine complex problems without any assumptions of what the correct solution should be. In a GA system, the following techniques are used:

- *Selection or survival of the fittest*—Gives preference or a higher weight to better outcomes.
- *Crossover*—Combines good portions of different outcomes to achieve a better outcome.
- *Mutation*—Tries combinations of different inputs randomly and evaluates the results.
- *Chromosome*—A set of parameters that defines a proposed solution to the problem the GA is trying to solve. It is usually represented as a simple string (sequence of characters).

Genetic algorithms are already used with neural networks and fuzzy logic systems to solve scheduling, engineering design, and marketing problems, among others. For example, a docking algorithm uses a neural network and fuzzy functions with a GA to find the best and shortest route for a robot to take to a docking bay.[29] In addition, researchers at General Electric and the Rensselaer Polytechnic Institute have used a GA to design a jet engine turbine in one-fourth the time it took to develop a model manually. It improved the design by 50 percent and better kept up with the many variables involved than an expert system did.[30]

Some hybrid products that combine AI technologies use GAs—for example, GeneHunter (*www.wardsystems.com/genehunter.asp*) and NeuroDimension (*www.nd.com/genetic*). You can find more information on the uses of GAs in robotics, telecommunications, computer games, and other fields at *http://brainz.org/15-real-world-applications-genetic-algorithms*.

13-8 NATURAL-LANGUAGE PROCESSING

Despite constant efforts to make information systems user friendly, they still require a certain degree of computer literacy and skills. As mentioned in Chapter 2, **natural-language processing (NLP)** was developed so users could communicate with computers in human language. Although GUI elements, such as menus and icons, have helped with communication problems between humans and computers, GUIs still involve some training, can be cumbersome to use, and often differ depending on the OS or application. An NLP system provides a question-and-answer setting that is more natural and easier for people to use. It is particularly useful with databases. Table 13.2 lists some currently available NLP systems.

At the time of this writing, these products are not capable of a dialogue that compares with conversations between humans. The size and complexity of the human language has made developing NLP systems difficult. However, progress has been steady, and NLP systems for tasks such as call routing, stock and bond trading, and banking by phone, among others, are already available.

NLP systems are generally divided into the following categories[31]:

- Interface to databases
- Machine translation, such as translating from French to English

TABLE 13.2 NLP SYSTEMS

NLP System	Use
Nuance Communications Dragon Speech Recognition Software (*www.nuance.com/naturallyspeaking*)	Business data retrieval, legal document processing, medical and ER applications, professional dictation systems
AT&T Natural Voices Wizzard Speech (*www.wizzardsoftware.com/att_nv_landing.php*)	Creates speech from computer-readable text
e-Speaking (*www.e-speaking.com*)	Voice and speech recognition for Windows

© Cengage Learning®

Genetic algorithms (GAs) are search algorithms that mimic the process of natural evolution. They are used to generate solutions to optimization and search problems using such techniques as mutation, [selec]tion, crossover, and chromosome.

[Natural]-language processing (NLP) was developed so [users could c]ommunicate with computers in human language.

- Text scanning and intelligent indexing programs for summarizing large amounts of text
- Generating text for automated production of standard documents
- Speech systems for voice interaction with computers

NLP systems usually perform two types of activities. The first is interfacing: accepting human language as input, carrying out the corresponding command, and generating the necessary output. The second is knowledge acquisition: using the computer to read large amounts of text and understand the information well enough to summarize important points and store information so the system can respond to inquiries about the content.

13-9 INTEGRATING AI TECHNOLOGIES INTO DECISION SUPPORT SYSTEMS

AI-related technologies, such as expert systems, natural-language processing, and artificial neural networks, can improve the quality of decision support systems (DSSs). They can add explanation capabilities (by integrating expert systems) and learning capabilities (by integrating ANNs) and create an interface that is easier to use (by integrating an NLP system). These systems are sometimes called integrated (or intelligent) DSSs (IDSSs), and the result is a more efficient, powerful DSS.[32,33] AI technologies, particularly expert systems and natural-language processing, can be integrated into the database, model base, and user interface components of a DSS.

The benefits of integrating an expert system into the database component of a DSS are[34]:

- Adding deductive reasoning to traditional DBMS functions
- Improving access speed
- Improving the creation and maintenance of databases
- Adding the capability to handle uncertainty and fuzzy data
- Simplifying query operations with heuristic search algorithms

Similarly, you can add AI technologies to a DSS's model base component. For example, expert systems can be added to provide reasons and explanations for output from the model base, to include heuristics in the

model base's analysis capabilities, to incorporate fuzzy sets in the model-building process, to reduce the time and cost of calculating data for models, and to select the best model for the problem.[35]

In addition, integrating expert system capabilities into the user interface component can improve the quality and user friendliness of a DSS. This integration can add features such as an explanation capability (explaining responses in more nontechnical terms). Integrating NLP can improve the effectiveness of an interface, too, by making it easier to use, particularly for decision makers who are not computer savvy.

13-10 CONTEXTUAL COMPUTING: MAKING MOBILE DEVICES SMARTER

We have been using GPSs on our smartphones for years. A GPS system delivers location-based services—for example, telling you which street to take when you're traveling from the eastern part of a city to the western part in order to arrive at a particular place. This is a great service for somebody who is not familiar with a city. Applications such as Apple's Siri and Google Now carry this idea a bit further. For example, Google Now uses information it has about a particular user to offer, say, weather forecasts, street directions, or sports scores for games that the user is interested in. It provides such information based on a user's previous behavior and the user's current location.

Contextual computing is expected to carry this idea much further still. For example, your smartphone may soon be able to predict with 80 percent probability that you will receive a job offer if you go to a particular job fair, based on information you have included on your various social media sites. For another example, what if you wanted to know the chances that you and your girlfriend will get married? Qualcomm, the chip maker for smartphones and other computing devices, has launched a line of Brain-Inspired Zeroth Processors to expedite contextual computing. According to the company, this line of software tools and technologies enables handheld devices to learn as they receive feedback from their environment.[36]

Humans make decisions based on what they know and how they feel about something, drawing on experiences they have accumulated throughout their lives.

For example, in a dark alley, when you hear a noise, you may quickly change directions. Alternatively, if you see a friend who looks sad, you ask them if there is something wrong. These scenarios would be difficult for computers to understand and perform a role in, but it is what contextual computing is designed to achieve.

Sometimes referred to as our sixth, seventh, and eighth senses, **contextual computing** refers to a computing environment that is always present, can feel our surroundings, and—based on who we are, where we are, and whom we are with—offer recommendations.[37]

The principle behind contextual computing is that computers can both sense and react to their environments similar to how human brains understand and interpret stimuli.[38] In essence, contextual computing allows for tailoring a course of action to a user in a particular situation and environment based on what it knows about the user. To achieve this, many of the information technologies discussed in this textbook may be used, including computer networks, software, hardware, database systems, and AI technologies.

The Industry Connection highlights Alyuda Research Company, a leading developer of neural network and trading software for business and personal use.

> **Contextual computing** refers to a computing environment that is always present, can feel our surroundings, and—based on who we are, where we are, and whom we are with—offer recommendations.

Industry Connection: Alyuda Research[39]

Alyuda Research is a major developer of neural networks and trading software for businesses and individuals. Its products and services include the following:

Tradecision: Provides tools to help investors and brokers make better decisions, such as advanced charting and automated trading. Includes modules for model building, strategies, alerts, simulations, and data analysis.

Scorto Credit Decision: Offers several methods for developing models for credit scoring, such as decision trees, neural networks, and fuzzy logic, and includes software for loan portfolio analysis.

NeuroIntelligence: Used to analyze and process data sets; find the best neural network architecture; train, test, and optimize a neural network; and apply the network to new data sets.

STUDY TOOLS 13

LOCATED AT BACK OF THE TEXTBOOK
☐ Rip out Chapter Review Card

LOCATED AT WWW.CENGAGE.COM/LOGIN
☐ Review Key Terms Flash Cards (Print or Online)
☐ Download audio and visual summaries to review

☐ Complete both Practice Quizzes to prepare for tests
☐ Play "Beat the Clock" and "Quizbowl" to master concepts
☐ Complete "Crossword Puzzle" to review key terms
☐ Watch video on "Method" for a real company example

KEY TERMS

artificial intelligence (AI) 263
artificial neural networks (ANNs) 274
backward chaining 268
case-based reasoning (CBR) 270
contextual computing 278
data-mining agents 272
expert systems 266
explanation facility 268
forward chaining 268
fuzzy logic 273
genetic algorithms (GAs) 276

inference engine 268
intelligent agents 270
knowledge acquisition facility 267
knowledge base 267
knowledge base management system (KBMS) 268
monitoring and surveillance agents 272
natural-language processing (NLP) 276
personal agents 272
robots 265
shopping and information agents 271

REVIEWS AND DISCUSSIONS

1. What is artificial intelligence?

2. What is the role of an inference engine in an expert system?

3. What is the difference between a case-based reasoning system and an expert system?

4. What are two advantages of expert systems?

5. What are three examples of intelligent agents?

6. What are two applications of fuzzy logic systems?

7. What are the three layers in a neural network configuration?

8. What are two applications of NLPs?

PROJECTS

1. After reading the information presented in this chapter and other sources, write a one-page paper that discusses the advantages and disadvantages of robotic surgery. When it comes to cost, which one is cheaper: human doctor or robot?

2. Contextual computing has generated a lot of excitement in the computing community. After reading the information presented in this chapter and other sources, write a one-page paper that discusses the advantages and disadvantages of this platform. In addition to Google Now, what other software applications currently offer this capability?

3. Nina, from Nuance.com, is being marketed as an intelligent agent. After reading the information presented in this chapter and other sources, write a one-page paper that explains the applications of this software. What types of businesses will benefit the most from it?

4. Case-based reasoning systems have been used in a variety of disciplines. In the future, they may also be used in the medical field. After reading the

information presented in this chapter and other sources, write a one-page paper that explains how case-based reasoning systems will be used in this area. How might that bring healthcare costs down?

5. Intelligent agents are being used in a variety of applications. After reading the information presented in this chapter and other sources, write a one-page paper that discusses the applications of these agents in the e-commerce environment. Identify three such agents and their specific applications in e-commerce.

6. Nuance Communication's Dragon Speech Recognition Software and AT&T's Natural VoicesWizzard Speech are two of the popular commercial NLPs currently on the market. After reading the information presented in this chapter and other sources, write a two-page paper that explains key features of each NLP. What are some of the business applications of each software? What types of businesses will benefit the most from these types of software?

1. An inference engine in an expert system is similar to the model base component of a decision support system. True or False?

2. In fuzzy logic systems, a degree of membership shows how irrelevant an item or object is to a set. True or False?

3. Artificial neural networks (ANNs) are networks that learn as they go forward. True or False?

4. Which of the following tasks are ANNs not used for?

   a. Bankruptcy prediction

   b. Credit rating

   c. Investment analysis

   d. Payroll processing

5. Which of the following techniques is not used in a GA system?

   a. Forward chaining

   b. Selection or survival of the fittest

   c. Crossover

   d. Mutation

6. Which of the following is not a category of intelligent agent?

   a. Data-mining agents

   b. Brainstorming agents

   c. Personal agents

   d. Shopping and information agents

CASE STUDY 13-1

Medical Robotics in Action

Robots have been used in the medical field for two decades. They have been used to train medical personnel (doctors, dentists, and nurses), they have assisted elderly patients during rehabilitation, and they have allowed surgeons to make smaller incisions for certain types of surgery. In addition, they have been used during the training process as dummies to mimic a live patient's feelings of pain.[40]

At 5 feet, 4 inches, and 140 pounds, the RP-VITA (Remote Presence Virtual + Independent Telemedicine Assistant) by iRobot Corporation is the first robot to receive FDA clearance for use in hospitals.[41]

To make the robot resemble humans as closely as possible, it has a video screen for a head, a microphone and speaker for a mouth, and two high-definition cameras for eyes. One of its functions is to assist doctors in making hospital rounds remotely. Controllable by a tablet computer using a wired or wireless network, it

© iStockphoto.com/EdStock

is able to move around and is intelligent enough to avoid obstacles.

RP-VITA does not offer medical advice, nor does it treat patients. Instead, it is used by doctors and nurses to communicate with their patients. It can also be used with InTouch Health's cloud service to provide doctors and nurses with real-time electronic medical record information. The robot is also able to connect with diagnostic devices such as otoscopes for examining the inside of the ears; it can also perform an ultrasound.[42]

Answer the following questions:

1. What are some of the typical applications of robots in the medical field?

2. What are some applications of RP-VITA?

3. Can RP-VITA treat patients?

4. Which area within the AI field might further improve the applications of robots in the medical field?

Natural-language processing makes more sense in a mobile environment than in a desktop environment because the users of mobile devices are on the go and want to use their hands as little as possible. NLP adds a user-friendly environment and enhances data entry and data input for mobile users. And the increased memory and speed of mobile devices (as well as the increased speed of mobile and wireless networks) make them good candidates for NLP. As a result, voice-activated functions, speech-to-text dictation, and voice-activated dialing are now available for most smartphones, and voice-driven apps are getting smarter. For example, instead of saying "Call 551-535-1922" to dial a phone number, users can now say "Dial Dad" or "Phone my father."

Nuance's Dragon Dictation, available as an iPhone app, allows users to dictate everything from memos and e-mails to Twitter updates; Dragon for E-mail offers similar capabilities for the BlackBerry. Also for the iPhone, Jibbigo (*http://jibbigo .com/index.html*) translates words, phrases, and simple sentences. Voice-driven apps such as Google Voice Search, Bing Voice

© iStockphoto.com/Nikada

Search, and Microsoft Tellme are among the more popular smartphone applications. Vlingo (*www.vlingo.com*), a multiplatform app, serves as a "virtual assistant" and is being used for such services as making restaurant reservations (OpenTable, *www.opentable.com/ info/aboutus.aspx*) and booking movie tickets (Fandango, *www.fandango.com*).

Apple's Siri is a new entrant in the fast-growing voice-activated mobile-device market, having been added to the iPhone 4S and beyond. One of the challenges with voice-activated mobile devices is to get the device to understand what you mean, not just what you say. Other challenges include the use of foreign names, accents, and maintaining accuracy in noisy environments.[43, 44]

Answer the following questions:

1. How can NLP make a smartphone smarter?
2. What are some examples of voice-based software used by iPhone?
3. What are some of the challenges that have to be overcome before achieving a full-featured voice-activated mobile device?

14 Emerging Trends, Technologies, and Applications

LEARNING OUTCOMES

After studying this chapter, you should be able to:

14-1 Summarize new trends in software and service distribution.

14-2 Describe virtual reality components and applications.

14-3 Discuss uses of radio frequency identification.

14-4 Explain quick response codes.

14-5 Summarize new uses of biometrics.

14-6 Describe new trends in networking, including grid, utility, and cloud computing.

14-7 Discuss uses of nanotechnology.

After you finish this chapter, go to **PAGE 300** for the **STUDY TOOLS**

This chapter discusses new trends in software and service distribution, including pull technology, push technology, and software as a service (SaaS). You also learn about virtual reality components and applications, including CAVE, and you learn how virtual worlds are becoming a new platform for communication and collaboration. Next, you learn about radio frequency identification (RFID) and quick response (QR) codes. Then, you learn about new uses of biometrics and new trends in networking, including grid, utility, and cloud computing. Finally, you get an overview of how nanotechnology is being used and its future applications.

14-1 TRENDS IN SOFTWARE AND SERVICE DISTRIBUTION

Recent trends in software and service distribution include pull technology, push technology, and application service providers, although pull technology has been around since the Web began. All these trends are discussed in the following sections.

14-1a Pull and Push Technologies

With **pull technology**, a user states a need before getting information, as when a URL is entered in a Web browser so the user can go to a certain Web site. However, this technology is not adequate for marketing certain products and services or for providing customized information. People rarely request marketing information, for example. With **push technology**, also known as "webcasting," a Web server delivers

information to users (who have signed up for the service) instead of waiting for them to request the information be sent to them. Webcasting is supported by many Web browsers and is also available from vendors (described later in this section). With push technology, your favorite Web content can be updated in real time and sent to your desktop or your mobile device. Push technology can be effective for business-to-consumer (B2C)

With **pull technology**, a user states a need before getting information, as when a URL is entered in a Web browser so the user can go to a certain Web site.

With **push technology**, also known as "webcasting," a Web server delivers information to users (who have signed up for this service) instead of waiting for users to request the information be sent to them.

and business-to-business (B2B) marketing as well. For example, a car manufacturer can send the latest information on new models, prices, and features to all its dealers in real time. Network administrators also use push technology when they need antivirus updates to be automatically downloaded on employees' workstations.

Push technology delivers content to users automatically at set intervals or when a new event occurs. For example, you often see notices such as "A newer version of Adobe Flash is available. Would you like to install it?" In this case, the vendor (Adobe) is pushing the updated product to you as soon as it is available, which is the event triggering the push. Of course, this assumes you have already downloaded a previous version of Adobe Flash; by doing so, you have signed up for pushed updates. The same process applies to content updates, such as news and movie releases. When users sign up, they specify what content they want (sports, stock prices, political news, etc.) and consent to the "push." They can also specify how often the content should be pushed. For example, if you have subscribed to an online news service and have indicated that you are interested in the latest economic news on China, this online service will send you such news as soon as it becomes available and will do so in the future as well. You do not need to make any further request.

Push technology streamlines the process of users getting software updates and updated content. It benefits vendors, too, because by keeping in constant touch with users, they build customer loyalty. This benefit often outweighs the costs of adding servers and other technology resources needed to use push technology. Push technology also improves business relations and customer service because users get the information they need in a more timely fashion.

Here are three examples of push technology:

- *Microsoft Direct Push (AT&T)*—Enables mobile professionals to stay connected to their Microsoft Outlook information while on the go. It helps users work more efficiently with full wireless synchronization of e-mail,

> **Push technology streamlines the process of users getting software updates and updated content.**

calendar, contacts, and tasks on a Windows Mobile-enabled smartphone. Users can receive and respond to e-mail quickly through a Microsoft Office Outlook Mobile interface.

- *Apple Push Notification*—Sends alerts, such as news updates or social networking status changes, to Apple devices.

- *BlackBerry*—Offers a push API (application programming interface) that allows software developers to push real-time content as well as alerts to its smartphone users. The content includes news, weather, banking, medical information, and games.

14-1b Application Service Providers

In Chapter 7, you learned about ISPs, which provide access to the Internet for a fee. Under a more recent business model, called **application service providers (ASPs)**, access to software or services is provided for a fee. **Software as a service (SaaS)**, also known as "on-demand software," is a model in which ASPs deliver software to users for a fee; the software is for temporary or long-term use. With this delivery model, users do not need to be concerned with new software versions and compatibility problems because the ASP offers the most recent version of the software. Users can also save all application data on the ASP's server so the software and data are portable. This flexibility is convenient for those who travel or work in different locations, but it can also create privacy and security issues. Saving data on the ASP's servers instead of

© Subbotina Anna/Shutterstock.com

Application service providers (ASPs) provide access to software or services for a fee.

Software as a service (SaaS), also known as "on-demand software," is a model for ASPs to deliver software to users for a fee; the software is for temporary or long-term use.

on users' own workstations might leave this data more exposed to theft or corruption by attackers.

Here is a simple example of how SaaS might work: Say you want to edit a document, Chapter14.doc, and you need word-processing software for this task. With SaaS, you do not need the software installed on your computer. You simply access it from the SaaS provider site. You can then run the software from the provider's server (and not take up your computing resources) or on your computer. The location of the Chapter14.doc file does not matter. You make use of the provider's SaaS service to edit the document, which stays on your hard drive (or wherever you had it stored—a flash drive, for example). The word-processing application is not stored on your computer, so the next time you access the word-processing software from the provider's SaaS site you might get a newer version of the word-processing software. SaaS deals only with software, not with data and document storage or with hardware resources, such as processing power and memory.

The SaaS model can take several forms, such as the following:

- Software services for general use, such as office suite packages
- A specific service, such as credit card processing
- A service in a vertical market, such as software solutions for doctors, accountants, and attorneys

Generally, the advantages of outsourcing—less expensive, delivers information more quickly—apply to the ASP model, too. However, ASPs have some specific advantages, including the following:

- The customer does not need to be concerned about whether software is current.
- Information systems (IS) personnel time is freed up to focus on applications (such as customer relationship management and financial information systems) that are more strategically important to the organization.
- Software development costs are spread over several customers, so vendors can absorb some expenses of software development and develop more improved software.
- Software is kept up to date, based on users' requests.
- The ASP contract guarantees a certain level of technical support.
- An organization's software costs can be reduced to a predictable monthly fee.

Here are some of the disadvantages of ASPs:

- Generally, users must accept applications as provided by ASPs; software customized to users' needs is not offered.

- Because the organization has less control over how applications are developed, there is the risk that applications might not fully meet the organization's needs.
- Integration with the customer's other applications and systems might be challenging.

Google, Microsoft, Salesforce, NetSuite, Basecamp, and Mint are all companies that offer SaaS. Google offers Google Apps for Business (*www.google.com/apps*), which includes several Google products similar to traditional office suites—Gmail for Business, Google Calendar, Google Talk, Google Docs, and Google Sites, among others. (The Standard Edition is free.) Microsoft Office 365 competes with Google by offering cloud-based office services called Microsoft Office Suite. Salesforce offers a variety of enterprise applications using the SaaS model, including CRM. NetSuite offers ERP and Financial Software Suite. Basecamp (*http://basecamphq.com*) is a Web-based project collaboration tool that allows users to share files, set deadlines, assign tasks, and receive feedback. Mint is a Web-based personal financial management service. SaaS is also common for human resources applications and has been used in ERP systems with vendors such as Workday (*www.workday.com*).

14-2 VIRTUAL REALITY

Virtual reality (VR) uses computer-generated, three-dimensional images to create the illusion of interaction in a real-world environment. Stereo sound and tactile sensations enhance the feeling of being immersed in a three-dimensional real world. In VR terminology, the everyday physical world is referred to as an "information environment."

Before VR technology, even the best graphics programs used a two-dimensional environment to illustrate a three-dimensional object. VR technology has added the third dimension so users can interact with objects in a way that has not been possible before. Thomas Furness, a notable VR pioneer, states, "The distinction between immersion in a VR world and analyzing the same information using blueprints, numbers, or text is the difference between looking at fish in an aquarium and putting on your scuba gear and diving in."[1]

Virtual reality (VR) uses computer-generated, three-dimensional images to create the illusion of interaction in a real-world environment.

> "The distinction between immersion in a VR world and analyzing the same information using blueprints, numbers, or text is the difference between looking at fish in an aquarium and putting on your scuba gear and diving in."

Virtual reality began with military flight simulations in the 1960s, but these VR systems were rudimentary compared with today's systems. In the 1990s, Japan's Matsushita built a virtual kitchen that enabled its customers to change fixtures and appliances and alter the design on a computer and then virtually walk around the kitchen space. A customer's preferences could then become the blueprint for the kitchen's final design. This was the first VR system designed not for games but for general public use.[2]

As you read through the following sections, you will want to be familiar with these terms:

- *Simulation*—Giving objects in a VR environment texture and shading for a 3D appearance.

- *Interaction*—Enabling users to act on objects in a VR environment, as by using a data glove to pick up and move objects.

- *Immersion*—Giving users the feeling of being part of an environment by using special hardware and software (such as a CAVE, discussed later in this section). The real world surrounding the VR environment is blocked out so users can focus their attention on the virtual environment.

- *Telepresence*—Giving users the sense that they are in another location (even one geographically far away) and can manipulate objects as though they are actually in that location. Telepresence systems use a variety of sophisticated hardware, discussed in a later section.

- *Full-body immersion*—Allowing users to move around freely by combining interactive environments with cameras, monitors, and other devices.

- *Networked communication*—Allowing users in different locations to interact and manipulate the same world at the same time by connecting two or more virtual worlds.

In an **egocentric environment**, the user is totally immersed in the VR world.

14-2a Types of Virtual Environments

There are two major types of user environments in VR: egocentric and exocentric. In an **egocentric environment**, the user is totally immersed in the VR world. The most common technology used with this environment is a head-mounted display (HMD). Another technology, a virtual retinal display (VRD), uses lasers. Exhibit 14.1 shows an example of each

Exhibit 14.1
Egocentric VR technologies

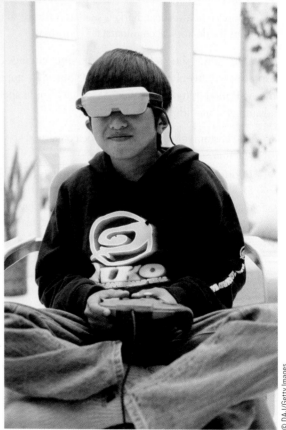

© DAJ/Getty Images

of these devices, which are discussed more in the next section.

In an **exocentric environment**, the user is given a "window view." Data is still rendered in 3D, but users can only view it on screen. They cannot interact with objects, as in an egocentric environment. The main technology used in this environment is 3D graphics.

14-2b Components of a Virtual Reality System

The following are the major components of a VR system:

- *Visual and aural systems*— These components allow users to see and hear the virtual world. HMDs, mentioned earlier, contain two small TV screens, one in front of each eye, along with a magnifying lens to generate the view. Sensing devices on top of the helmet determine the orientation and position of the user's head. The information is then transmitted to the computer, which generates two pictures so each eye has a slightly different view, just as humans' eyes do. HMDs can also incorporate stereo sound into a VR environment to make the environment more convincingly real. With VRDs, a very low-power laser beam carrying an image is projected onto the back of the user's eyes. As with an HMD, users can move their heads in any direction without losing sight of the image.

- *Manual control for navigation*—This component allows the user to navigate in the VR environment and control various objects. The most commonly used device is the data glove (see Exhibit 14.2). With it, users can point to, "grab," and manipulate objects and experience limited tactile sensations, such as determining an object's shape, size, and hardness or softness. A data glove can also be used as an input device, much like a mouse. Users can use a data glove with software to open a dialog box or pull down a menu, for example. A data glove is covered with optical sensors that send information to a computer that reconstructs the user's movements graphically. The agent representing the user's hand in the virtual world duplicates the user's hand movements.

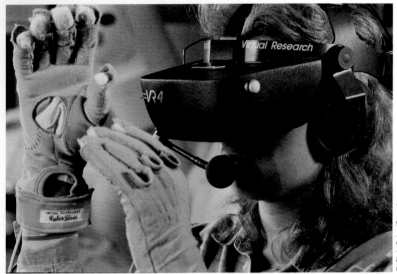

Exhibit 14.2
VR components

© Chris Salvo/Getty Images

- *Central coordinating processor and software system*— This component generates and manipulates high-quality graphics in real time, so it needs a very fast processor. To display images in real time, 3D image graphics must be rendered rapidly, and the screen's refresh rate has to be extremely fast.

- *Walker*—This input device captures and records movements of the user's feet as the user walks or turns in different directions.

14-2c CAVE

A **cave automatic virtual environment (CAVE)** is a virtual environment consisting of a cube-shaped room in which the walls are rear-projection screens. It uses holographic devices that create, capture, and display images in true 3D form (see Exhibit 14.3). People can enter other CAVEs in other locations as well, no matter how far away they are geographically, and interact with the other users. High-speed digital cameras capture one

In an **exocentric environment**, the user is given a "window view." Data is still rendered in 3D, but users can only view it on screen. They cannot interact with objects, as in an egocentric environment.

A **cave automatic virtual environment (CAVE)** is a virtual environment consisting of a cube-shaped room in which the walls are rear-projection screens. CAVEs are holographic devices that create, capture, and display images in true 3D form.

Exhibit 14.3
Example of a CAVE

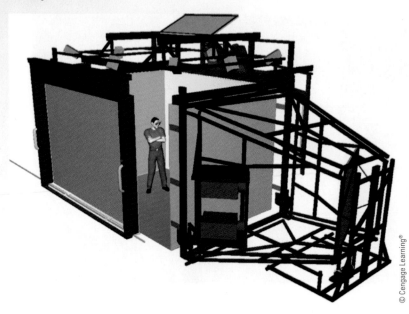

© Cengage Learning®

user's presence and movements and then re-create and send these images to users in other CAVEs. In this way, people can carry on a conversation as though they are all in the same room.

CAVEs are used for research in many fields, including archaeology, architecture, engineering, geology, and physics. Some engineering companies use CAVEs to improve product design and development. With a CAVE, they can create and test prototypes, develop interfaces, and simulate factory layouts and assembly lines, for example, without investing in physical equipment and layouts. Many universities, including Brown University, the University of Illinois at Urbana-Champaign, and Duke University, use CAVEs for geological research, architectural studies, and anatomy studies.

14-2d Virtual Reality Applications

The U.S. military uses VR systems for flight simulations. In medicine, they are used for "bloodless" surgery. In the entertainment industry, they are used in games and theaters. Eventually, they will be used for user interfaces in information systems. You might have seen an example of this in the movie *Minority Report*, in which Tom Cruise uses a 3D user interface to examine documents, graphics, and video files in crime reports. This technology is here and has been used in many real-life applications.

VR systems can be used for many other business applications, too. For example, a VR system could be used for site selection when a company wants to open a new plant. A simulation model combined with VR capabilities would allow a virtual walk-through of the potential site, a more realistic view than is possible with maps and blueprints. The following are some current business applications of VR systems:

- *Assistance for the disabled*— Virtual reality helps extend the capabilities of the disabled. For example, quadriplegics can use an HMD to get a 360-view of their surroundings, or people with cerebral palsy can learn how to operate a motorized wheelchair in a VR environment.[3]

- *Architectural design*—Architects and engineers use VR systems to create blueprints and build prototypes to show to clients. With a VR system, several versions of a design can be created to demonstrate to clients the outcome of modifying different factors. Architects and engineers can also use VR systems to safely test different conditions (such as wind shear) without the expense of using physical materials.

- *Education*—VR systems are used in educational games and simulations, such as VR "flash cards" for teaching math skills. Incorporating visuals, sound, and touch into a game can help improve the learning process. For example, in a world geography class, a VR globe could be used with touch technology that displays different facts about a country—language, population, political system, and so forth—when a student touches it.

- *Flight simulation*—Commercial airlines and the military have been using flight simulators for many years. Flight simulators are used for training pilots to handle new equipment or unusual operating conditions. Training in a VR environment is safer and less expensive than training on actual equipment.

VR systems can also be used in videoconferencing and group support systems. Current technologies using TV screens cannot fully capture the sense of other people being physically present, and people cannot shake hands or engage in direct eye contact effectively. VR systems could help overcome these obstacles by giving participants the impression of

VIRTUAL REALITY AT LOCKHEED MARTIN

Lockheed Martin Aeronautics Company, with headquarters in Bethesda, Maryland, builds some of the most sophisticated military aircraft in the world. A virtual reality and simulation laboratory that the company built in Littleton, Colorado is used to simulate and test new products and processes before introducing them into the market. The lab can be used for testing space systems, satellites, launch vehicles, and missile defense systems. According to Lockheed Martin, the virtual reality technology brings production costs down by making models of products and testing those before physically manufacturing them. Naturally, modifications on a model are a lot simpler and cheaper than modifying an actual product. One of the lab's first products in which VR was used was the Air Force's GPS III system (a $1.46 billion project), also referred to as "next-generation" GPS. According to the company, the lab may also be used for NASA's Orion project, a space vehicle that will take astronauts to the International Space Station and beyond. Lockheed Martin spokesman Michael Friedman says that the principle behind using virtual reality technology is that "it is easier to move electrons than it is to move molecules".[4]

being in the same room, which makes achieving true interaction more possible. With data gloves, they can even shake hands, even though they may be thousands of miles apart. This scenario might sound like science fiction, but the technology already exists. It gives a new meaning to the old AT&T slogan "Reach out and touch someone."

The information box above highlights the applications of VR at Lockheed Martin.

14-2e Obstacles in Using VR Systems

One major obstacle to using VR technology is that not enough fiber-optic cables are currently available to carry the data transmissions needed for a VR environment capable of re-creating a conference. With people in different geographical locations, high-speed transmission capabilities are necessary for participants to interact in real time. Without them, having to wait several seconds every time you want to act in a VR environment would be frustrating.

VR systems have generated a lot of excitement in recent years, but the following problems must be solved before this technology's potential can be realized:

- *Confusion between the VR environment and the real environment*—The possibility of users becoming unable to distinguish reality from virtual reality is a potential danger, particularly if people come to

believe that anything they do in the virtual environment is acceptable in the real world. This risk is especially a concern in computer games that allow users to torture or kill others.

- *Mobility and other problems with HMDs*—With current technology, users are "tethered" to a limited area while wearing an HMD and cannot switch to performing tasks outside the virtual world without removing the gear. In addition, refresh rates in HMDs still are not quite fast enough, so a degree of visual distortion can happen while wearing an HMD.

- *Difficulty representing sound*—Representing sound in a 3D environment is difficult if the sound needs to move, such as a plane passing overhead. Creating stationary sound is easy, but re-creating the effect of sound fading or getting louder as an object moves away or closer is much harder with current technology.

- *Need for additional computing power*—VR systems require a lot of memory and speed to manipulate large graphics files and provide the instantaneous response needed to give the impression of a real world. Drawing and refreshing frames continuously and rapidly requires extremely fast computers with a lot of memory.

With the rapid pace of technology, however, these problems should be solved in the near future so that VR systems can be used more widely.

With data gloves, users can shake hands, even though they may be thousands of miles apart.

14-2f Virtual Worlds

A **virtual world** is a simulated environment designed for users to interact via avatars. An **avatar** is a 2D or 3D graphical representation of a person in the virtual world, used in chat rooms and online games (see Exhibit 14.4). The market research firm Strategy Analytics predicts that 640 million people worldwide will inhabit virtual worlds by 2015, an increase of 244 percent over the 186 million who did so in 2009.[5]

Users can manipulate objects in the simulated world and experience a limited telepresence that gives them the feeling of being in another location. Communication between users can take the form of text, graphical icons, and sound. For example, you can shop while you are at Second Life, a virtual world platform developed by Linden Lab; there are also fan sites, blogs, forums, news sites, and classified ads. Currently, virtual worlds are used most often for gaming, social networking, and entertainment. However, they are beginning to be used in business and education. For example, IBM is using virtual worlds for client training. Other organizations use virtual worlds to conduct a variety of business activities, such as marketing and sales, product development, recruiting, and team meetings. For example, Northrop Grumman Corporation uses Second Life for displaying prototypes, performing simulations, and training employees in situations that would be dangerous, expensive, or unfeasible in the physical world. The National Oceanic and Atmospheric Administration (NOAA) also uses Second Life as a marketing channel to reach new customers.[6] Widely used virtual worlds include the following:

- ActiveWorlds (3D virtual reality platform, *https://www.activeworlds.com/index.html*)
- Club Penguin (online game, *www.clubpenguin.com*)
- EGO (social networking game, *www.kongregate.com/games/PunchEnt/ego-city*)
- Entropia Universe (multiplayer virtual universe, *www.entropiauniverse.com*)
- Habbo (social networking site, *www.habbo.com*)
- RuneScape (multiplayer role-playing game, *www.runescape.com*)
- Second Life (virtual world platform, *http://secondlife.com*)

> A **virtual world** is a simulated environment designed for users to interact with one another via avatars.
>
> An **avatar** is a 2D or 3D graphical representation of a person in the virtual world, used in chat rooms and online games.

Exhibit 14.4
Avatars in Second Life

© Moneca/Shutterstock.com, Hywit Dimyadi/Shutterstock.com, Moneca/Shutterstock.com

14-3 RADIO FREQUENCY IDENTIFICATION: AN OVERVIEW

A **radio frequency identification (RFID)** tag is a small electronic device consisting of a microchip and an antenna (see Exhibit 14.5). This device performs the same task as bar codes, universal product codes (UPCs), and magnetic strips on credit and debit cards: It provides a unique identification for the card or the object carrying the tag.

Unlike bar codes and other systems, RFID devices do not have to be in contact with the scanner to be read. Exhibit 14.6 shows an RFID reader. Because of its embedded antenna, it can be read from a distance of about 20 feet. The RFID tag's advantages, along with its decreasing price (less than 10 cents per tag), have made this device more popular with the retail industry and other industries.

There are two types of RFID tags: passive and active. Passive RFID tags have no internal power supply, so they can be very small. Typically, they absorb the signal from the receiver device, convert this signal into energy, and use this energy to respond to the receiver. Passive tags usually last longer than active

Exhibit 14.6
An RFID reader

© Tilborg Jean-Pierre/Shutterstock.com

tags; the best ones have about 10 years of life. Active RFID tags have an internal power source and are usually more reliable than passive tags and can broadcast signals over a much wider range. These tags can also be embedded in a sticker or under the skin (human or animal).

Despite RFID's advantages, there are some technical problems and some privacy and security issues. On the technical level, signals from multiple readers can overlap, signals can be jammed or disrupted, and the tags are difficult to remove. Privacy and security issues include being able to read a tag's contents after an item has left the store, tags being read without

Exhibit 14.5
RFID tags

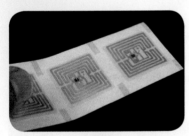

© Albert Lozano/Shutterstock.com, Albert Lozano/Shutterstock.com, Lawrence Wee/Shutterstock.com

A **radio frequency identification (RFID)** tag is a small electronic device consisting of a small chip and an antenna. This device provides a unique identification for the card or the object carrying the tag.

COCA-COLA COMPANY USES RFID-BASED DISPENSERS FOR GENERATING BI

Coca-Cola is implementing a system that will offer more than 100 varieties of soda, juice, tea, and flavored water. The system is being tested in California, Georgia, and Utah, and the company has plans to implement it nationwide. The "Freestyle" dispensers not only give customers many choices of soft drinks by allowing them to mix their own flavor combinations, they collect valuable business intelligence that Coca-Cola can use to improve the efficiency and effectiveness of its soft drink production and distribution. The dispensers contain cartridges that are tagged with RFID chips, and each dispenser contains an RFID reader. The system collects data on which drinks customers buy and how much they purchase. This information is then transmitted through a wireless network to a data warehouse system in Atlanta, Georgia. Coca-Cola analyzes the data and generates reports on how new drinks are performing in the marketplace.[8]

© Katherine Welles/Shutterstock.com

the customer's knowledge, and tags with unique serial numbers being linked to credit card numbers.

14-3a RFID Applications

RFID devices have been used by many organizations in the public and private sectors, including Walmart, the U.S. Department of Defense, Toyota, and The Gap. Table 14.1 lists some common applications of RFID, divided into five major categories.[7]

The information box above describes how the Coca-Cola Company uses RFID.

> A **QR (quick response) code** is a matrix barcode consisting of black modules arranged in a square pattern on a white background.

14-4 QUICK RESPONSE CODES

A **QR (quick response) code** is a matrix barcode consisting of black modules arranged in a square pattern on a white background. It offers a larger storage capacity compared to standard UPC barcodes (see Exhibit 14.7). Although they have been around for many years, QR codes have recently grown in popularity, particularly as a marketing tool.

Compared to conventional bar codes, QR codes have the following features[9]:

- High storage capacity
- Small printout size
- Dirt and dust resistance
- Readable from any direction
- Compatible with the Japanese character set

TABLE 14.1 RFID APPLICATIONS

Category	Examples
Tracking and identification	Railway cars and shipping containers, livestock and pets, supply-chain management (tracking merchandise from manufacturers to retailers to customers), inventory control, retail checkout and POS systems, recycling and waste disposal
Payment and stored-value systems	Electronic toll systems, contactless credit cards (require no swiping), subway and bus passes, casino tokens, concert tickets
Access control	Building access cards, ski-lift passes, car ignition systems
Anticounterfeiting	Casino tokens, high-denomination currency notes, luxury goods, prescription drugs
Health care	Tracking medical tools and patients (particularly newborns and patients with Alzheimer's), process control, monitoring patient data

© Cengage Learning®

QR CODES IN ACTION

Sacre Bleu Wine, headquartered in Prior Lake, Minnesota, employs social media—including Facebook, Twitter, and YouTube—for its advertising campaigns. And now the company has added a QR code to the labels on its two types of wines: Cabernet Sauvignon and Sauvignon Blanc. The QR code enables the company to deliver important information to its consumers when they purchase its wines. Customers who scan the QR code with their smartphones are transferred to a Web site that includes information on promotions, special offers, and even recommendations, advice, and tips from top chefs regarding mixing and matching the wine with various foods. This is an inexpensive way for the company to promote its products and create customer loyalty.[12,13]

Here are other examples of companies that are using QR codes in an effective way:
- Best Buy uses QR codes to keep a record of what its customers are scanning in its stores.[14]
- Calvin Klein used QR codes for billboards in 2010.[15]
- Dell Computer has used QR codes for an online contest in which entrants could win a new laptop.[16]
- McDonald's uses QR codes to display nutritional information.[17]
- Pepsi uses QR codes to push video content.[18]
- Ralph Lauren uses QR codes to draw consumers into its store locations.[19]
- Starbucks uses QR codes as a payment method.[20]

Exhibit 14.7
A QR code

© rangizzz/Shutterstock.com

QR codes can be read by smartphones that are equipped with cameras. The scanner app must first be downloaded to the smartphone in order for the camera to be able to read the QR code. This eliminates the need for bulky handheld scanners, which is one of the reasons for the QR code's growing popularity over bar codes and RFID tags.

You may soon see QR codes in magazine advertisements and on billboards, restaurant menus, Web pages, blogs, and social networking sites—even on t-shirts and sporting gear. In fact, a major reason for their popularity is that they can store (and digitally present) much more data, including URL links, geo coordinates, and text. Businesses, too, may start using more QR codes—on their business cards, coffee mugs, Web sites, pop-up banners, and so forth—to convey information to their business partners. In other words, business employees will no longer need to carry paper documents around with them. QR codes could also be used in trade shows to attract customers to a business's Web site.[10]

You can create a QR code online for free. One way would be to use the Google URL shortener (*http://goo.gl*), which automatically creates a QR code for a Web page each time a URL is shortened.[11] Other Web sites create QR codes for free as well—for example, *invx.com*. The information box above highlights several companies that are using QR codes as a marketing tool.

 BIOMETRICS: A SECOND LOOK

Since September 11, 2001, biometrics have become more widespread in forensics and related law enforcement fields, such as criminal identification, prison security, and airport security. Because biometrics offer a high degree of accuracy that is not possible with other security measures, they have the potential to be used in many civilian fields, too. They are already used in e-commerce and banking by phone, for example, using voice synthesizers and customers' voices as the biometric element that identifies them remotely. Here are some current and future applications of biometrics:

- *ATM, credit, and debit cards*—Even if users forget their PINs, they can use their cards if their biometric attributes are stored. Biometrics make ATM and credit/debit cards more secure, too, because they cannot be used by others if they are lost or stolen.

- *Network and computer login security*—Other security measures, such as ID cards and passwords, can be copied or stolen, which is not likely with biometric security measures. Fingerprint readers, for example, are already available at moderate prices.

- *Web page security*—Biometric measures could add another layer of security to Web pages to guard against attacks and eliminate or reduce defacing (a kind of electronic graffiti). For example, a stock-trading Web site might require customers to log on with a fingerprint reader.

- *Voting*—Biometrics could be used to make sure people do not vote more than once, for example, and could be useful for authentication purposes in voting via the Internet.

- *Employee time clocks*—Biometrics could uniquely identify each employee and verify clock-in and clock-out times. This technology could also prevent one coworker from checking in for another.

- *Member identification in sport clubs*—Fingerprint scanners are used in some sport clubs to allow members admittance. This technology improves convenience and security.

- *Airport security and fast check-in*—Israeli airports have been using biometrics for this purpose for years. Ben Gurion International Airport in Tel Aviv has a frequent flyer's fast check-in system based on smart cards that store information on users' hand geometry. With this system, travelers can pass through check-in points in less than 20 seconds.[21]

- *Passports and highly secured government ID cards*—A biometrically authenticated passport or ID card can never be copied and used by an unauthorized person. German citizens over the age of 24 can apply for an ePass, a passport containing a chip that stores a digital photograph and fingerprints. Other biometric identifiers, such as iris scans, could be added.

- *Sporting events*—Germany used biometric technology at the 2004 Summer Olympic Games in Athens, Greece, to protect its athletes. NEC, Inc. created an ID card containing an athlete's fingerprints, which was used for secure access.

- *Cell phones and smart cards*—Biometrics can be used to prevent unauthorized access to cell phones and smart cards and prevent others from using them if they are lost or stolen. An example is Apple's iPhone 5s, which features a fingerprint identity sensor.

14-6 TRENDS IN NETWORKING

The following sections discuss recent trends in networking technologies. Many are already used, such as wireless technologies and grid computing. Others, such as WiMAX and cloud computing, are newer but are attracting a lot of attention.

14-6a Wi-Fi

Wireless Fidelity (Wi-Fi) is a broadband wireless technology that allows computers and other devices to communicate over a wireless signal. Information can be transmitted over short distances—typically 120 feet indoors and 300 feet outdoors—in the form of radio waves. You can connect computers, mobile phones and smartphones, MP3 players, PDAs, and game consoles to the Internet with Wi-Fi. Some restaurants, coffee shops, and university campuses provide Wi-Fi access; these are called "hotspots." Wi-Fi connections are easy to set up, they have fast data transfer rates, and they offer mobility and flexibility. However, they are susceptible to interference from other devices and to being intercepted, which raises security concerns. In addition, there is a lack of support for high-quality media streaming.

14-6b WiMAX

Worldwide Interoperability for Microwave Access (WiMAX) is a broadband wireless technology based on the IEEE 802.16 standards. It is designed for wireless metropolitan area networks (WMANs, discussed in Chapter 6) and usually has a range of about 30 miles for fixed stations and 3–10 miles for mobile stations. Compared with Wi-Fi, WiMAX theoretically has faster data transfer rates and a longer range. In addition,

---

Wireless Fidelity (Wi-Fi) is a broadband wireless technology. Information can be transmitted over short distances—typically 120 feet indoors and 300 feet outdoors—in the form of radio waves.

Worldwide Interoperability for Microwave Access (WiMAX) is a broadband wireless technology based on the IEEE 802.16 standards. It is designed for wireless metropolitan area networks and usually has a range of about 30 miles for fixed stations and 3–10 miles for mobile stations.

it is fast and easy to install and enables devices using the same frequency to communicate. A single station can serve hundreds of users.

Disadvantages of WiMAX include interference from other wireless devices, high costs, and interruptions from weather conditions, such as rain. This technology also requires a lot of power; and when bandwidth is shared among users, transmission speed decreases.

14-6c Bluetooth

Bluetooth is a wireless technology that allows fixed and mobile devices to transfer data over short distances (usually within 30 feet). It can also be used to create a personal area network (PAN) for communication among computerized devices, including telephones, PDAs, and game consoles. (PANs can be used for communication among the devices themselves or for connecting to the Internet.) Used with mobile headsets, Bluetooth has become popular as a safer method of talking on cell phones while driving. Bluetooth uses a radio technology called Frequency Hopping Spread Spectrum (FHSS), which separates data into chunks and transmits each chunk on a different frequency, if needed. Bluetooth is also used to connect devices such as computers, global positioning systems (GPSs), mobile phones, laptops, printers, and digital cameras. Unlike infrared devices, Bluetooth has no line-of-sight limitations. Similar to other wireless devices, its susceptibility to interception is a security concern.

Bluetooth can also be used in the following ways:

- Video game consoles, such as Nintendo and Sony PlayStation, can use it in their wireless controllers.

- Companies can use it to send short advertisements to Bluetooth-enabled devices. For example, a restaurant can send announcements of dinner specials.

- A wireless device, such as a mouse, keyboard, printer, or scanner, can be connected via Bluetooth.

- Computers that are close together can network via Bluetooth.

- Contact information, to-do lists, appointments, and reminders can be transferred wirelessly between devices with Bluetooth and OBject EXchange (OBEX), a communication protocol for transmitting binary data.

14-6d Grid Computing

Generally, **grid computing** involves combining the processing powers of various computers (see Exhibit 14.8). With this configuration, users can make use of other computers' resources to solve problems involving large-scale, complex calculations, such as circuit analysis or mechanical design—problems that a single computer is not capable of solving in a timely manner. Each participant in a grid is referred to as a "node." Cost saving is a major advantage of grid computing because companies do not have to purchase additional equipment. In addition, processing on

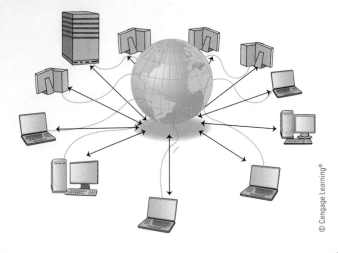

Exhibit 14.8
A grid computing configuration

© Cengage Learning®

© Darren Baker/Shutterstock.com

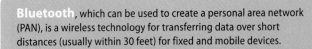

Bluetooth, which can be used to create a personal area network (PAN), is a wireless technology for transferring data over short distances (usually within 30 feet) for fixed and mobile devices.

Grid computing involves combining the processing powers of various computers. With this configuration, users can make use of other computers' resources to solve problems involving large-scale, complex calculations, such as circuit analysis or mechanical design—problems that a single computer is not capable of solving in a timely manner.

overused nodes can be switched to idle servers and even desktop systems. Grid computing has already been used in bioinformatics, oil and gas drilling, and financial applications.

Other advantages of grid computing include:

- *Improved reliability*—If one node in the grid fails, another node can take over.

- *Parallel processing*—Complex tasks can be performed in parallel, which improves performance. In other words, a large complex task can be split into smaller tasks that run simultaneously on several nodes.

- *Scalability*—If needed, more nodes can be added for additional computing power without affecting the network's operation. Upgrades can also be managed by segmenting the grid and performing the upgrade in stages without any major effect on the grid's performance.

Grid computing does have some drawbacks. Some applications cannot be spread among nodes, so they are not suitable for grid computing, and applications requiring extensive memory that a single node cannot

14-6e Utility (On-Demand) Computing

Utility (on-demand) computing is the provision of IT services on demand. Users pay for computing or storage resources on an as-needed basis, similar to the way one pays for utilities such as heat and water. Convenience and cost savings are two main advantages of utility computing, but this service does have drawbacks in the areas of privacy and security. Because the service is outside the company's location, theft or corruption of data is a concern.

Utility computing can work with the SaaS model you learned about earlier. Returning to the example of editing a Word document, suppose the Chapter14.doc file is very large because it contains a lot of images. You notice that your computer is running slowly because it has an older CPU and does not have enough RAM to handle the file size adequately. With utility computing, you can request computing power and memory from the provider. It is like leasing a more powerful computer just for the period of time you need it. Whereas SaaS deals with software, utility computing handles hardware resources, such as CPU processing and memory.

> With utility computing, you can request computing power and memory from the provider. It is like leasing a more powerful computer just for the period of time you need it.

provide cannot be used on a grid. In addition, licensing agreements can be challenging, and synchronizing operations in several different network domains can be difficult and require sophisticated network management tools. Finally, some organizations are resistant to sharing resources, even if doing so benefits them.

Universities and research centers take advantage of utility computing to run complex programs for which they do not have the necessary resources. Sun Microsystems (now a part of Oracle) and IBM offer this service in the form of storage and virtual servers. Other companies offer virtual data centers that allow users to combine memory, storage, and computing capabilities—Liquid Computing's LiquidIQ (*www.liquidweb.com*), for example. Other vendors include ENKI (*http://enki.co*), Joyent (*www.joyent.com*), and Layered Tech (*www.layeredtech.com*). Even NASA leases its supercomputer for a fee, which both brings in income and ensures that the supercomputer is used.

14-6f Cloud Computing

Cloud computing incorporates, under one platform, many recent technologies, including the SaaS model, Web 2.0, grid computing, and utility computing; hence,

Utility (on-demand) computing is the provision of IT services on demand. Users pay for computing or storage resources on an as-needed basis, similar to the way one pays for utilities such as heat and water.

Cloud computing incorporates, under one platform, many recent technologies, including the SaaS model, Web 2.0, grid computing, and utility computing; hence, a variety of resources can be provided to users over the Internet. Business applications are accessed via a Web browser, and data is stored on the providers' servers.

a variety of resources can be provided to users over the Internet. Business applications are accessed via a Web browser, and data is stored on the providers' servers.[22] In addition, cloud providers, such as Amazon, set up an environment that enables you to subscribe to SaaS, utility, grid, and other services you need and coordinates all these services for you.

Nearly all tech vendors are involved in cloud computing. BTC Logic, an IT consulting firm, has classified cloud computing into seven areas and has identified some of the top players in each category (see Table 14.2).[23]

Going back to the example of editing the Chapter14 .doc file, say you are using your iPhone instead of your computer. Clearly, your iPhone does not have the storage

resources are like a cloud that surrounds you wherever you go and are available whenever you need them.

Generally, cloud computing includes components in the form of infrastructure as a service (IaaS), platform as a service (PaaS), and software as a service (SaaS). See Exhibit 14.9.

Cloud computing has many of the advantages and disadvantages of distributed computing. With this platform, users can request services, applications, and storage. For small and medium businesses, it means they do not have to invest in expensive equipment to compete effectively with large companies and can concentrate on the services and products they provide. Cloud computing services typically require a fee, although some are free. Google Apps,

> The document, the software, and the computing resources are like a cloud that surrounds you wherever you go and are available whenever you need them.

space to save such a large file, and it does not have the necessary computing power or Word software installed. With cloud computing, you can subscribe to Word at the provider's SaaS site, store the document on an external storage unit provided by the vendor, and run Word on a multiprocessor system the vendor provides. You might even get extra RAM from another computer available in the cloud, and the cloud provider coordinates all these tasks for you. Your iPhone is simply the device for viewing the document while you are editing it, and because it is a mobile device, you can do your work anywhere. In other words, the document, the software, and the computing

which includes Gmail, Google Talk, and Google Docs, provides commonly used applications accessed via a Web browser; software and data are stored on Google's servers, not on the user's computer. The standard edition for personal use is free.[24] The information box on the next page looks at cloud computing services currently being used.

Public, Private, Hybrid, and Community Clouds: Which One to Choose

There are four options when using cloud computing: public, private, hybrid, and community clouds. Organizations usually choose which one to use based on

TABLE 14.2 CLOUD COMPUTING CATEGORIES AND THE TOP PLAYERS

Categories	Top Players
Foundations (tools and software that make it possible to build cloud infrastructure)	VMware, Microsoft, Red Hat
Infrastructure	Amazon, IBM
Network services (the communication components that combine with cloud foundation and infrastructure to form cloud architecture)	Level 3 Computing Services, Amazon, Cisco, Citrix
Platforms	Amazon, IBM
Applications	Google, Salesforce.com, Oracle, DROPBOX
Security	EMC/RSA, Symantec, IBM
Management	IBM, Amazon

© Cengage Learning®

Exhibit 14.9
Cloud Computing

Cloud Computing

Servers
Laptop
Desktop

Application — Monitoring · Content · Collaboration · Communication · Finance

Platform — Object Storage · Identity · Runtime · Queue · Database

Infrastructure — Compute · Block Storage · Network

Phones
Tablets

© Cengage Learning®

their security needs and the level of involvement that their IT managers require.

With a public cloud, users connect with an off-site infrastructure over the Internet. Because the public cloud services and the infrastructure are shared by a large number of users, this option offers the highest cost savings and is the most popular of the four. However, it also carries higher security and privacy risks. Therefore, it is more suitable for organizations that need scalability (the ability to add or drop resources), do collaborative projects over the Web, and/or offer standard applications over the Web, such as e-mail. Public cloud computing requires the least amount of involvement by IT managers.

Overall, security and reliability are the main concerns with this option. Examples of public cloud services are Amazon Elastic Compute Cloud (EC2), IBM Blue Cloud, Google AppEngine, and Windows Azure Services Platform.

In a private cloud, the services and the infrastructure are run on a private network. Naturally, this option offers higher security and privacy than a public cloud. Because a participating organization purchases and maintains the software and infrastructure itself, this option offers less cost savings than a public cloud. It is recommended for organizations that operate on highly secure data.

It is important to mention that some providers of public cloud services offer private versions of their public clouds. Also, some providers of private cloud services offer public versions of their private cloud. A private cloud achieves cost savings by integrating fragmented infrastructures, automating common data center tasks, and providing financial accuracy and responsibility. In addition, an organization gains a greater degree of automation through the standardization of previously custom-configured services into predefined infrastructure products offered in a controlled self-service manner. Major private cloud services are Eucalyptus, Elastra Enterprise Cloud Server, VMware Private-Cloud Architecture, and Microsoft Azure.

Organizations that operate on both private and public data may choose a hybrid cloud. This is a collection of at least one private and at least one public cloud. In a hybrid cloud environment, there are a variety of public and private options, with multiple providers. An organization may run its sensitive data on a private cloud and its public information on the public cloud (with less security and privacy). A hybrid cloud allows an organization to take advantage of the scalability and cost-effectiveness

CLOUD COMPUTING IN ACTION

Jeff Bezos, the founder of Amazon says, "You don't generate your own electricity. Why generate your own computing?"[25] Amazon has established a computing platform that companies can use, regardless of their locations. This platform provides storage and processing power on demand, and companies pay only for the resources they use. By using this service, companies do not have to invest in technology that might become obsolete quickly.[26]

Google Apps, introduced in February 2007, competes with Microsoft's Office Suite, and many companies use it now, including universities such as Arizona State and Northwestern.[27] In response to Google, Microsoft now offers Office 365 that is basically the Office suite in the cloud. Other vendors offering cloud computing services include IBM and Salesforce.com.

that a public cloud computing environment offers without exposing mission-critical applications and data to the outside world. However, this option may require multiple security platforms, and making sure that all systems communicate with each other in an efficient manner might be a challenge.

With the community option, the cloud infrastructure is designed for exclusive use by a specific community of users from organizations that share common concerns (e.g., security requirements, policy, mission, and compliance considerations). This infrastructure may be owned, managed, and operated by one or more of the organizations in the community, a third party, or some combination; and the infrastructure may exist on or off premises.

With this alternative, the costs are spread over fewer users than with a public cloud (but more than with a private cloud) to realize its full cost-saving potential.[28] Allocation of costs and responsibilities, governance, and the implementation of a tight security system are among the challenges that must be overcome when choosing this option.

Cloud Computing Security

Most experts believe that security is a concern when using a cloud computing platform, and users play an important role in its success. The organization that uses cloud computing should provide end-user education, force software updates, and work with the cloud computing provider to spot unusual activities.

In a cloud computing environment, there are two types of security issues: client (the user) side and server (the provider) side. The organization using the cloud services does not have much control over the server-side security issues; the provider is responsible for that. However, the client-side security is the responsibility of the organization using the cloud services. Many of the security measures discussed in Chapters 5 and 6 can be used to improve client-side security. Gartner, an analyst firm, has identified seven cloud-computing security risks[29]:

- *Privileged user access*—Who has access to your data
- *Regulatory compliance*—Availability of external audits and security certifications
- *Data location*—Specific jurisdictions and commitment to local privacy
- *Data segregation*—How your data is kept separate from other data in the cloud
- *Recovery*—What will happen to your data in case of a disaster

- *Investigative support*—Contractual commitment to support specific forms of investigation
- *Long-term viability*—What will happen to your data if the provider goes out of business

Before choosing one, an organization must make sure a cloud provider has a clear policy regarding these security risks and has indicated in writing how it will deal with each of them. Additionally, some level of trust between the provider and the user is needed. Without it, using cloud computing becomes a risky venture.

14-7 NANOTECHNOLOGY

Nanotechnology incorporates techniques that involve the structure and composition of materials on a nanoscale. Based on the nanometer, which is one billionth of a meter, it has become an exciting development in many fields. For example, scientists are working on miniature devices that can unclog arteries, detect and eradicate cancer cells, filter water pollution, and more. Nanotechnology is also being developed to make computers much faster and smaller, with more memory. However, nanotechnology is currently too expensive to justify its use in many applications. Further research and development should reduce its cost in the future.

In the field of information systems, the current technology for miniaturizing transistors and other microprocessor components might reach its limit in the next decade, so new technologies, including nanotechnology, will be necessary.[30] Nanotechnology might also play a role in the following areas:

- Energy (reduction of energy consumption, increase in the efficiency of energy production, more environmentally friendly energy systems)

- Information and communication (larger and faster storage devices, faster and cheaper computers, display monitors with low energy consumption)

- Heavy industry (aerospace, construction, refineries, vehicle manufacturing)

Some consumer goods incorporating nanotechnology are already on the market. They use what is called

Nanotechnology incorporates techniques that involve the structure and composition of materials on a nanoscale.

"nanomaterials." Nanomaterials have been added to sports gear, such as tennis and golf balls and tennis rackets, to make them more durable and improve their responsiveness; tennis balls incorporating nanomaterials bounce better. Nanomaterials have also been applied as coatings on eyewear for increased comfort and durability, and they are used in clothing and footwear to cut down bacteria growth and improve stain resistance.[31] In addition, IBM has developed the scanning tunneling microscope (STM), which is capable of imaging atoms and incorporating nanomaterial layers into hard disk recording heads and magnetic disk coatings. This technology might also improve electronic circuits and data storage devices.[32] As mentioned in Chapter 2, IBM is replacing silicon with carbon nanotubes (CNTs) in its computer chips.

The Industry Connection highlights Mechdyne Corporation and its virtual reality products.

Industry Connection: Mechdyne Corporation[33]

Mechdyne developed the first commercial CAVE system with rear-projection screens and offers a wide variety of VR hardware, software, and services, including the following:

- *CAVELib*—An application programming interface (API) with tools for creating interactive three-dimensional applications

- *Trackd*—Enhanced VR software for the immersive displays industry; capable of incorporating input from a variety of devices

- *Conduit for Google Earth Pro*—Includes features such as stereoscopic viewing for true depth perception to emulate real-life settings and real-time changes in viewing perspectives as users navigate through the environment to simulate a real-world experience

STUDY TOOLS 14

LOCATED AT BACK OF THE TEXTBOOK
- ☐ Rip out Chapter Review Card

LOCATED AT WWW.CENGAGE.COM/LOGIN
- ☐ Review Key Terms Flash Cards (Print or Online)
- ☐ Download audio and visual summaries to review

- ☐ Complete both Practice Quizzes to prepare for tests
- ☐ Play "Beat the Clock" and "Quizbowl" to master concepts
- ☐ Complete "Crossword Puzzle" to review key terms
- ☐ Watch video on "Method" for a real company example

KEY TERMS

application service providers (ASPs) 284
avatar 290
Bluetooth 295
cave automatic virtual environment (CAVE) 287

cloud computing 296
egocentric environment 286
exocentric environment 287
grid computing 295

nanotechnology 299
pull technology 283
push technology 283
QR (quick response) code 292
radio frequency identification (RFID) 291
software as a service (SaaS) 284

utility (on-demand) computing 296
virtual reality (VR) 285
virtual world 290
Wireless Fidelity (Wi-Fi) 294
Worldwide Interoperability for Microwave
    Access (WiMAX) 294

REVIEWS AND DISCUSSIONS

1. What is the difference between pull and push technologies?

2. What are two advantages of the SaaS model?

3. What are the two types of VR environments?

4. What are two business applications of VR systems?

5. What are two types of RFID? Which type lasts longer?

6. What are two marketing applications of QR codes?

7. What are four types of cloud computing options?

8. What is the impact of nanotechnology on information systems?

PROJECTS

1. After reading the information presented in this chapter and other sources, write a one-page paper that describes two companies (in addition to those mentioned in this book) that have been using VR. What are two advantages and two disadvantages of using VR?

2. Virtual worlds are being increasingly used for communication and collaboration. After reading the information presented in this chapter and other sources, write a one-page paper that describes two companies (in addition to those mentioned in this book) that have been using virtual worlds for these purposes. What are two advantages and two disadvantages of using virtual worlds?

3. RFID is playing a major role in increasing the efficiency and effectiveness of supply chain management. After reading the information presented in this chapter and other sources, write a one-page paper that describes the experience of a company that has been using RFID for this purpose. What are two advantages and two disadvantages of using RFID in supply chain management?

4. Most experts believe that security and privacy risks are two of the drawbacks of the cloud computing environment. After reading the information presented in this chapter and other sources, write a one-page paper that outlines five recommendations for improving the security and privacy of information in cloud computing.

5. After reading the information presented in this chapter and other sources, write a one-page paper that describes two companies (in addition to those mentioned in this book) that have been using cloud computing. Explain the advantages these companies have gained by using cloud computing and their major reasons for adopting this technology. Is cost-saving the only reason, or there were other reasons?

6. After reading the information presented in this chapter and other sources, write a one-page paper that identifies two companies (in addition to those mentioned in this book) that have been using biometrics. Explain the advantages of using biometrics in these companies and their major reasons for adopting this technology.

1. Push technology and "webcasting" are the same thing. True or False?

2. "SaaS" stands for "software as a system." True or False?

3. Nanotechnology is being developed to make computers much faster and smaller, with more memory. True or False?

4. The SaaS model can take all of the following forms except:

   a. Software services for general use, such as office suite packages

   b. Hardware services for personal computers

   c. A specific service, such as credit card processing

   d. A service in a vertical market, such as software solutions for doctors, accountants, and attorneys

5. Which of the following is not a type of cloud computing?

   a. Public

   b. Private

   c. Community

   d. Closed

6. Which of the following wireless technologies has line-of-sight limitations?

   a. Infrared

   b. Bluetooth

   c. WiMAX

   d. Wi-Fi

CASE STUDY 14-1

Cloud Computing at Intercontinental Hotels Group (IHG)

InterContinental Hotels Group (IHG) is an international hotel company that operates over 4,000 hotels in 100 countries under seven brands, including Holiday Inn and InterContinental. It is also the largest Western supplier of hotel rooms in China.[34]

IHG uses several types of cloud computing platforms to improve the efficiency and effectiveness of its operations. In addition to cost savings, the company says, it moved into cloud computing because it wanted a faster and more flexible way of getting a new system up and running and because of the cloud's ease of integrating various applications under a single platform. According to Bryson Koehler, IHG's senior VP of revenue and guest information, the savings also come in other ways, "such as improved flexibility and the ability to deliver a more tailored experience to end users." IHG is building a private cloud for its CRM applications using Salesforce.com's CRM. It is also building a private cloud called Camelot for its loyalty program. Camelot

© iStockphoto.com/stevegeer

performs several major tasks. First, it analyzes current guest activity and maintains historical records for future use. Second, it is used for promotions tailored to individual guests. Third, it runs revenue management and room-yield operations, from which room rates are calculated.

The public cloud at IHG is used for application development and testing. The platform chosen for this operation is Amazon Web Services' Elastic Compute Cloud infrastructure. At least three cloud delivery models are in use at IHG, including SaaS, Infrastructure as a Service (IaaS), and a private cloud.[35]

Answer the following questions:

1. What are some of the reasons that IHG is moving into cloud computing?

2. Which platform is used for IHG CRM applications?

3. Which applications run in IHG's private cloud?

4. What are the three cloud delivery models used at IHG?

RFID at Macy's Department Store

Macy's opened its first store in 1858 in New York City and has now grown to over 800 department stores in 45 states, the District of Columbia, Guam, and Puerto Rico.

Increasing sales and profitability is the goal of any retail business, including Macy's. To achieve this goal, Macy's has launched an omnichannel initiative, which seeks to integrate the physical stores, the Internet, and mobile technologies. Using this strategy, a sales associate who cannot find a product at his or her particular store will be able to quickly find it somewhere in the Macy's operation and have it sent to the customer. The strategy is designed to create a seamless communication among all sales channels. Similarly, the Macy's online Web site can draw products from any of Macy's physical stores.[36]

© Tyler Olson/Shutterstock.com

RFID is the key technology behind the Macy's omnichannel strategy, allowing the company to replenish merchandise accurately and in a timely manner. Macy's also deploys RFID tags in all of its shoe departments' sample shoes. This helps the stores eliminate or reduce the incidence of missing samples. Before RFID was implemented, keeping the samples together was a time-consuming task that caused lost sales.[37]

Answer the following questions:

1. What is the goal of Macy's omnichannel initiative?
2. With the omnichannel initiative, which three operations or technologies are integrated?
3. What is the role of RFID in the success of this initiative?
4. For which purpose is Macy's using RFID tags within its shoe departments?

ENDNOTES

1

1. Reinhard, C-G. "YouTube Brands: 5 Outstanding Leaders in YouTube Marketing." *Mashable.com*. Accessed 5 July 2010 @ *http://mashable.com/2009/06/01/youtube-brands*.

2. Evans, M. "34 Ways to Use YouTube for Business." 28 July 2009. *Gigaom.com*. Accessed 20 January 2012 @ *http://gigaom.com/collaboration/34-ways-to-use-youtube-for-business*.

3. "Number of Active Users at Facebook Over the Years." *The Associated Press*. Accessed 2 January 2014 @ *http://news.yahoo.com/number-active-users-facebook-over-230449748.html*.

4. Gaudin, S. "Half of Social Networkers Post Risky Information, Study Finds." *InfoWorld*. 4 May 2010. Accessed 5 July 2010 @ *www.infoworld.com/d/security-central/half-social-networkers-post-risky-information-study-finds-623?source=IFWNLE_nlt_wrapup_2010-05-04*.

5. "Making Pizza Since 1960." *Dominosbiz.com*. Accessed 21 January 2013 @ *www.dominosbiz.com/Biz-Public-EN/Site+Content/Secondary/About+Dominos/History*.

6. "Domino's Pizza Hits Unprecedented $1 Billion in U.S. Digital Sales in One Year, Thanking Customers with Weeklong 50% Off Pizza Online Offer." *Dominosbiz.com*. Accessed 21 January 2013 @ *http://www.prnewswire.com/news-releases/dominos-pizza-hits-unprecedented-1-billion-in-us-digital-sales-in-one-year-thanking-customers-with-weeklong-50-off-pizza-online-offer-158409165.html*.

7. Bidgoli, H. *Handbook of Management Information Systems: A Managerial Perspective*. San Diego, CA: Academic Press, 1999.

8. Bednarz, A. "The Home Depot's Latest Project: XML, Web Services." *NetworkWorldFusion*. 21 January 2002. Accessed 20 October 2010 @ *http://www.highbeam.com/doc/1G1-82328410.html*.

9. Songini, M. L. "Home Depot Kicks Off Big Data Warehouse Effort." *Computerworld*. 30 September 2002. Accessed 20 October 2010 @ *www.computerworld.com/action/article.do?command=viewArticleBasic&articleId=74751*.

10. "The Home Depot Transitions to Fujitsu U-Scan Retail Self-Checkout Software." *Fujitsu.com*. 13 January 2010. Accessed 20 October 2010 @ *www.fujitsu.com/ca/en/news/pr/fci_20100113.html*.

11. Sarah, F. G. "Small Companies Can Benefit from an HRIS." 26 September 2012. Accessed 2 January 2014 @ *www.workforce.com/articles/small-companies-can-benefit-from-an-hris*.

12. Osterhaus, E. "Compare Web-Based Human Resources (HR) Software." 17 December 2013. Accessed 2 January 2014 @ *www.softwareadvice.com/hr/web-based-hr-software-comparison*.

13. "UPS Uses Technology and Operational Efficiencies to Reduce Fuel Consumption and Emissions." *UPS Pressroom*. Accessed 20 January 2012 @ *www.pressroom.ups.com/Fact+Sheets/ci.UPS+Uses+Technology+and+Operational+Efficiencies+to+Reduce+Fuel+Consumption+and+Emissions.print*.

14. "Why UPS Trucks Never Turn Left." *YouTube*. Accessed 20 January 2012 @ *www.youtube.com/watch?v=HV_bJxkdNDE&feature=relmfu*.

15. Terdiman, D. "UPS Turns Data Analysis into Big Savings." *CNET*. 2 April 2010. Accessed 20 January 2012 @ *http://news.cnet.com/8301-13772_3-20001576-52.html*.

16. "UPS Changes the Delivery Game with New Intercept Service." *UPS Pressroom*. 26 March 2007. Accessed 20 January 2012 @ *http://www.businesswire.com/news/home/20070326005256/en/UPS-Delivery-Game-Intercept-Service#.U8lUKzNOXb0*.

17. Porter, M. "How Competitive Forces Shape Strategy." *Harvard Business Review*. (March–April 1979): 57.

18. Hays, C. "What Wal-Mart Knows About Customers' Habits." *The New York Times*. 14 November, 2004. Accessed 28 January 2013 @ *www.nytimes.com/2004/11/14/business/yourmoney/14wal.html?_r=1&*.

19. Porter, M. "How Competitive Forces Shape Strategy." *Harvard Business Review*. (March–April 1979): 57.

20. "By the Numbers: Netflix Subscribers." *Yahoo! News*. 22 July 2013. Accessed 2 January 2014 @ *http://news.yahoo.com/numbers-netflix-subscribers-205626746.html*.

21. McEntee, K. "Cloud Connect Keynote: Complexity and Freedom." *Netflix Tech Blog*. 8 March 2011. Accessed 9 August 2011@ *http://techblog.netflix.com/2011/03/cloud-connect-eynote-complexity-and.html*.

22. Thompson, C. "If You Liked This, You're Sure to Love That." *New York Times Magazine*. 21 November 2008. Accessed 9 August 2011 @ *www.nytimes.com/2008/11/23/magazine/23Netflix-t.html?pagewanted=all*.

23. Borrelli, C. "How Netflix Gets Your Movies to Your Mailbox So Fast." *Chicago Tribune*. 4 August 2009. Accessed 9 August 2011 @ *http://articles.chicagotribune.com/2009-08-04/entertainment/0908030313_1_dvd-by-mail-warehouse-trade-secrets*.

24. Linthicum, D. "The Cloud Job Market: A Golden Opportunity for IT Pros." *InfoWorld*. 1 July 2010. Accessed 3 July 2010 @ *http://www.infoworld.com/d/cloud-computing/the-cloud-job-market-golden-opportunity-it-pros-817*.

25. "10 Jobs That Didn't Exist 10 Years Ago." *Kiplinger*. 15 February 2011. Accessed 20 January 2012 @ *http://finance.yahoo.com/career-work/article/111973/jobs-that-didnt-exist-10-years-ago*.

26. Strohmeyer, R. "The 6 Hottest New Jobs in IT." *InfoWorld*. 14 June 2011. Accessed 20 January 2012 @ *www.infoworld.com/t/information-technology-careers/the-6-hottest-new-jobs-in-it-052?page=0,0&source=IFWNLE_nlt_daily_2011-06-18*.

27. This information has been gathered from the Microsoft Web site and other promotional materials. For detailed information and updates, visit *www.microsoft.com*.

28. Babcock, C. "FedEx Locks in Customers by Tying Shipping Data to Back-Office Apps." *InformationWeek*. 12 September 2006. Accessed 3 July 2010 @ *www.informationweek.com/news/management/showArticle.jhtml?articleID=192700339*.

29. Gage, D. "FedEx: Personal Touch." *Baseline*. 13 January 2005. Accessed 3 July 2010 @ *www.baselinemag.com/c/a/Projects-Supply-Chain/FedEx-Personal-Touch*.

30. Zimmerman, A. "Check Out the Future of Shopping." *Wall Street Journal Online*. 18 May 2011. Accessed 20 January 2012 @ *http://online.wsj.com/article/SB100014240527487034212045763292535050637400.html?mod=djem_jiewr_IT_domainid*.

2

1. Alabaster, J. "IBM Moving to Replace Silicon with Carbon Nanotubes in Computer Chips." *Computerworld*. 29 October 2012. Accessed 21 January 2013 @ *www.computerworld.com/s/article/9232997/IBM_moving_to_replace_silicon_with_carbon_nanotubes_in_computer_chips*.

2. Paul, I. "IBM Watson Wins Jeopardy, Humans Rally Back." *PCWorld*. 17 February 2011. Accessed 9 August 2011 @ *www.pcworld.com/article/219900/ibm_watson_wins_jeopardy_humans_rally_back.html*.

3. Hamm, S. "Watson on Jeopardy! Day Three: What We Learned About How Watson Thinks." *Building a Smarter Planet*. 16 February 2011. Accessed 9 August 2011 @ *http://asmarterplanet.com/blog/2011/02/watson-on-jeopardy-day-three-what-we-learned-about-how-watson-thinks.html*.

4. IBM. "This Is Watson." *IBM.com*. Accessed 8 February 2012 @ *http://www.ibm.com/smarterplanet/us/en/ibmwatson/*.

5. "The Disruptors: TouchlessComputing." *CNBC.com*. Accessed 2 January 2014 @ *http://video.cnbc.com/gallery/?play=1&video=3000178509*.

6. Ibid.

7. Welch, Chris. "Reading News Through Gestures: New York Times to Release Leap Motion App." *The Verge*. 18 July 2013. Accessed 2 January 2014 @ *www.theverge.com/2013/7/18/4535284/new-york-times-to-release-leap-motion-gesture-app*.

8. Shah, A. "SanDisk SD Card Can Store Data for 100 Years." *InfoWorld*. 23 June 2010. Accessed 31 July 2010 @ *www.infoworld.com/d/storage/sandisk-sd-card-can-store-data-100-years-161?source=IFWNLE_nlt_storage_2010-06-30*.

9. Prigge, M. "Cloud Storage: The Final Nail in Tape's Coffin." *InfoWorld*. 24 September 2012. Accessed 21 January 2013 @ *www.infoworld.com/d/data-explosion/cloud-storage-the-final-nail-in-tapes-coffin-202891?source=IFWNLE_nlt_daily_2012-09-24*.

10. "Apple Sells Three Million iPads in 80 Days." *Apple Press Info*. 22 June 2010. Accessed 5 July 2010 @ *www.apple.com/pr/library/2010/06/22ipad.html*.

11. Fleck, C. "Top 10 Business Uses for the iPad." 1 April 2010. Accessed 5 July 2010 @ *http://community.citrix.com/display/ocb/2010/04/01/Top+10+Business+Uses+For+the+iPad*.

12. Thompson, C. "Google, Nike, Jawbone and the Fight to Win Wearable Computing." *Cnet.com*. 2 May, 2013. Accessed 2 January 2014 @ *www.entrepreneur.com/article/230555*.

13. Greene, Jay. "Google, Nike, Jawbone and the Fight to Win Wearable Computing." *CNET.com*. 2 May 2013. Accessed 2 January 2014 @ *http://news.cnet.com/8301-11386_3-57582475-76/google-nike-jawbone-and-the-fight-to-win-wearable-computing/*.

14. "Google Docs." *Google.com*. Accessed 24 January 2012 @ *www.google.com/google-d-s/intl/en/tour1.html*.

15. "Go Google: Google Drive." *Google.com*. Accessed 21 January 2013 @ *www.google.com/intl/en_US/drive/start/index.html*.

16. Kirk, J. "Security Analyst Spots Three Flaws in Google Docs." *ITWorld*. 27 March 2009. Accessed 31 July 2010 @ *www.itworld.com/saas/65211/security-analyst-spots-three-flaws-google-docs*.

17. "iOS 7." *Apple.com*. Accessed 2 January 2014 @ *www.apple.com/ios/whats-new*.

18. This information has been gathered from the IBM Web site and other promotional materials. For detailed information and updates, visit *www.ibm.com*.

19. Sherr, I. "Manage Your Money." *Wall Street Journal*. 27 August 2012. Accessed 21 January 2013 @ *http://online.wsj.com/article/SB10000872396390444405804577559160257519518.html?mod=djem_jiewr_IT_domainid*.

20. Gaudin, S. "Tablets, Other Devices to Overshadow PCs by 2013." *Computerworld*. 12 August 2011. Accessed 14 August 2011 @ *www.computerworld.com/s/article/9219151*.

21. Ullal, M. "Laptop vs. Tablet Computers (iPad), Advantages and Disadvantages." *TekNoise*. 6 April 2011. Accessed 10 August 2011 @ *http://teknoise.com/2011/04/06/laptop-vs-tablet-computers-ipad-advantages-and-disadvantages*.

3

1. Smalltree, H. "Business Intelligence Case Study: Gartner Lauds Police for Crime-Fighting BI." *Data Management*. 5 April 2007. Accessed 5 July 2010 @ *http://searchbusinessanalytics.techtarget.com/news/1507220/Business-intelligence-case-study-Gartner-lauds-police-for-crime-fighting-BI*.

2. Kim, S. H., B. Yu, and J. Chang. "Zoned-Partitioning of Tree-Like Access Methods." *Information Systems* 33(3) (2008): 315–331.

3. Storey, V. C., and R. C. Goldstein. "Knowledge-Based Approaches to Database Design." *MIS Quarterly* (March 1993): 25–32.

4. Rauch-Hindin, W. "True Distributed DBMSs Presage Big Dividends." *Mini-Micro Systems* (May 1987): 65–73.

5. Brueggen, D., and S. Lee. "Distributed Database Systems: Accessing Data More Efficiently." *Journal of Information Systems Management* (Spring 1995): 15–20.

6. Inmon, W. H. *Building the Data Warehouse*. 4th ed. New York, NY: Wiley, 2005.

7. "About Marriott International: A World of Opportunity." *Marriott.com*. Accessed 27 January 2013 @ *www.marriott.com/marriott/aboutmarriott.mi*.

8. Swanborg, R. "CRM: How Marriott Broke Down Customer Data Siloes." *CIO (USA)*. 16 November 2009. Accessed 27 January 2013 @ *www.cio.com.au/article/325944/crm_how_marriott_broke_down_customer_data_siloes*.

9. Tremblay, M. C., R. Fuller, D. Berndt, et al. "Doing More with More Information: Changing Healthcare Planning with OLAP Tools." *Decision Support Systems* 43(4) (2007): 1305–1320.

10. Nicas, J. "How Airlines Mine Personal Data In-Flight." *The Wall Street Journal*.

8 November 2013. Accessed 2 January 2014 @ *http://online.wsj.com/news/article_email/SB10001424052702304384104579139923818792360-lMyQjAxMTAzMDEwMjExNDIyWj?mod=djem_jiewr_IT_domainid*.

11. Vijayan, J. "Time Has Come for Chief Analytics Officers." *Computerworld*. 11 October 2012. Accessed 18 December 2012 @ *www.computerworld.com/s/article/9232283/Time_has_come_for_chief_analytics_officers*.

12. Vijayan, J. "How Predictive Analytics Can Deliver Strategic Benefits" *Computerworld*. 20 September 2011. Accessed 11 November 2011 @ *www.computerworld.com/s/article/9220131/How_predictive_analytics_can_deliver_strategic_benefits*.

13. "eBay Study: How to Build Trust and Improve the Shopping Experience." *knowWPC*, Arizona State University. 8 May 2012. Accessed 2 January 2014 @ *http://knowwpcarey.com/article.cfm?aid=1171*.

14. Laney, D. "3D Data Management: Controlling Data Volume, Velocity, and Variety." *META Group*. 6 February 2001.

15. Hagen, C., K. Khan, M. Ciobo, J. Miller, D. Wall, H. Evans, and A Yadav. "Big Data and the Creative Destruction of Today's Business Models." *ATKearney*. January 2013. Accessed 2 January 2014 @ *www.atkearney.com/paper/-/asset_publisher/dVxv4Hz2h8bS/content/big-data-and-the-creative-destruction-of-today-s-business-models/10192*.

16. "How Will Big Data Revolutionize Retail?" *SAS*. Accessed 2 January 2014 @ *www.sas.com/en_us/news/sascom/2012q3/big-data-in-retail.html*.

17. Jennings, R. "The Future of Retail: It's All in Big Data." *Forbes.com*. 26 November 2013. Accessed 2 January 2014 @ *www.forbes.com/sites/netapp/2013/11/26/future-of-retail-big-data*.

18. Nemschoff, M. "Big Ways Big Data Is Changing PR." *DashBurst*. 19 October 2013. Accessed 2 January 2014 @ *http://dashburst.com/big-data-changing-pr*.

19. Brandon, A. "Why Big Data Analytics Could Spell Big Energy Savings." *TIBCO Spotfire*. 27 February 2013. Accessed 2 January @ *http://spotfire.tibco.com/blog/?p=17868*.

20. Kennedy, K. "Analysis of Huge Data Sets Will Reshape Health Care." *USA Today*. 24 November 2013. Accessed 2 January 2014 @ *www.usatoday.com/story/news/nation/2013/11/24/big-data-health-care/3631211*.

21. King, J. "Data Analytics: Eye-Popping Results from Intel, UPS and Express Scripts." *Computerworld*. 15 July 2013. Accessed 2 January 2014 @ *www.computerworld.com/s/article/9240746/Data_analytics_Eye_popping_results_from_Intel_UPS_and_Express_Scripts*.

22. Ibid.

23. This information has been gathered from the company Web site and other promotional materials. For detailed information and updates, visit *www.oracle.com*.

24. Parry, M. "Big Data on Campus." *The New York Times*. 18 July 2012. Accessed 14 December 2012 @ *www.nytimes.com/2012/07/22/education/edlife/colleges-awakening-to-the-opportunities-of-data-mining.html?pagewanted=all&_r=0*.

25. Fiegerman, S. "Pandora Now Has 200 Million Registered Users." *Mashable*. 9 April 2013. Accessed 2 January 2014 @ *http://mashable.com/2013/04/09/pandora-200-million-users*.

26. "About Pandora." *Pandora.com*. Accessed 11 November 2011 @ *www.pandora.com/corporate/*.

27. "Data Mining: Because You're Worth It." *Nabil's Brainstorm*. Accessed 11 November 2011 @ *www.laoudji.com/?p=306*.

28. Lardinois, F. "Four Approaches to Music Recommendations: Pandora, Mufin, Lala, and eMusic." *ReadWriteWeb*. 26 January 2009. Accessed 11 November 2011 @ *www.readwriteweb.com/archives/music_recommendations_four_approaches.php*

4

1. Samson, T. "Want a Job? This Site Reveals If You Have a Chance." *InfoWorld*. 11 November 2011. Accessed 28 January 2013 @ *www.infoworld.com/t/information-technology-careers/want-job-site-reveals-if-you-have-chance-178745*.

2. Read, B. "A MySpace Photo Costs a Student a Teaching Certificate." *Chronicle of Higher Education*. 27 April 2007. Accessed 6 August 2010 @ *http://chronicle.com/blogs/wiredcampus/a-myspace-photo-costs-a-student-a-teaching-certificate/2994*.

3. Ante, S., and L. Weber. "Memo to Workers: The Boss Is Watching." *The Wall Street Journal (wsj.com)*. 22 October 2013. Accessed 2 January 2014 @ *http://online.wsj.com/news/articles/SB10001424052702303672404579151440488919138?mod=djem_jiewr_IT_domainid*.

4. Pace, J., and K. Dozier. "White House Task Force Urges Limit on NSA Snooping." *Associated Press*. 18 December 2013. Accessed 2 January 2014 @ *http://finance.yahoo.com/news/white-house-task-force-urges-234420576.html;_ylt=AwrTWf0s0LxSwxEAaz2TmYlQ*.

5. "NSA Spying on Americans." *Electronic Frontier Foundation*. Accessed 2 January 2014 @ *www.eff.org/nsa-spying/timeline*.

6. "Internet 2012 in Numbers." *Pingdom*. 16 January 2013. Accessed 6 February 2013 @ *http://royal.pingdom.com/2013/01/16/internet-2012-in-numbers*.

7. Erlanger, L. "Email Users Still Running Amok Despite Years of Warnings." *InfoWorld*. 25 April 2011. Accessed 10 August 2011 @ *www.infoworld.com/t/data-loss-prevention/email-users-still-running-amok-despite-years-warnings-906?source=IFWNLE_nlt_wrapup_2011-04-25*.

8. Chattopadhyay, D. "Where Ethics Rule: A Global Ranking of the World's Most Ethical Companies Could Be a Source of Worry for Asian Firms." *Business Today* 8. March 2009.

9. Hills, S. *News of the World* Bosses Ordered Mass Email Deletion in Bid to Cover Up Phone Hacking." *Mail Online*. 24 February 2012. Accessed 28 January 2013 @ *www.dailymail.co.uk/news/article-2105810/News-World-bosses-ordered-mass-email-deletion-bid-cover-phone-hacking.html*.

10. Kirk, J. "Identity Fraud in U.S. Reaches Highest Level in Three Years." *InfoWorld*. 21 February 2013. Accessed 2 January 2014 @ *http://akamai.infoworld.com/d/security/identity-fraud-in-us-reaches-highest-level-in-three-years-213199?source=IFWNLE_nlt_sec_2013-02-21*.

11. "General Moral Imperatives." *Association for Computing Machinery* (Web site). Accessed 31 January 2012 @ *www.acm.org/about/code-of-ethics*.

12. Barquin, R. "Ten Commandments of Computer Ethics." *Computer Ethics Institute*. Accessed 5 December 2011 @ *http://computerethicsinstitute.org/publications/tencommandments.html*.

13. Morris, S., and B. Lennett. "Whither Net Neutrality?" *Slate.com*. 23 December 2013. Accessed

2 January 2014 @ *http://akamai.infoworld.com/d/security/identity-fraud-in-us-reaches-highest-level-in-three-years-213199?source=IFWNLE_nlt_sec_2013-02-21*.

14. "What Is Intellectual Property?" *World Intellectual Property Organization.* Accessed 6 August 2010 @ *www.wipo.int/about-ip/en*.

15. "What Is Copyright Protection?" *WhatIsCopyright.org.* 4 March 2007. Accessed 6 August 2010 @ *www.whatiscopyright.org*.

16. Pavento, M. "Patent Protection for E-Commerce Business Models." *Jones & Askew, LLP.* April 1999. Accessed 6 August 2010 @ *http://www.inventorfraud.com/Benefits.aspx*.

17. Tetzeli, R. "Getting Your Company's Internet Strategy Right." *Fortune.* 18 March 1996: 72–78.

18. "Verizon Wins $33 Million in Suit Over Domain Names." *Bloomberg News.* 24 December 2008. Accessed 6 August 2010 @ *www.nytimes.com/2008/12/25/technology/companies/25verizon.html*.

19. Milone, M. N. Jr., and J. Salpeter. "Technology and Equity Issues." *Technology & Learning.* January 1996: 39–47.

20. Skyrme, D. "The Virtual Organization." *KMKnowledge.com.* 24 March 2011. Accessed 2 February 2012 @ *www.skyrme.com/kmbriefings/2virtorg.htm*.

21. Hindle, T. "The Virtual Organisation." *Economist.* 23 November 2009. Accessed 2 February 2012 @ *www.economist.com/node/14301746*.

22. Tessler, F. "The Hidden Danger of Touchscreens." *InfoWorld.* 11 January 2012. Accessed 2 February 2012 @ *www.infoworld.com/t/laptops/the-hidden-danger-touchscreens-181774?source=IFWNLE_nlt_daily_2012-01-11*.

23. Becker, D. "When Games Stop Being Fun." *CNET News.* 12 April 2002. Accessed 6 August 2010 @ *http://news.cnet.com/2100-1040-881673.html*.

24. Edwards, D. "Man Dies After Playing Computer Games for 27 Days in a Row." *The Beijinger.* 23 February 2001. Accessed 26 December 2011 @ *www.thebeijinger.com/blog/2011/02/23/Man-Dies-After-Playing-Computer-Games-for-27-Days-in-a-Row*.

25. "Xbox Addict Dies from Blood Clot." *AFP.* 1 August 2011. Accessed 26 December 2011 @ *http://games.yahoo.com/blogs/plugged-in/xbox-addict-dies-blood-clot-212851422.html*.

26. "Explaining Green Computing." *ExplainingComputers.com.* Accessed 4 July 2010 @ *www.youtube.com/watch?v=350Rb2sOc3U*.

27. Murugesan, S. "Harnessing Green IT: Principles and Practices." *IT Professional* 10.1 (2006): 24–33.

28. "IBM Project Big Green." *ibm.com.* 10 May 2007. Accessed 4 July 2010 @ *www-03.ibm.com/press/us/en/photo/21514.wss*.

29. This information has been gathered from the company Web site (*http://anonymizer.com*) and other promotional materials. For more information and updates, visit the Web site.

30. Shellenbarger, S. "Working from Home Without Slacking Off." *The Wall Street Journal.* 11 July 2012. Accessed 28 January 2013 @ *http://online.wsj.com/article/SB100014240527023036840045775089534830 21234.html?mod=djem_jiewr_IT_domainid*.

31. "Employee Monitoring." *Awareness Technologies.* Accessed 28 January 2013 @ *www.awarenesstechnologies.com/products_employee.html*.

32. Shellenbarger, S. "Working from Home Without Slacking Off." *The Wall Street Journal.* 11 July 2012. Accessed 28 January 2013 @ *http://online.wsj.com/article/SB100014240527023036840045775089534830 21234.html?mod=djem_jiewr_IT_domainid*.

33. Parfeni, L. "Google Street View WiFi Lawsuit to Move Ahead." *softpedia.com.* 1 July 2011. Accessed 10 August 2011 @ *http://news.softpedia.com/news/Google-Street-View-WiFi-Lawsuit-to-Move-Ahead-209339.shtml*.

34. Zetter, K. "Lawsuits Pour in Over Google's Wi-Fi Data Collection." *wired.com.* 26 May 2010. Accessed 10 August 2011 @ *www.wired.com/threatlevel/2010/05/google-sued*.

35. Axon, S. "Google Maps Lawsuit: Woman Follows Directions, Gets Run Over." *Mashable.* 30 May 2010. Accessed 10 August 2011 @ *http://mashable.com/2010/05/30/google-maps-lawsuit*.

36. Mueller, K. P. "Google Map Brings Recreation-Seeking Motorists in Droves to Clinton Township Family's Driveway." *nj.com.* 11 July 2011. Accessed 10 August 2011 @ *www.nj.com/business/index.ssf/2011/07/google_map_brings_recreation-s.html*.

37. Brito, J. "Explaining the Google Books Case Saga." *Time.* 23 March 2011. Accessed 10 August 2011 @ *http://techland.time.com/2011/03/23/explaining-the-google-books-case-saga*.

5

1. Corbin, K. "Cyber Crime Costs U.S. Economy $100 Billion and 500,000 Jobs." *InfoWorld.* 24 July 2013. Accessed 2 January 2014 @ *www.infoworld.com/d/security/cyber-crime-costs-us-economy-100-billion-and-500000-jobs-223352?source=IFWNLE_nlt_sec_2013-07-25*.

2. Lederman, J. "IRS Missing Billions in ID Theft." *Associated Press.* 2 August 2012. Accessed 14 December 2012 @ *http://news.yahoo.com/irs-missing-billions-id-theft-164707999.html*.

3. Vijayan, J. "90 Percent of Companies Say They've Been Hacked." *Computerworld.* 23 June 2011. Accessed 11 August 2011 @ *www.infoworld.com/d/security/90-percent-companies-say-theyve-been-hacked-118*.

4. Saunders, S. "Putting a Lock on Corporate Data." *Data Communications.* January 1996: 78–80.

5. McCumber, J. *Assessing and Managing Security Risk in IT Systems.* Boca Raton, FL: Auerbach, 2004.

6. Cooney, M. "U.S. Needs to Be On Guard for a Big Cyberattack." *Computerworld.* 14 September 2011. Accessed 12 November 2011 @ *www.computerworld.com/s/article/9220018/U.S._needs_to_be_on_guard_for_a_big_cyberattack*.

7. Perlroth, N., and Q. Hardy. "Bank Hacking Was the Work of Iranians, Officials Say." *New York Times.* 8 January 2013. Accessed 2 January 2014 @ *www.nytimes.com/2013/01/09/technology/online-banking-attacks-were-work-of-iran-us-officials-say.html?_r=0*.

8. Kitten, T. "DDoS Attacks on Banks: No Break in Sight." *BankInfoSecurity.com.* 24 April 2013. Accessed 2 January 2014 @ *www.bankinfosecurity.com/ddos-attacks-on-banks-no-break-in-sight-a-5708/op-1*.

9. Samson, T. "Cyber Criminals Tying Up Emergency Phone Lines Through TDoS Attacks." *InfoWorld.* 1 April 2013. Accessed 2 January 2014 @ *www.infoworld.com/t/cyber-crime/cyber-criminals-tying-emergency-phone-lines-through-tdos-attacks-215585?source=IFWNLE_nlt_sec_2013-04-02*.

10. "Prevent Data Theft Using Removable Devices." *Get Safe Online.* 2009. Accessed 10 August 2010 @ *www.getsafeonline.org/nqcontent.cfm?a_id=1103*.

11. Anderson, H. "Case Study: The Motivation for Biometrics." *HealthCareInfoSecurity.com.* 24 June 2010. Accessed 13 July 2010 @ *www.healthcareinfosecurity.com/articles.php?art_id=2686*.

12. Bueb, F., and P. Fife. "Line of Defense: Simple, Complex Security Measures Help Prevent Lost and Stolen Laptops." *California Society of Certified Public Accountant and Gale Group* (2006). Accessed 10 August 2010 @ *www.thefreelibrary.com/Line+of+defense:+simple,+complex+security+measures+help+prevent+lost...-a0155477162*.

13. "The Sarbanes–Oxley Act of 2002." Accessed 10 August 2010 @ *www.soxlaw.com*.

14. Bidgoli, H., ed. *Global Perspectives in Information Security: Legal, Social and International Issues.* Hoboken, NJ: John Wiley & Sons, 2008.

15. This information was gathered from the company Web site (*www.mcafee.com*) and other promotional materials. For more information and updates, visit the Web site.

16. Talbot, D. "Computer Viruses Are 'Rampant' on Medical Devices in Hospitals." *MIT Technology Review.* 17 October 2012. Accessed 2 January 2014 @ *www.technologyreview.com/news/429616/computer-viruses-are-rampant-on-medical-devices-in-hospitals*.

17. Weaver, C. "Patients Put at Risk by Computer Viruses." *Wall Street Journal.* 13 June 2013. Accessed 2 January 2014 @ *http://online.wsj.com/news/articles/SB10001424127887324188604578543162744943762?mod=djem_jiewr_IT_domainid*.

18. Stuart, K. "PlayStation 3 Hack—How It Happened and What It Means." *Theguardian.com.* 7 January 2011. Accessed 11 August 2011 @ *www.guardian.co.uk/technology/gamesblog/2011/jan/07/playstation-3-hack-ps3*.

19. Kuchera, B. "Sony Admits Utter PSN Failure: Your Personal Data Has Been Stolen." *ars technica.* Accessed 11 August 2011 @ *http://arstechnica.com/gaming/news/2011/04/sony-admits-utter-psn-failure-your-personal-data-has-been-stolen.ars*.

20. Schiesel, S. "PlayStation Security Breach a Test of Consumers' Trust." *New York Times.* 27 April 2011. Accessed 11 August 2011 @ *www.nytimes.com/2011/04/28/arts/video-games/sony-playstation-security-flaw-tests-consumer-trust.html*.

6

1. "GoToMeeting." *gotomeeting.com.* Accessed 5 February 2013 @ *www3.gotomeeting.com/m/g2mab_49sideLMlp.tmpl*.

2. Miners, Z. "70% of US Residents Have Broadband Access, Study Finds." *Computerworld.* 26 August 2013. Accessed 2 January 2014 @ *www.computerworld.com/s/article/9241917/70_of_US_residents_have_broadband_access_study_finds#disqus_thread*.

3. Keizer, G. "Forrester: iPad and Tablet Rivals Will Kill Netbooks." *InfoWorld.* 17 June 2010. Accessed 31 July 2010 @ *www.infoworld.com/d/mobilize/forrester-ipad-and-tablet-rivals-will-kill-netbooks-633*.

4. Lopez, S. "Mobile Apps and Their Global Growth." *Andovar Global Solutions.* 13 November 2013. Accessed 2 January 2014 @ *http://www.andovar.com/mobile-apps-and-their-global-growth/*.

5. Tweney, D. "Mobile App Growth Exploding and Shows No Signs of Letting Up." *VentureBeat.* 10 July 2013. Accessed 2 January 2014 @ http://venturebeat.com/2013/07/10/state-of-the-apposphere.

6. Whalen, J. "Health-Care Apps That Doctors Use." *Wall Street Journal.* 17 November 2013. Accessed 2 January 2014 @ http://online.wsj.com/news/articles/SB10001424052702303376904579137683810827104?mod=djem_jiewr_IT_domainid.

7. Tan, K. "20 Mobile Apps for Shopping Discounts and Deals." *Hongkiat.com.* Accessed 2 January 2014 @ www.hongkiat.com/blog/mobile-shopping-apps.

8. "Americans Rank the Top 10 Best and Worst Mobile Banking Apps." *The Financial Brand.* 22 August 2013. Accessed 2 January 2014 @ http://thefinancialbrand.com/32967/americans-rate-the-best-and-worst-u-s-mobile-banking-apps.

9. Prindle, D. "Best Music Apps for Your Smartphone (iPhone, Android, or Windows)." *Digital Trends.* 19 March 2014. Accessed 2 January 2014 @ www.digitaltrends.com/mobile/best-music-apps.

10. Baldwin, H. "Wireless Bandwidth: Are We Running Out of Room?" *Computerworld.* 25 January 2012. Accessed 14 March 2013 @ www.infoworld.com/d/mobile-technology/wireless-bandwidth-are-we-running-out-of-room-184999?source=IFWNLE_nlt_wrapup_2012-01-25.

11. "iPhone 5s." *Apple.com.* Accessed 2 January 2104 @ www.apple.com/iphone-5s/features/?cid=wwa-us-kwg-iphone-com&siclientid=5894&sessguid=d3ac023e-509d-4795-9525-6df9956a3a30&userguid=d3ac023e-509d-4795-9525-6df9956a3a30&permguid=d3ac023e-509d-4795-9525-6df9956a3a30.

12. "iPhone 5c." *Apple.com.* Accessed 2 January 2014 @ www.apple.com/iphone-5c/features/?cid=wwa-us-kwg-iphone-com&siclientid=5894&sessguid=d3ac023e-509d-4795-9525-6df9956a3a30&userguid=d3ac023e-509d-4795-9525-6df9956a3a30&permguid=d3ac023e-509d-4795-9525-6df9956a3a30.

13. Apple. "iPhone in Business." *Apple.com.* Accessed 17 July 2010 @ http://www.apple.com/iphone/business/.

14. Gralla, P. "Preston Gralla: IT Is Caught in Crossfire When It Comes to Smartphone Privacy." *Computerworld.* 9 May 2011. Accessed 7 December 2011 @ www.computerworld.com/s/article/356168/Smartphone_Privacy_IT_Caught_in_the_Crossfire.

15. Robertson, J. "Smart-Phone Tracking Raises Privacy Issues." *NorthJersey.com.* 23 April 2011. Accessed 7 December 2011 @ www.northjersey.com/news/120560334_Smart-phone_tracking_raises_privacy_issues.html.

16. Cisco. "TelePresence: In-Person Experiences for All." Accessed 17 February 2012 @ www.cisco.com/web/telepresence/customers/retail.html.

17. This information has been gathered from the company Web site (www.cisco.com) and from other promotional materials. For more information and updates, visit the Web site.

18. "Data, Data Everywhere." *The Economist* 25 February 2010. Accessed 16 February 2012 @ www.economist.com/node/15557443?story_id=15557443.

19. "What Is Walmart's Retail Link Program?" Petersen, Chris. "Wal-Mart's secret sauce: How the largest survives and thrives." 27 March 2013.

20. "Wal-Mart Takes on Industrial Vehicle Management." *Food Manufacturing.* 29 September 2009. Accessed 10 August 2010 @ http://www.foodmanufacturing.com/articles/2009/09/wal-mart-takes-industrial-vehicle-management.

21. Gralla, P. "Preston Gralla: IT Is Caught in Crossfire When It Comes to Smartphone Privacy." *Computerworld.* 9 May 2011. Accessed 7 December 2011 @ www.computerworld.com/s/article/356168/Smartphone_Privacy_IT_Caught_in_the_Crossfire.

22. Gahran, A. "Mobile Phone Security: What Are the Risks?" *CNN.com.* 17 June 2011. Accessed 11 August 2011 @ http://edition.cnn.com/2011/TECH/mobile/06/17/mobile.security.gahran.

7

1. Greene, T. "Cisco: Is the Internet as Important as Food, Water, Air, Shelter?" *Network World.* 23 September 2011. Accessed 25 November 2011 @ www.infoworld.com/d/networking/cisco-the-internet-important-food-water-air-shelter-173844?source=IFWNLE_nlt_networking_2011-09-27.

2. Gaudin, S. "Internet Hits Major Milestone, Surpassing 1 Billion Monthly Users." *Computerworld.* 26 January 2009. Accessed 10 August 2010 @ www.computerworld.com/action/article.do?command=viewArticleBasic&articleid=9126796.

3. Elgin, J. "New Generic Top-Level Domains Nearing Launch Date." *Mass Media Headlines.* January 2011. Accessed 16 July 2011 @ www.wileyrein.com/publications.cfm?sp=articles&id=6728.

4. "Get Ready for the Next Big .Thing." *ICANNnews.* 2 July 2011. Accessed 16 July 2011 @ www.youtube.com/watch?v=AybZsS3NmFo.

5. Said, S. "Top 10 Countries with the Fastest Internet Connections Available." *TheRichest.com.* 14 July 2013. Accessed 2 January 2014 @ www.therichest.com/business/technology/top-10-countries-with-the-fastest-internet-connections-available.

6. Carlson, N. "This Is What Facebook Graph Search Looks Like." *Business Insider.* 15 January 2013. Accessed 5 February 2013 @ www.businessinsider.com/this-is-what-facebook-graph-search-looks-like-2013-1.

7. Sullivan, D. "How the New Facebook Search Is Different and Unique from Google Search." *Search Engine Land.* 15 January 2013. Accessed 5 February 2013 @ http://searchengineland.com/facebook-search-not-google-search-145124.

8. Sullivan, D. "Google Launches Knowledge Graph to Provide Answers, Not Just Links." *Third Door Media.* 16 May 2012. Accessed 5 February 2013 @ http://searchengineland.com/google-launches-knowledge-graph-121585.

9. Magid, L. "What Is Snapchat and Why Do Kids Love It and Parents Fear?" *Forbes.com.* May 2013. Accessed 5 January 2014 @ www.forbes.com/sites/larrymagid/2013/05/01/what-is-snapchat-and-why-do-kids-love-it-and-parents-fear-it.

10. Gross, D. "Millions of Accounts Compromised in Snapchat Hack." *CNN Tech.* 2 January 2014. Accessed 5 January 2014 @ www.cnn.com/2014/01/01/tech/social-media/snapchat-hack.

11. Mossberg, W. "A Real-Estate App When You're Buying or Just Nosy." *The Wall Street Journal.* 9 May 2012. Accessed 5 February 2013 @ http://online.wsj.com/article/SB10001424052702304451104577392112842506378.html?mod=djem_jiewr_IT_domainid.

12. Callow, R. "The Advantages of Buying Software Online vs Purchasing Programs in a Box." *Bright Hub.* 6 May 2010. Accessed 24 February 2012 @ www.brighthub.com/environment/green-computing/articles/8187.aspx.

13. Ruiz, R. "How the Internet Is Changing Health Care." *Forbes.com.* 20 July 2009. Accessed 24 February 2012 @ www.forbes.com/2009/07/30/health-wellness-internet-lifestyle-health-online-facebook.html.

14. Snyder, B. "How Kaiser Bet $4 Billion on Electronic Records—and Won." *InfoWorld.* 2 May 2013. Accessed 2 January 2014 @ www.infoworld.com/d/the-industry-standard/how-kaiser-bet-4-billion-electronic-health-records-and-won-217731?source=IFWNLE_nlt_daily_am_2013-05-02.

15. Marsan, C. "10 Fool-Proof Predictions for the Internet in 2020." *Network World.* 4 January 2010. Accessed 13 July 2010 @ www.networkworld.com/news/2010/010410-outlook-vision-predictions.html?page=1.

16. "Internet Users in the World: Distribution by World Regions—2011." *Internet World Stats.* Accessed 24 February 2012 @ www.internetworldstats.com/stats.htm.

17. Bidgoli, H. *Electronic Commerce: Principles and Practice.* San Diego, CA: Academic Press, 2002.

18. Mista, M. "Extranet vs. Intranet in Company Use." *chron.com.* Accessed 24 February 2012 @ http://smallbusiness.chron.com/extranet-vs-intranet-company-use-11878.html.

19. Weston, R. "Business Value in Communication." *InformationWeek.* Accessed 24 February 2012 @ www.informationweek.com/720/commun.htm.

20. "Intranets & Extranets: Understanding Their Business Application and Value." *Lyris HQ.* Accessed 24 February 2012 @ www.lyris.com/us-en/web-cms/57-Intranets-Extranets-Understanding-their-business-application-and-value.

21. O'Reilly, T. "What Is Web 2.0: Design Patterns and Business Models for the Next Generation of Software." *O'Reilly.* 30 September 2005. Accessed 10 August 2010 @ www.oreillynet.com/pub/a/oreilly/tim/news/2005/09/30/what-is-web-20.html.

22. Cho, A. "What Is Web 3.0? The Next Generation Web: Search Context for Online Information." *Suite101.com.* 22 July 2008. Accessed 10 August 2010 @ http://internet.suite101.com/article.cfm/what_is_web_30.

23. King, R. "No Rest for the Wiki." *Bloomberg Businessweek.* 12 March 2007. Accessed 10 August 2010 @ www.businessweek.com/technology/content/mar2007/tc20070312_740461.htm.

24. Gaudin, S. "Americans Spend 16 Minutes of Every Hour Online on Social Nets." *Computerworld.* 17 April 2013. Accessed 5 January 2014 @ www.computerworld.com/s/article/9238469/Americans_spend_16_minutes_of_every_hour_online_on_social_nets.

25. Van Grove, J. "LinkedIn Member Base Climbs Almost 40 Percent in Year." *CNET.* 29 October 2013. Accessed 5 January 2014 @ http://news.cnet.com/8301-1023_3-57609878-93/linkedin-member-base-climbs-almost-40-percent-in-year.

26. Kawasaki, G. "Ten Ways for Small Businesses to Use LinkedIn." *LinkedIn Blog.* 12 April 2010. Accessed 23 February 2012 @ http://blog.linkedin.com/2010/04/12/linkedin-small-business-tips.

27. King, R. "How Companies Use Twitter to Bolster Their Brands." *Bloomberg BusinessWeek.* 6 September 2008. Accessed 10 August 2010 @ www.businessweek.com/technology/content/sep2008/tc2008095_320491.htm.

28. Corbett, Karina. "Twitter Fastest Growing Social Network in the World." *Business & Leadership.* 31 January 2013. Accessed 5 February 2013 @ www.businessandleadership.com/marketing/item/39391-twitter-fastest-growing.

29. "The Fastest, Simplest Way to Stay Close to Everything You Care About." *Twitter.com*. Accessed 5 February 2013 @ *http://twitter.com/about*.

30. "Case Studies." *Twitter.com*. Accessed 5 February 2013 @ *https://business.twitter.com/optimize/case-studies*.

31. Roques, G. "IoT: 50 Billion Things and Their Developer Kings." *Developer Tech*. Accessed 2 January 2014 @ *www.developer-tech.com/news/2013/sep/11/iot-50-billion-things-and-their-developer-kings*.

32. Epps, S. "There Is No Internet of Things." *Forbes.com*. 17 October 2013. Accessed 2 January 2014 @ *www.forbes.com/sites/forrester/2013/10/17/there-is-no-internet-of-things*.

33. Evans, D. "The Internet of Everything." *Cisco.com*. Accessed 2 January 2014 @ *www.google.com/url?sa=t&rct=j&q=&esrc=s&frm=1&source=web&cd=2&sqi=2&ved=0CEwQFjAB&url=http%3A%2F%2Fwww.cisco.com%2Fweb%2Fabout%2Fac79%2Fdocs%2Finnov%2FIoE.pdf&ei=MqRhUt_oHejPigL8iIDQDg&usg=AFQjCNGNL-F_2G4BLBLKQ-9KwTe8sFkUIw*.

34. Ibid.

35. Stephenson, D. "Why Collaboration Will Replace 'Zero Sum Game' for Internet of Things Success." *Inex Advisors*. 3 September 2013. Accessed 2 January 2014 @ *http://inexadvisors.com/2013/09/03/why-zero-sum-game-is-not-the-answer-to-internet-of-things-success*.

36. Evans, D. "The Internet of Everything." *Cisco.com*. Accessed 2 January 2014 @ *www.google.com/url?sa=t&rct=j&q=&esrc=s&frm=1&source=web&cd=2&sqi=2&ved=0CEwQFjAB&url=http%3A%2F%2Fwww.cisco.com%2Fweb%2Fabout%2Fac79%2Fdocs%2Finnov%2FIoE.pdf&ei=MqRhUt_oHejPigL8iIDQDg&usg=AFQjCNGNL-F_2G4BLBLKQ-9KwTe8sFkUIw*.

37. This information has been gathered from the company Web site (*www.google.com*) and other promotional materials. For more information and updates, visit the Web site.

38. "The Scotts Miracle-Gro Company." *thescottsmiraclegrocompany.com*. Accessed 6 February 2013 @ *www.thescottsmiraclegrocompany.com*.

39. Manchester, A. "Intranet Award Winners Scotts Miracle-Gro Get Some Great Recognition #Intranet." *alexmanchester.com*. 14 February 2012. Accessed 6 February 2013 @ *www.alexmanchester.com/alexmanchester/2012/02/intranet-innovation-award-winners-scottsmiracle-gro-get-some-great-recognition-intranet.html*.

40. Kass, K. "How 'The Garden' Grew Into an Award-Winning Intranet for Scotts Miracle-Gro." *Simplycommunicate.com*. Accessed 6 February 2013 @ *www.simply-communicate.com/case-studies/company-profile/how-%E2%80%98-garden%E2%80%99-grew-award-winning-intranet-scottsmiracle-gro*.

41. Balwani, S. "5 Easy Social Media Wins for Your Small Business." *Mashable Business*. 28 July 2009. Accessed 11 August 2011 @ *http://mashable.com/2009/07/28/social-media-small-business*.

1. Porter, M. *Competitive Advantage: Creating and Sustaining Superior Performance*. New York, NY: Free Press, 1985.

2. Zetlin, M. "Use Twitter to Find Customers." *Inc*. 24 July 2009. Accessed 14 July 2010 @ *www.inc.com/news/articles/2009/07/twitter.html*.

3. Hill, R. "Ecommerce 2020." *ECRM*. 8 January 2014. Accessed 8 January 2014 @ *www.ezinepost.com/articles/article-237942.html*.

4. Brandweiner, N. "Mobile Forecast to Drive 25% of Retail Sales by 2020". *MyCustomer.com*. 5 March 2013. Accessed 8 January 2014 @ *www.mycustomer.com/topic/customer-experience/mobile-forecast-drive-25-retail-sales-2020/162691*.

5. Lei, R. "E-Commerce Boom on the Way: Alibaba." *Global Times*. 9 May 2013. Accessed 8 January 2014 @ *www.globaltimes.cn/content/780600.shtml#.UsnyvoKA3b0*.

6. O'Malley, G. "E-Commerce Forecast to Hit $279B by 2015." *MediaPost*. 28 February 2011. Accessed 8 January 2014 @ *www.mediapost.com/publications/article/145775/ecommerce-forecast-to-hit-279b-by-2015.html*.

7. "Best of the Web and Digital Government Achievement Awards 2012 Results." *Center for Digital Government*. 23 August 2012. Accessed 13 February 2013 @ *www.centerdigitalgov.com/survey/88/2012*.

8. Towns, S. "What Makes the Best Government Website?" *Government Technology*. 1 October 2012. Accessed 8 February 2013 @ *www.govtech.com/e-government/What-Makes-the-Best-Government-Website.html*.

9. Ovans, A. "E-procurement at Schlumberger." *Harvard Business Review*. (May–June 2000): 21–22.

10. Caswell, S. "Voice-Based E-Commerce Looms Large." *E-Commerce Times*. 28 March 2000. Accessed 14 July 2010 @ *www.ecommercetimes.com/story/2838.html*.

11. "Speak It, Get It: New Hound App Kicks Off Next Paradigm in User Behavior on Mobile." *Business Wire*. 16 May 2011. Accessed 3 March 2012 @ *www.soundhound.com/index.php?action=s.press_release&pr=30*.

12. "Are You Talking to Me?" *The Economist*. 7 June 2007. Accessed 3 March 2012 @ *www.economist.com/node/9249338*.

13. Flacy, M. "Google Launches Micropayments Service for Web Content." *Digital Trends*. 2 October 2012. Accessed 11 February 2013 @ *www.digitaltrends.com/web/google-launches-micropayments-for-web-content*.

14. Elgan, M. "Are We Giving Up on Mobile Payments Already?" *Computerworld*. 23 November 2013. Accessed 2 January 2014 @ *www.computerworld.com/s/article/9244285/Are_we_giving_up_on_mobile_payments_already*.

15. Stewart, C., and S. Vranica. "Phony Web Traffic Tricks Digital Ads." *The Wall Street Journal*. 30 September 2013. Accessed 2 January 2014 @ *http://online.wsj.com/article_email/SB10001424052702303464504579107082064962434-lMyQjAxMTAzMDAwMTEwNDEyWj.html?mod=djem_jiewr_IT_domainid*.

16. This information has been gathered from the company Web site (*www.amazon.com*) and other promotional materials. For more information and updates, visit the Web site.

17. Smith, B. "Yahoo's FareChase: The Stealth Disruptor?" *Search Engine Watch.com*. 27 April 2006. Accessed 14 July 2010 @ *http://searchenginewatch.com/3601971*.

18. Minyanville. "Qualcomm's Android-Capable Answer to Apple's iBeacon." *Nasdaq.com*. 11 December 2013. Accessed 2 January 2014 @ *www.nasdaq.com/article/qualcomms-androidcapable-answer-to-apples-ibeacon-cm310857*.

19. "With iBeacon, Apple Aims to Guide You Inside Its Stores and, Soon, Everywhere." *Associated Press*. 7 December 2013. Accessed 2 January 2014 @ *http://gadgets.ndtv.com/mobiles/news/with-ibeacon-apple-aims-to-guide-you-inside-its-stores-and-soon-everywhere-455739*.

20. Danova, Tony. "Apple's iBeacon Is Ready to Take Over the Retail Sector." *Business Insider*. 9 December 2013. Accessed 2 January 2014 @ *www.onenewspage.us/n/Business/74w637lqo/Apple-iBeacon-Is-Ready-To-Take-Over.htm*.

1. Ives, B., and S. Jarvenpaa. "Application of Global Information Technology: Key Issues for Management." *Management Information Systems Quarterly* 15 (1991): 33–49.

2. Eom, S. B. "Transnational Management Systems: An Emerging Tool for Global Strategic Management." *Advanced Management Journal* 59.2 (1994): 22–27.

3. Lucas, H. *Information Systems Concepts for Managers*. San Francisco, CA: McGraw-Hill; 1994: 137–152, 289–315.

4. Harrington, L. H. "The Information Challenge: The Dream of Global Information Visibility Becomes a Reality." *IndustryWeek* 246.7 (1997): 97–100.

5. "World Internet Usage and Population Statistics: June 30, 2012." *Miniwatts Marketing Group*. Accessed 15 February 2013 @ *www.internetworldstats.com/stats.htm*.

6. Schreiber Translations, Inc. "Globalizing Your Website." *schreibernet.com*. Accessed 4 December 2011 @ *www.schreibernet.com/site.nsf/1/globalizing-your-website-1.htm*.

7. Eom, S. B. "Transnational Management Systems: An Emerging Tool for Global Strategic Management." *Advanced Management Journal* 59.2 (1994): 22–27.

8. Ives, B., and S. Jarvenpaa. "Application of Global Information Technology: Key Issues for Management." *Management Information Systems Quarterly* 15 (1991): 33–49.

9. Edwards, R., A. Ahmad, and S. Moss. "Subsidiary Autonomy: The Case of Multinational Subsidiaries in Malaysia." *Journal of International Business Studies* 33.1 (2002): 183–191.

10. Karimi, J., and B. R. Konsynski. "Globalization and Information Management Strategies." *Journal of Management Information Systems* 7.4 (1991): 7–26.

11. Bar, F., and M. Borrus. "Information Networks and Competitive Advantage: Issues for Government Policy and Corporate Strategy." *International Journal of Technology Management* 7.6-8 (1992): 398–408.

12. Klein, K. "Globalization, Small Biz-Style." *Bloomberg Business*. 23 January 2008. Accessed 14 July 2010 @ *www.businessweek.com/smallbiz/content/jan2008/sb20080123_487864.htm*.

13. Yadav, V., M. Adya, V. Sridhar, and D. Nath. "Flexible Global Software Development (GSD): Antecedents of Success in Requirements Analysis." *Journal of Global Information Management* 17.1 (2009): 1–30.

14. Global Industry Analysts, Inc. "Globalization of Businesses and Workforce Mobility to Drive Global Market for Video Conferencing Systems."*PRWEB*. 19 February 2013. Accessed 2 January 2014 @ *www.prweb.com/releases/video_conferencing_system/enterprise_ISDN_IP_system/prweb10444764.htm*.

15. Kong, L. "Data Protection and Transborder Data Flow in the European and Global Context." *European Journal of International Law* 21.2 (2010): 441–456. Accessed 13 February 2013 @ *http://ejil.oxfordjournals.org/content/21/2/441.full*.

16. Mun, J. "Barriers for Globalization." *Yahoo! Voices* 27. October 2008. Accessed 14 February 2013 @ http://voices.yahoo.com/barriers-globali zation-2070111.html?cat=3.

17. Haeckel, S., and R. Nolan. "Managing by Wire (Enterprise Models Driving Strategic Information Technology)." *Harvard Business Review* 71 (1993): 122–132.

18. Passino, J., and D. Severance. "Harnessing the Potential of Information Technology for Support of the New Global Organization." *Human Resource Management* 29.1 (1990): 69–76.

19. Karimi, J., and B. R. Konsynski. "Globalization and Information Management Strategies." *Journal of Management Information Systems* 7.4 (1991): 7–26.

20. Schatz, W. "Scatter-Shot Systems: There Are Lots of Ways to Buy and Manage Technology in a Decentralized Environment." *Computerworld* 30. August 1993: 75–79.

21. Kettinger, W., V. Grover, S. Guha, and A. Segars. "Strategic Information Systems Revisited: A Study in Sustainability and Performance." *Management Information Systems Quarterly* 18.1 (1994): 31–58.

22. Ives, B., and S. Jarvenpaa. "Application of Global Information Technology: Key Issues for Management." *Management Information Systems Quarterly* 15 (1991): 33–49.

23. Fleenor, D. "The Coming and Going of the Global Corporation." *Columbia Journal of World Business* 28.4 (1993): 6–10.

24. Laudon, K., and J. Laudon. *Management Information Systems: Organization and Technology.* Upper Saddle River, New Jersey: Prentice Hall, 1996. 668–692.

25. Lynch, C. "How General Motors Uses Socialtext for Its R&D." *Socialtext.* March 2010. Accessed 14 February 2013 @ www.socialtext.com/ blog/2010/03/how-general-motors-uses-social.

26. Kettinger, W., V. Grover, S. Guha, and A. Segars. "Strategic Information Systems Revisited: A Study in Sustainability and Performance." *Management Information Systems Quarterly* 18.1 (1994): 31–58.

27. "Technological Innovation at FedEx." *FedEx .com.* Accessed 19 February 2013 @ www.fedex .com/ma/about/overview/innovation.html.

28. Ibid.

29. "FedEx Global Trade Manager." *Fedex.com.* Accessed 19 February 2013 @ www.fedex.com/ GTM?cntry_code=us.

30. "FedEx Innovation." *Fedex.com.* Accessed 19 February 2013 @ http://about.van.fedex.com/ fedex-innovation.

31. "Gartner Says Alternative Offshore Locations for IT Services Are Emerging as Global Delivery Models Evolve." *Gartner, Inc.* 27 September 2012. Accessed 14 February 2013 @ www.gartner.com/ newsroom/id/1500514.

32. Biehl, M. "Success Factors for Implementing Global Information Systems." *Communications of the ACM* 50.1 (2007).

33. Gelfand, M. J., M. Erez, and Z. Aycan. "Cross-Cultural Organizational Behavior." *Annual Review of Psychology* 58 (2007): 479–514.

34. Huff, S. "Managing Global Information Technology." *Business Quarterly* 56.2 (1991): 71–75.

35. Ambrosio, J. "Global Software: When Does It Make Sense to Share Software with Offshore Units?" *Computerworld* (2 August 1993): 74–77.

36. Spiegler, R. "Globalization: Easier Said Than Done." *The Industry Standard.* 9 October 2000: 136–155.

37. Huff, S. "Managing Global Information Technology." *Business Quarterly* 56.2 (1991): 71–75.

38. Aune, S. "Two Persian Gulf Nations Planning to Block Features of BlackBerry Phones." *TechBlorge.* 2 August 2010. Accessed 17 August 2010 @ http:// tech.blorge.com/Structure:%20/2010/08/02/two -persian-gulf-nations-planning-to-block-features-of -blackberry-phones.

39. *Yahoo v. LICRA.* U.S. Court of Appeals (9th Cir., 2004, D.C. No. CV-00-21275-JF).

40. Business Software Alliance. "Fifth Annual BSA and IDC Global Software Piracy Study." 2007. Accessed 17 August 2010 @ http://global.bsa.org/ idcglobalstudy2007/studies/2007_global_piracy _study.pdf.

41. Huff, S. "Managing Global Information Technology." *Business Quarterly.* 56.2 (1991): 71–75.

42. This information has been gathered from the company Web site (www.sap.com) and other promotional materials. For more detailed information and updates, visit the Web site.

43. Norris, R. "An Evaluation of Toyota Motor Company (TMC) Information Systems." *Yahoo! Voices.* 23 May 2007. Accessed 15 February 2013 @ http://voices.yahoo.com/an-evaluation -toyota-motor-company-tmc-information-354371 .html?cat=27.

44. "Toyota Production System." *Toyota-global .com.* Accessed 15 February 2013 @ www .toyota-global.com/company/vision_philosophy/ toyota_production_system.

45. Duvall, M. "Toyota: Daily Dealings." *Baseline.* 5 September 2006. Accessed 15 February 2013 @ www.baselinemag.com/c/a/Infrastructure/ What-Is-Driving-Toyota/2.

46. "Toyota Motor Europe Upgrades Vehicle Order Management System with Oracle E-Business Suite 12.1 to Help Improve Customer Allocation and Lead Times." *Oracle.* 5 October 2011. Accessed 15 February 2013 @ www.oracle.com/us/corporate/ press/513279.

47. Baldwin, H. "Supply Chain 2013: Stop Playing Whack-a-Mole with Security Threats." *InfoWorld.* 30 April 2013. Accessed 2 January 2014 @ www.infoworld.com/d/security/supply-chain -2013-stop-playing-whack-mole-security-threats -217547?source=IFWNLE_nlt_sec_2013-05-02.

48. Ibid.

10

1. Wood, J., and D. Silver. *Joint Application Design.* New York, NY: John Wiley and Sons, 1989.

2. Bensaou, M., and M. Earl. "The Right Mind-Set for Managing Information Technology." *Harvard Business Review* 76.5 (1998): 109.

3. Duvall, M. "Airline Reservation System Hits Turbulence." *Baseline.* 25 July 2007. Accessed 16 July 2010 @ http://www.baselinemag.com/c/a/ Business-Intelligence/Airline-Reservation-System -Hits-Turbulence/.

4. Sun, L., and S. Wilson. "Health Insurance Exchange Launched Despite Signs of Serious Problems." *Washington Post.* 21 October 2013. Accessed 2 January 2014 @ www.washingtonpost .com/national/health-science/health-insurance -exchange-launched-despite-signs-of-serious -problems/2013/10/21/161a3500-3a85-11e3-b6a9 -da62c264f40e_story.html.

5. Britt, R. "Five Key Things That Are Wrong with Obamacare's HealthCare.gov Site." *Wall Street Journal.* 7 November 2013. Accessed 2 January 2014 @ http://blogs.marketwatch.com/ health-exchange/2013/11/07/five-key-things-that -are-wrong-with-obamacares-healthcare- gov-site.

6. Kanaracus, C. "The Worst IT Project Disasters of 2013." *Computerworld.* 10 December 2013. Accessed 2 January 2014 @ www.computerworld .com/s/article/9244679/The_worst_IT_project _disasters_of_2013.

7. Storm, D. "HealthCare.gov More Vulnerable to Hacking and Privacy Breaches After 'Fix'." *Computerworld.* 3 December 2013. Accessed 2 January 2014 @ http://blogs.computerworld.com/ application-security/23226/healthcaregov-more -vulnerable-hacking-privacy-breaches-after-fix.

8. Unknown. "A Software Developer's View on the HealthCare.gov Glitches." *Daily Kos.* 17 October 2013. Accessed 2 January 2014 @ www.dailykos .com/story/2013/10/17/1248260/-A-software -developer-s-view-on-the-HealthCare-gov -glitches.

9. Storm, D. "HealthCare.gov More Vulnerable to Hacking and Privacy Breaches After 'Fix'." *Computerworld.* 3 December 2013. Accessed 2 January 2014 @ http://blogs.computerworld .com/application-security/23226/healthcaregov -more-vulnerable-hacking-privacy-breaches -after-fix.

10. Doke, E. "An Industry Survey of Emerging Prototyping Methodologies." *Information and Management* 18.4 (1990): 169–176.

11. Benson, D. "A Field Study of End User Computing: Findings and Issues." *MIS Quarterly* 7.4 (1983): 35–40.

12. Tayntor, C. "New Challenges or the End of EUC." *Information Systems Management* 11.3 (1994): 86–88.

13. Overby, S. "IT Outsourcing to China Increases Despite Drawbacks." *InfoWorld.* 6 December 2011. Accessed 8 December 2011 @ www.infoworld .com/d/the-industry-standard/it-outsourcing -china-increases-despite-drawbacks-181027?source =IFWNLE_nlt_wrapup_2011-12-06.

14. McKendrick, J. "Ten Examples of SOA at Work, Right Now." *ZDNet.* 3 January 2006. Accessed 15 July 2010 @ www.zdnet.com/blog/service-oriented/ ten-examples-of-soa-at-work-right-now/508.

15. Goldstein, A. "Hand-to-Hand Teamwork. Sabre Airline Solutions." *Dallas Morning News.* 23 February 2003. Accessed 20 March 2012 @ http://www.sabre-holdings.com/newsroom/ inTheNews/02_23_03.html.

16. Ibid.

17. "Principles Behind the Agile Manifesto." Accessed 16 July 2010 @ www.agilemanifesto.org/ principles.html.

18. "About HomeAway, Inc." *Homeaway.com.* Accessed 19 February 2013 @ www.homeaway .com/info/about-us/company-info.

19. Krill, P. "Agile: How It Became a Way of Life at HomeAway." *InfoWorld.* 1 May 2012. Accessed 18 February 2013 @ www.infoworld.com/d/ application-development/agile-how-it-became-way -of-life-homeaway-192102?source=IFWNLE_nlt _daily_2012-05-01.

20. This information has been gathered from the company Web site (www.ca.com) and other promotional materials. For more information and updates, visit the Web site.

21. "Latvia Builds a Comprehensive Business Platform on IBM Lotus Notes and Domino." *IBM.* 16 December 2008. Accessed 16 July 2010 @ ftp://ftp.software.ibm.com/common/ssi/pm/ab/n/ loc14088lven/LOC14088LVEN.PDF.

22. York, M. "Crowdsourcing: It's Not Just for Logos and Web Design." *About.com Entrepreneurs.* Accessed 14 August 2011 @ *http://entrepreneurs. about.com/od/casestudies/a/Crowdsourcing-Its-Not -Just-For-Logos-And-Web-Design.htm.*

11

1. "Our Company." *Coca-colacompany.com.* Accessed 20 February 2013 @ *www.coca -colacompany.com/our-company.*

2. "Coca-Cola Supply Chain Management Success Story." *CSC.* Accessed 20 February 2013 @ *www.csc.com/application_services/success _stories/78846-coca_cola_supply_chain _management_success_story.*

3. Payne, A. *Handbook of CRM: Achieving Excellence Through Customer Management.* Oxford, MA: Butterworth-Heinemann, 2005.

4. Wailgum, T. "CRM Definition and Solutions." *CIO.com.* Accessed 26 July 2010 @ *www.cio.com/ article/40295/Customer_Relationship_Management.*

5. "Team Productivity Increases Up to 10% After Time Warner Cable Business Class Implements Salesforce." *Salesforce.com.* Accessed 16 July 2010 @ *www.salesforce.com/showcase/stories/ timewarnercable.jsp.*

6. Caitlin. "Personalization vs. Customization." *RapLeaf Blog.* 22 February 2012. Accessed 26 March 2012 @ *http://blog.rapleaf.com/blog/2012/ 02/22/personalization-vs-customization-2.*

7. Ray, R. "Selling with Personalization: Why Amazon.com Succeeds and You Can Too." *Smallbiztechnology.com.* 2 August 2008. Accessed 16 July 2010 @ *smallbiztechnology.com/ archive/2008/08/selling-with-personalization-w.html.*

8. Sherman, C. "Google Personalized Search Leaves Google Labs." *SearchEngineWatch.com.* 9 November 2005. Accessed 26 March 2012 @ *http://searchenginewatch.com/3563036.*

9. Aaronson, J. "Personalization Technologies: A Primer." *ClickZ.com.* 24 August 2007. Accessed 26 March 2012 @ *www.clickz.com/3626837.*

10. Rao, L. "How Amazon Is Tackling Personalization and Curation for Sellers on Its Marketplace." *TechCrunch.* 31 August 2013. Accessed 2 January 2014 @ *http://techcrunch .com/2013/08/31/how-amazon-is-tackling -personalization-and-curation-for-sellers-on-its -marketplace.*

11. Levinson, M. "Knowledge Management Definition and Solutions." *CIO.com.* Accessed 16 July 2010 @ *www.cio.com/article/40343/ Knowledge_Management_Definition_and_Solutions.*

12. D'Auria, J. "Goodwin Procter Makes Strong Case for Knowledge Management." *CIO.com.* 1 August 2008. Accessed 3 December 2011 @ *www .cio.com/article/440726/Goodwin_Procter_Makes _Strong_Case_for_Knowledge_Management?page=1 &taxonomyId=3154.*

13. Vormittag Associates. "Saudi-Based Auto and Electronics Distributor Streamlines Operations with VAI's S2K Software." *VAI.net.* Accessed 17 July 2010 @ *www.vai.net/company/success -stories/naghi-group.html.*

14. This information has been gathered from the company Web site (*www.salesforce.com*) and other promotional materials. For more information and updates, visit the Web site.

15. IBM. "Johns Hopkins Unites Its Businesses with SAP and IBM." *IBM.com.* 26 June 2008. Accessed 16 July 2010 @ *www-01.ibm.com/ software/success/cssdb.nsf/CS/STRD-7FYDM6?Ope nDocument&Site=gicss67sap&cty=en_us.*

16. Starbucks. "Starbucks Company Profile." Accessed 14 August 2011 @ *http://assets.starbucks.com/assets/ aboutuscompanyprofileq12011final13111.pdf.*

17. Salesforce. "Powered by Salesforce, My Starbucks Idea Brews Customer Feedback and Community Engagement." *Salesforce.com.* Accessed 14 August 2011 @ *www.salesforce.com/ showcase/stories/starbucks.jsp.*

12

1. Simon, H. *The New Science of Management Decision.* Englewood Cliffs, NJ: Prentice-Hall, 1977.

2. Sprague, R., and E. Carlson. *Building Effective Decision Support Systems.* Englewood Cliffs, NJ: Prentice-Hall, 1982.

3. Ibid.

4. Keen, P. "Value Analysis: Justifying Decision Support Systems." *MIS Quarterly* 5(1) (1981): 1–15.

5. Alter, S. *Decision Support Systems Current Practice and Continuing Challenges.* Reading, MA: Addison-Wesley, 1979.

6. Sauter, V. *Decision Support Systems for Business Intelligence,* 2nd Edition. Hoboken, NJ: Wiley, 2011.

7. Turban, E., R. Sharda, and D. Delen. *Decision Support and Business Intelligence Systems,* 9th Edition. Englewood Cliffs, NJ: Prentice-Hall, 2011.

8. "Family Dollar Accelerates Growth in Down Times." *SAS.* Accessed 2 January 2014 @ *www .sas.com/offices/europe/uk/industries/retail/family -dollar-case-study.html.*

9. Watson, H., and K. Rainer Jr. "A Manager's Guide to Executive Support Systems." *Business Horizons* 34(2) (1991): 44–50.

10. Liang, L., and R. Miranda. "Dashboards and Scorecards: Executive Information Systems for the Public Sector." *Government Finance Review.* December 2001. Accessed 16 July 2012 @ *www .inst-informatica.pt/v20/ersi/13_ersi/documentos/ Dashboards%20and%20Scorecards%20Executive %20Information%20Systems%20for%20the%20 Public%20Sector.pdf.*

11. Glover, H., H. Watson, and R. Rainer. "20 Ways to Waste an EIS Investment." *Information Strategy: The Executive's Journal* (Winter 1992): 11–17.

12. Watson, H., and J. Satzinger. "Guidelines for Designing EIS Interfaces." *Information Systems Management* (Fall 1994): 46–52.

13. "Technology Drives Decisions at Hyundai." *SAS.* Accessed 2 January 2014 @ *www.sas.com/ success/pdf/hyundai.pdf.*

14. Cronk, R. "EISes Mine Your Data." *Byte* 18(7) (1993): 121–128.

15. Overton, K., M. Frolick, and R. Wilkes. "Politics of Implementing EISs." *Information Systems Management* 13(3) (1996): 50–57.

16. "What Is SharePoint." *Microsoft.com.* Accessed 2 January 2014 @ *http://office.microsoft.com/ en-us/sharepoint-server-help/what-is-sharepoint -HA010378184.aspx.*

17. Kibbe, D. "Why Clinical Groupware May Be the Next Big Thing in Health IT." *e-Care Management* (8 February 2009). Accessed 18 July 2010 @ *http://e-caremanagement.com/why -clinical-groupware-may-be-the-next-big-thing-in -health-it.*

18. "Skype Overview." *Downloadinfo.* Accessed 20 February 2013 @ *http://allthingsd.com/20120821/a -chance-to-call-15-friends-to-video-chat-in-high-def/.*

19. "FaceTime: Be in Two Places at Once." *Apple .com.* Accessed 20 February 2013 @ *www.apple .com/ios/facetime/?cid=wwa-us-kwg-features -00001&siclientid=6381&sessguid=e5147ce0-ea29 -47ed-914d-e1058f57bd71&userguid=e5147ce0-ea29 -47ed-914d-e1058f57bd71&permguid=e5147ce0 -ea29-47ed-914d-e1058f57bd71.*

20. "Tango." *Tango.me.* Accessed 20 February 2013 @ *tango.me.*

21. "Get Started. About Hangouts." *Google.com.* Accessed 20 February 2013 @ *http://support.google. com/plus/bin/answer.py?hl=en&answer=1215273.*

22. "What Is ooVoo?" *Oovoo.com.* Accessed 20 February 2013 @ *http://cdn01.oovoo.com/ cdnhelp/en/content/what_is_oovoo.htm.*

23. "Unified Meeting Experience." *Zoom.us.* Accessed 20 February 2013 @ *zoom.us.*

24. "A Chance to Call 15 Friends to Video Chat in High Def." *The Wall Street Journal.* 21 August 2012. Accessed 20 February 2013 @ *http://online.wsj.com/article/SB1000087 2396390444435045776033832 38616426. html?mod=djem_jiewr_IT_domainid.*

25. eSpatial. "GIS for Insurance." *eSpatial.com.* Accessed 1 April 2012 @ *http://www.esri.com/ library/whitepapers/pdfs/gis-for-insurance-claims -process.pdf.*

26. Caliper Corporation. "GIS Software for Insurance Mapping." *caliper.com.* Accessed 1 April 2012 @ *www.caliper.com/maptitude/insurance/ default.htm.*

27. Romeo, J. "Target Marketing with GIS." *Geospatial Solutions.* May 2005. Accessed 1 April 2012 @ *http://tetrad.com/pub/documents/geospatial.pdf.*

28. eSpatial. "Mapping Software for Sales & Marketing." *eSpatial.com.* Accessed 1 April 2012 @ *https://www.espatial.com/info/mapping -software-for-online?utm_source=Google&utm _medium=PPC&kw=%2Bmapping%20% 2Bsoftware&Network=Search&Placement =&Creative=25617727576&utm _campaign=US_Mapping&gclid=CKCHj8 _Mz78CFQORfgodzkQAvQ.*

29. ESRI. "GIS for Real Estate." *ESRI.com.* February 2007. Accessed 1 April 2012 @ *www.esri .com/library/bestpractices/real-estate.pdf.*

30. Geographic Information Systems. "Why Use GIS?" *GIS.com.* Accessed 1 April 2012 @ *www.gis .com/content/why-use-gis.*

31. O'Cionnaith, F. "GIS on the Front Lines in Fighting Disease." *gpsworld.com.* 19 December 2007. Accessed 1 April 2012 @ *http://www.esri .com/news/arcnews/fall12articles/can-gis-help-fight -the-spread-of-malaria.html.*

32. Dempsey, C. "Monitor the Swine Flu Real Time with Google Maps." *GIS Lounge.* 27 April 200. Accessed 16 July 2010 @ *http://gislounge.com/ monitor-the-swine-flu-real-time-with-google-maps.*

33. This information has been gathered from the company Web site (*www.sas.com*) and other promotional materials. For more information and updates, visit the Web site.

34. "UPS Speeds ORION Deployment and Takes Routing Optimization to New Heights." *USP.com.* 30 October 2013. Accessed 2 January 2014 @ *www .pressroom.ups.com/Press+Releases/Archive/2013/ Q4/UPS+Speeds+ORION+Deployment+And+Take s+Routing+Optimization+To+New+Heights.*

35. Mearian, L. "Healthcare IT Spending to Hit $40B This Year." *InfoWorld.* 26 May 2011. Accessed 30 November 2011 @ *www.infoworld .com/d/the-industry-standard/healthcare-it-spending -hit-40b-year-323?source=IFWNLE_nlt_ wrapup_2011-05-28.*

36. Mearian, L. "500 Million People Will Be Using Mobile Health Apps by 2015." *Computerworld.* 16 November 2010. Accessed 30 November 2011 @ *www.computerworld.com/s/article/9196823/500_million_people_will_be_using_mobile_health_apps_by_2015.*

13

1. McCarthy, J. "What Is Artificial Intelligence?" *Stanford University.* 12 November 2007. Accessed 16 July 2010 @ *www.formal.stanford.edu/jmc/whatisai/whatisai.html.*

2. Gaudin, S. "Researchers Enable Computers to Teach Themselves Common Sense." *Computerworld.* 27 November 2013. Accessed 2 January 2014 @ *www.computerworld.com/s/article/9244403/Researchers_enable_computers_to_teach_themselves_common_sense.*

3. Ibid.

4. Schwartz, E. "Smart Programs Go to Work." *BusinessWeek.* 2 March 1992.

5. Lesaint, D, C. Voudouris, and N. Azarmi. "Dynamic Workforce Scheduling for British Telecommunications." *Interfaces* 30(1) (2000): 45–56.

6. Hall, O. "Artificial Intelligence Techniques Enhance Business Forecasts." *Graziadio Business Review* 5(2) (2002). Accessed 15 November 2009 @ *http://gbr.pepperdine.edu/022/intelligence.html.*

7. Bidgoli, H. "Integration of Technologies: An Ultimate Decision-Making Aid." *Industrial Management & Data Systems* 93(1) (1993): 10–17.

8. Ives, B. "Expert System Puts More Meaning into Semantic Search." *The AppGap.* 2 June 2008. Accessed 16 July 2010 @ *www.theappgap.com/expert-system-puts-more-meaning-into-semantic-search.html.*

9. Kneale, D. "How Coopers & Lybrand Put Expertise into Its Computers." *Wall Street Journal.* 14 November 1986.

10. Roby, R. "Expert Systems Help Labs Process DNA Samples." *NIJ Journal.* July 2008. Accessed 16 July 2010 @ *www.ojp.usdoj.gov/nij/journals/260/expert-systems.htm.*

11. Wai, K., A. Rahman, M. Zaiyadi, and A. Aziz. "Expert System in Real-World Applications." *Generation5.* Accessed 16 July 2010 @ *www.generation5.org/content/2005/Expert_System.asp.*

12. Adderley, R., and P. Musgrove. "Police Crime Recording and Investigation Systems—A User's View." *Policing: An International Journal of Police Strategies and Management* 24(1) (2001): 100–114.

13. Lewis, D. "Intelligent Agents and the Semantic Web." *IBM developerWorks.* 28 October 2008. Accessed 16 July 2010 @ *https://www14.software.ibm.com/webapp/iwm/web/signup.do?source=swg-BA_WebOrganic&S_PKG=ov7816&form=336&S_TACT=101KR81W&cmp=101KR&ct=101KR81W&cr=google&cm=k&csr=BI-Cognos_Enterprise_Demo-Branded-Search&ccy=us&ck=cognos%20powerplay&cs=e&mkwid=sqXkhASmC-dc_33445287668_43246d30503&gclid=CLvpxsrPz78CFZSEfgod8kIAkQ.*

14. Dobbs, D., K. Boehle, D. Goldwasser, and J. Stamps. "The Return of Artificial Intelligence." *Training.* November 2000.

15. Gaydos, M. "The Debate Room: Virtual Agents Will Replace Live Customer Service Reps." *Bloomberg Businessweek.* Accessed 5 March 2013 @ *www.businessweek.com/debateroom/archives/2010/07/virtual_agents_will_replace_live_customer_service_reps.html.*

16. "Louise." *Chatbots.org.* Accessed 5 March 2013 @ *www.chatbots.org/virtual_agent/louise.*

17. Claburn, T. "Google Search App Vs. Apple Siri: 8 Questions." *InformationWeek.* 1 November 2012. Accessed 5 March 2013 @ *www.informationweek.com/internet/google/google-search-app-vs-apple-siri-8-questi/240012715.*

18. Snyder, B. "Beyond 'Jeopardy': How IBM Will Make Billions from Watson." *InfoWorld.* 10 May 2012. Accessed 5 March 2013 @ *www.infoworld.com/d/the-industry-standard/beyond-jeopardy-how-ibm-will-make-billions-watson-192814?source=IFWNLE_nlt_daily_2012-05-10.*

19. Dorrier, J. "IBM's Watson Expands Commercial Applications, Aims to Go Mobile." *SingularityHub.* 14 October 2012. Accessed 5 March 2013 @ *http://singularityhub.com/2012/10/14/ibms-watson-jeopardy-champ-expands-commercial-applications-aims-to-go-mobile.*

20. "Web Channel: Provide Instant, Accurate Answers to Your Customers' Questions Right from Your Website." *Intelliresponse.com.* Accessed 12 March 2013 @ *http://www.intelliresponse.com/products/intelliresponse-virtual-agent/.*

21. Zadeh, L. "Yes, No and Relatively, Part 2." *Chemtech.* July 1987.

22. Foster, Z. "Guesstimates, Fuzzy Logic and Believe it or Not…Better Decision Making." blogspot.com. 9 March 2011. Accessed 3 December 2011 @ *http://talk-technology.blogspot.com/2011/03/guesstimates-fuzzy-logic-and-believe-it.html.*

23. Klingenberg, B. "Fuzzy Logic Applications." *Getting Started with Fuzzy Logic.* Accessed 12 March 2013 @ *http://www.calvin.edu/~pribeiro/othrlnks/Fuzzy/home.htm.*

24. Sherman, C. "In Praise of Fuzzy Searching." *Search Engine Watch.* 5 June 2001. Accessed 12 March 2013 @ *http://searchenginewatch.com/article/2064695/In-Praise-of-Fuzzy-Searching.*

25. Feng, G. "A Survey on Analysis and Design of Model-Based Fuzzy Control Systems." *IEEE Transactions on Fuzzy Systems* 14(5) (2006): 676–697.

26. California Scientific. "Maximize Returns on Direct Mail with BrainMaker Neural Network Software." *CalSci.com.* Accessed 20 July 2010 @ *www.calsci.com/DirectMail.html.*

27. California Scientific. "Neural Network Red-Flags Police Officers with Potential for Misconduct." *CalSci.com.* Accessed 20 July 2010 @ *www.calsci.com/Police.html.*

28. Holland, J. "Genetic Algorithms." *Scientific American.* July 1992.

29. Wiggins, R. "Docking a Truck: A Genetic Fuzzy Approach." *AI Expert* 7(5) (1992): 28–35.

30. Marczyk, A. "Genetic Algorithms and Evolutionary Computation." *The Talk Origins Archive.* Accessed 9 May 2012@ *www.talkorigins.org/faqs/genalg/genalg.html.*

31. Obermeir, K. "Natural Language Processing." *Byte.* December 1987.

32. Liu, S., A. Duffy, R. Whitfield, and I. Boyle. "Integration of Decision Support Systems to Improve Decision Support Performance." In *Knowledge and Information Systems.* London, UK: Springer, 2009.

33. Basu, R., N. Archer, and B. Mukherjee. "Intelligent Decision Support In Healthcare." *Analytics.* January, February 2012. Accessed 5 March 2013 @ *www.analytics-magazine.org/januaryfebruary-2012/507-intelligent-decision-support-in-healthcare.*

34. Turban, E., and P. Watkins. "Integrating Expert Systems and Decision Support Systems." *MIS Quarterly* 10(2) (1986): 121–136.

35. Ibid.

36. Mortensen, P. "The Future of Technology Isn't Mobile, It's Contextual." *Fast Company.* 24 May 2013. *Fast Company.* Accessed 2 January 2014 @ *www.fastcodesign.com/1672531/the-future-of-technology-isnt-mobile-its-contextual.*

37. Ibid.

38. Cohen, R. "Contextual Computing: Our Sixth, Seventh and Eighth Senses." *Forbes.com.* 18 October 2013. Accessed 2 January 2013 @ *www.forbes.com/sites/reuvencohen/2013/10/18/contextual-computing-our-sixth-seventh-and-eighth-senses.*

39. This information has been gathered from the company Web site (*www.alyuda.com*) and other promotional materials. For more information and updates, visit the Web site.

40. Alison, D. "12 Advances in Medical Robotics." *InformationWeek.* 29 January 2011. Accessed 5 March 2013 @ *www.informationweek.com/healthcare/patient/12-advances-in-medical-robotics/229100383.*

41. "FDA Clears First Autonomous Telemedicine Robot for Hospitals." *iRobot.com.* 24 January 1 2013. Accessed 5 March 2013 @ *www.irobot.com/us/Company/Press_Center/Press_Releases/Press_Release.aspx?n=012413.*

42. Mearian, L. "Physician Robot to Begin Making Rounds." *Computerworld.* 25 July 2012. Accessed 5 March 2013 @ *www.computerworld.com/s/article/9229616/Physician_robot_to_begin_making_rounds.*

43. Yegulalp, S. "Speech Recognition: Your Smartphone Gets Smarter." *Computerworld.* 16 March 2011. Accessed 14 August 2011 @ *www.computerworld.com/s/article/9213925/Speech_recognition_Your_smartphone_gets_smarter?taxonomyId=15&pageNumber=1.*

44. Cringely, R. "Apple's Siri Is AI with an Attitude." *InfoWorld.* 17 October 2011. Accessed 1 December 2011 @ *www.infoworld.com/t/cringely/apples-siri-ai-attitude-176323?source=IFWNLE_nlt_blogs_2011-10-18.*

14

1. Hamilton, J. "Virtual Reality: How a Computer-Generated World Could Change the Real World." *BusinessWeek.* 5 October 1992. Accessed 16 July 2010 @ *www.businessweek.com/1989-94/pre88/b328653.htm.*

2. Marquez, J. "Introduction to Virtual Reality: Humans and Automation Seminar." Spring 2002. Accessed 11 April 2012 @ *http://web.mit.edu/16.459/www/VR1.pdf.*

3. D'Angelo, M., and S. Narayaran. "A Virtual Reality Environment to Assist Disabled Individuals." Accessed 11 April 2012 @ *www.wright.edu/lwd/documents/virtual_rehabilitation_dangelo.pdf.*

4. Schneider, R. "Lockheed Starts Using Virtual Reality to Test New Products." *Talking Points Memo.* 26 January 2011. Accessed 2 January 2014 @ *http://talkingpointsmemo.com/dc/lockheed-starts-using-virtual-reality-to-test-new-products—2.*

5. Reuters. "Strategy Analytics: Virtual Worlds Forecast to Grow at 23% Through 2015." *Reuters.com.* 15 June 2009. Accessed 9 March 2012 @ *www.reuters.com/article/2009/06/15/idUS165902+15-Jun-2009+BW20090615.*

6. "Virtual NOAA." *ESRL Quarterly Newsletter.* Summer 2009. Accessed 9 March 2010 @ *www.esrl.noaa.gov/news/quarterly/summer2009/virtual_noaa.html.*

7. Weis, S. "RFID: Technical Considerations." *The Handbook of Technology Management.* Edited

by H. Bigdoli. Hoboken, NJ: John Wiley & Sons, 2010.

8. Weier, M. "Coke's RFID-Based Dispensers Redefine Business Intelligence." *InformationWeek*. 6 June 2009. Accessed 7 October 2010 @ *www.informationweek.com/news/mobility/RFID/showArticle.jhtml?articleID=217701971*.

9. "QR Code Features." *QR Code.com*. Accessed 7 December 2011 @ *www.denso-wave.com/qrcode/qrfeature-e.html*.

10. Campbell, A. "QR Codes, Barcodes and RFID: What's the Difference?" *Small Business Trends*. 21 February 2011. Accessed 2 December 2011 @ *http://smallbiztrends.com/2011/02/qr-codes-barcodes-rfid-difference.html*.

11. Sprague, M. "Close-Up with Google's New QR Code Generator." *Search Engine Land*. 7 October 2010. Accessed 2 December 2011 @ *http://searchengineland.com/close-up-with-googles-new-qr-code-generator-52248*.

12. Kotz, L. "Sacre Bleu Wine Adapts QR Codes to Labels." *DataMart Direct*. 8 August 2012. Accessed 6 March 2013 @ *www.datamartdirect.com/index.php/2012/08/sacre-blue-wine-adapts-qr-codes-to-labels*.

13. Ankeny, J. "How One Small Company Is Using QR Codes." *Entrepreneur*. 3 October 2011. Accessed 6 March 2013 @ *www.entrepreneur.com/article/220359*.

14. Goldberg, J. "Best Buy Deploys QR Codes to Enhance Shopping Experience." *Retailgeek*. Accessed 6 March 2013 @ *http://retailgeek.com/best-buy-deploys-qr-codes-to-enhance-shopping-experience*.

15. Tsirulnik, G. "Calvin Klein Activates Billboards with QR Codes Pushing Mobile Video Ad." *Mobile Marketer*. 29 July 2010. Accessed 6 March 2013 @ *www.mobilemarketer.com/cms/news/advertising/6933.html*.

16. Sprague, M. "QR Codes: Are You Ready for Paper-Based Hyperlinks?" *Search Engine Land*. 13 September 2010. Accessed 6 March 2013 @ *http://searchengineland.com/qr-codes-are-you-ready-for-paper-based-hyperlinks-49684*.

17. Kats, R. "McDonald's Continues Mobile Reign with QR Code Push." *Mobile Marketer*. 30 July 2012. Accessed 6 March 2013 @ *www.mobilemarketer.com/cms/news/software-technology/13410.html*.

18. "Pepsi Uses QR Codes to Push Video Content." *MediaPoondi*. 30 October 2012. Accessed 6 March 2013 @ *www.mediapoondi.com/2012/10/30/pepsi-uses-qr-codes-to-push-video-content*.

19. Lamb, R. "Ralph Lauren Steps up Mobile Game with Customized QR Codes." *Luxury Daily*. 20 September 2011. Accessed 6 March 2013 @ *www.luxurydaily.com/ralph-lauren-steps-up-mobile-game-with-customized-qr-codes*.

20. "Starbucks Customers Pay for Drinks with QR Codes." *BeQRious*. Accessed 6 March 2013 @ *http://beqrious.com/starbucks-customers-pay-for-drinks-with-qr-codes*.

21. Mesenbrink, J. "Biometrics Plays Big Role with Airport Security." *Security*. 4 February 2002. Accessed 9 March 2012 @ *www.securitymagazine.com/articles/biometrics-plays-big-role-with-airport-security-1*.

22. Knorr, E., and G. Gruman. "What Cloud Computing Really Means." *InfoWorld*. 7 April 2008. Accessed 9 March 2012 @ *www.infoworld.com/article/08/04/07/15FE-cloud-computing-reality_1.html*.

23. Brodkin, J. "Amazon and IBM Are the Cloud's Biggest Players." *InfoWorld*. 15 July 2010. Accessed 31 July 2010 @ *www.infoworld.com/d/cloud-computing/amazon-and-ibm-are-the-clouds-biggest-players-484? page=0,0&source=IFWNLE_nlt_wrapup_2010-07-16*.

24. Knorr, E., and G. Gruman. "What Cloud Computing Really Means." *InfoWorld*. 7 April 2008. Accessed 21 August 2010 @ *www.infoworld.com/article/08/04/07/15FE-cloud-computing-reality_1.html*.

25. McFedries, P. "The Cloud Is the Computer." *IEEE Spectrum*. August 2008. Accessed 21 August 2010 @ *www.spectrum.ieee.org/computing/hardware/the-cloud-is-the-computer*.

26. Dignan, L. "Amazon's Cloud Computing Will Surpass Its Retailing Business." *ZDNet*. 14 April 2008. Accessed 21 August 2010 @ *http://blogs.zdnet.com/BTL/?p=8471*.

27. Brodkin, J. "10 Cloud Computing Companies to Watch." *NetworkWorld*. 18 May 2009. Accessed 21 August 2010 @ *www.networkworld.com/supp/2009/ndc3/051809-cloud-companies-to-watch.html?page=1*.

28. Mell, P., and M. Grance. "The NIST Definition of Cloud Computing." *NIST*. September 2011. Accessed April 11 2012 @ *http://csrc.nist.gov/publications/nistpubs/800-145/SP800-145.pdf*.

29. Brodkin, J. "Gartner: Seven Cloud-Computing Security Risks." *InfoWorld*. 2 July 2008. Accessed 3 December 2011 @ *www.infoworld.com/d/security-central/gartner-seven-cloud-computing-security-risks-853?page=0,0*.

30. Wood, S., R. Jones, and A. Geldart. "Commercial Applications of Nano-Technology in Computing and Information Technology." *Azonano.com*. 29 October 2008. Accessed 21 August 2010 @ *www.azonano.com/details.asp?ArticleID=1057*.

31. "New Report on Nanotechnology in Consumer Goods." *AZoM.com*. 26 May 2009. Accessed 21 August 2010 @ *www.azom.com/news.asp?NewsID=17263*.

32. IBM. "Nanotech Defined." *IBM.com*. Accessed 11 April 2012 @ *http://domino.research.ibm.com/comm/research.nsf/pages/r.nanotech.defined.html*.

33. This information has been gathered from the company Web site (*www.mechdyne.com*) and other promotional materials. For more information and updates, visit the Web site.

34. "About IHG." *InterContinental Hotels Group*. Accessed 6 March 2013 @ *www.ichotelsgroup.com/ihg/hotels/us/en/global/support/about_ihg*.

35. Babcock, C. "4 Companies Getting Real Results from Cloud Computing." *InformationWeek*. 15 January 2011. Accessed 6 March 2013 @ *www.informationweek.com/hardware/data-centers/4-companies-getting-real-results-from-cl/229000706*.

36. Prince, P. "RFID a 'Very Big Part of Macy's Future.'" *RFID Journal*. 21 June 2013. Accessed 2 January 2014 @ *www.rfidjournal.com/articles/view?10783*.

37. Ibid.

INDEX

knowledge base management system (KBMS), **268**
user interface, 268
uses, 269
explanation facility, **268**
explicit knowledge, 235–236
Expression Web, 18
Express Scripts, 61, 62
Extended ASCII, 28
Extended Binary Coded Decimal Interchange Code (EBCDIC), 28
Extensible Markup Language (XML), 39, 155–156, 228
Extensible Stylesheet Language (XSL), 39
external bus, 25
external data, 8, 47, 57, 250
external users, **204**
extraction, transformation, and loading (ETL), **57**
extranets, 13, **151**, 232
business-to-business (B2B) e-commerce, 171
competitive advantage, 151
coordination, 152
electronic funds transfer (EFT), 151
e-mail, 151
virtual private networks (VPNs), 100
extreme programming (XP), **218**–219

F

Facebook, 4, 6, 66, 87, 139, 142–143, 155, 161, 168, 176, 241, 293
facial recognition, 94
Fair Credit Reporting Act (1970), 69
Fair Use Doctrine, 75
Family Dollar, 249
Fandango Web site, 281
Farms.com, 229
fault-tolerant systems, **90**
fax servers, 35
feasibility study, **205**–207
Federal Express (FedEX), 20–21, 165, 193, 199
Federal Express Tracking System, 151
FedEx.com Web site, 20, 232
FedEx Global Trade Manager, 193
FedEx INTrade, Traders Information Exchange System, 193
fiber optics, 114–115, 118
fields, 45, 50
fifth-generation computers, 26
fifth-generation languages (5GLs), **39**
Fifth Third Bank, 92
File Forces Model, 15–16
FileMaker Pro, 9, 37
files, 48
file servers, 35
File Transfer Protocol (FTP), 149, 187
financial apps, 41–42
financial information system (FIS), 13
financial institutions Web sites, 147
financial planning and accounting software, 38
fingerprints, 94
Firefox, 71, 142
fire protection systems, 104
firewalls, **95**–97, 104
first-generation computers, 26

first normal form (1NF), 50
First Phones, 21
Five Forces Model, **14**–16
flat files, 46
flight simulation, 288
Flikr, 153
Flurry, 127
Food and Drug Administration, 107
Force.com Builder, 239
Ford, 193
Ford Models, 5
forecasting model, 10
foreign key, **50**
forensics lab work, 269
formal information, 10
Forrester Research, 113
forward chaining, **268**
4G networks, 131, 174
Foursquare, 4, 6, 168
fourth-generation computers, 26
fourth-generation languages (4GLs), 27, **38**–39, 211
"Four Things You Can Do to Help Protect Kids Online" (Microsoft), 74
Fox television network, 155
fragmentation, **54**
FreeScan, 105
Frequency Hopping Spread Spectrum (FHSS), 295
front-end processors (FEP), 120
front-end servers, 89
Fujitsu U-Scan self-checkout software, 11
full-body immersion, 286
full-duplex transmission, 187
Functional Robotics (FROB), 266
Furness, Thomas, 285
fuzzy logic, **273**–274

G

Galaxy, 4
Galaxy Tab, 42
gallium arsenide chips, 26
game consoles, 113
Gantt charts, 212–213
The Gap, 168, 292
The Garden intranet, 161
Gartner, Inc., 83, 299
GeneHunter, 276
General Electric, 189
General Motors, 155, 191
generic top-level domains (gTLDs), 140
generic word TLDs, 140
genetic algorithms (GAs), **276**
geographic domains, 140
geographic information systems (GISs), **256**–258
Gigabit Ethernet, 131
Gigandet, Christy, 12
gigapops, **157**
GIGO (garbage in, garbage out), 24
Gimbal proximity beacons, 181
global companies, 190
global database, 186
global environment, 188
global information systems (GIS), **185**
components, 186–188
cultural differences, 195–196
diverse regulatory practices, 196

e-business, 184–185
global structure, **191**–192
growth of Internet, 185
implementing, 189–190
information sharing technologies, 186–188
intellectual property laws, 196
international structure, **192**
Internet, 187
jurisdiction issues, 196
migration to, 190
multinational structure, **190**–191
network architecture, 187
obstacles to using, 194–196
offshore outsourcing, **194**
operational support, 189
reasons to use, 183–184
requirements, 188–189
time zones, 195
transborder data flow (TDF), **188**
transmission technologies, 187
transnational structure, **192**–193
global structure, **191**–192
Global Times, 79
Gmail, 174
Goldman Sachs, 168
Goodwin Procter, 237
Google+, 4, 6
Google, Inc., 16, 35, 83, 127, 142–143, 158–159, 167, 174, 176, 178, 234, 271
Google AdSense, 153
Google AppEngine, 298
Google Apps, 297
Google Books, 83
Google Chat, 144
Google.com/flights/, 145
Google Docs, 35
Google Drive, 35
Google Glass, 34
Google Gmail, 143
Google Hangouts, 255
Google Maps, 83, 93, 257–258
Google Now, 277
Google Play, 127
Google Search, 153
Google Shopper, 127
Google URL shortener, 293
Google Voice, 174
Google Voice Search, 281
Gore, Al, 139
GoToMeeting, 112, 236
government, 257
e-commerce, 169
Government Technology magazine, 170
government-to-business (G2B) e-commerce, 169
government-to-citizen (G2C) e-commerce, 169
government-to-employee (G2E) e-commerce, 169
government-to-government (G2G) e-commerce, 169
GPS, 12, 261
graphical user interfaces (GUIs), 9–10, 29, 251
graphics cards, 31
graphics software, 36, 37
Graph Search, 142–143
green computing, **79**–80
grid computing, **295**–296
Groove, 153

LEARNING OUTCOMES

2-1 Define a computer system, and describe its components.

A computer is a machine that accepts data as input, processes data without human intervention by using stored instructions, and outputs information. The instructions, also called a "program," are step-by-step directions for performing a specific task, written in a language the computer can understand.

A computer system consists of hardware and software. Hardware components are physical devices, such as keyboards, monitors, and processing units. The software component consists of programs written in computer languages.

2-2 Discuss the history of computer hardware and software.

Major developments in hardware have taken place over the past 60 years. To understand these developments better, computers are often categorized into "generations" to mark technological breakthroughs.

2-3 Explain the factors distinguishing the computing power of computers.

Computers draw their power from three factors that far exceed human capacities: speed, accuracy, and storage and retrieval capabilities.

Every character, number, or symbol on the keyboard is represented as a binary number in computer memory. A binary system consists of 0s and 1s, with a 1 representing "on" and a 0 representing "off," similar to a light switch.

Computers and communication systems use data codes to represent and transfer data between computers and network systems.

Three data codes are: ASCII, Extended ASCII, and Unicode.

2-4 Summarize computer operations.

Computers can perform three basic tasks: arithmetic operations, logical operations, and storage and retrieval operations. All other tasks are carried out by one or a combination of these operations.

Computers can add, subtract, multiply, divide, and raise numbers to a power. They also perform comparison operations.

Computers can store massive amounts of data in very small spaces and locate a particular item quickly.

2-5 Discuss the types of input, output, and memory devices.

Input devices send data and information to the computer. Examples include keyboards, a mouse, touch screens, light pens, a trackball, data tablet, bar code reader, optical character readers, magnetic ink character recognition, and optical mark recognition.

Output devices are available for both mainframe and personal computers. Output devices are either soft copy (plasma display, LCD) or hard copy (printers).

Memory is considered either main or secondary.

2-6 Explain how computers are classified.

Computers are classified based on cost, amount of memory, speed, and sophistication.

Computers can be classified as: subnotebook; notebook; personal computer; minicomputer; mainframes; or supercomputer.

2-7 Describe the two major types of software.

Software is all of the programs that run a computer system.

The two types of software include: system software and application software.

System software is run by operating systems (OS). Examples of OS include Microsoft Windows, Mac OS, and Linux.

Application software is used to perform a variety of tasks on a personal computer, including word processing, spreadsheet, database, presentation, graphics, etc. Examples of application software suites include Microsoft Office, OpenOffice, and Corel.

KEY TERMS

2-1

A **computer** is a machine that accepts data as input, processes data without human intervention by using stored instructions, and outputs information.

The **central processing unit (CPU)** is the heart of a computer. It is divided into two components: the arithmetic logic unit (ALU) and the control unit.

The **arithmetic logic unit (ALU)** performs arithmetic operations (+, −, *, /) as well as comparison or relational operations (<, >, =); the latter are used to compare numbers.

The **control unit** tells the computer what to do, such as instructing the computer which device to read or send output to.

A **bus** is a link between devices connected to the computer. It can be parallel or serial, internal (local) or external.

A **disk drive** is a peripheral device for recording, storing, and retrieving information.

A **CPU case** is also known as a computer chassis or tower. It is the enclosure containing the computer's main components.

A **motherboard** is the main circuit board containing connectors for attaching additional boards. It usually contains the CPU, Basic Input/Output System (BIOS), memory, storage, interfaces, serial and parallel ports, expansion slots, and all the controllers for standard peripheral devices, such as the display monitor, disk drive, and keyboard.

2-5

Input devices send data and information to the computer. Examples include a keyboard and mouse.

An **output device** is capable of representing information from a computer. The form of this output might be visual, audio, or digital; examples include printers, display monitors, and plotters.

Main memory stores data and information and is usually volatile; its contents are lost when electrical power is turned off. It plays a major role in a computer's performance.

Secondary memory, which is nonvolatile, holds data when the computer is off or during the course of a program's operation. It also serves as archival storage.

Random access memory (RAM) is volatile memory, in which data can be read from and written to; it is also called read-write memory.

KEY TERMS

cache RAM, resides on the processor. Because memory access from main RAM storage takes several clock cycles (a few nanoseconds), cache RAM stores recently accessed memory so the processor is not waiting for the memory transfer.

read-only memory (ROM) is nonvolatile; data cannot be written to ROM.

Magnetic tape, made of a plastic material, resembles a cassette tape and stores data sequentially.

A **magnetic disk**, made of Mylar or metal, is used for random-access processing. In other words, data can be accessed in any order, regardless of its order on the surface.

Optical discs use laser beams to access and store data. Examples include CD-ROMs, WORM discs, and DVDs.

Used for online storage and backup, **cloud storage** involves multiple virtual servers that are usually hosted by third parties. Customers buy or lease storage space from third parties based on their current or future needs.

A **redundant array of independent disks (RAID) system** system is a collection of disk drives used for fault tolerance and improved performance, and is typically found in large network systems.

A **storage area network (SAN)** is a dedicated high-speed network consisting of both hardware and software used to connect and manage shared storage devices, such as disk arrays, tape libraries, and optical storage devices.

Network-attached storage (NAS) is essentially a network-connected computer dedicated to providing file-based data storage services to other network devices.

2-6
A **server** is a computer and all the software for managing network resources and offering services to a network.

2-7
Application software can be commercial software or software developed in house and is used to perform a variety of tasks on a personal computer.

An **operating system (OS)** is a set of programs for controlling and managing computer hardware and software. It provides an interface between a computer and the user and increases computer efficiency by helping users share computer resources and by performing repetitive tasks for users.

2-8
Machine language, the first generation of computer languages, consists of a series of 0s and 1s representing data or instructions. It is dependent on the machine, so code written for one type of computer does not work on another type of computer.

2-8 List the generations of computer languages.

There are four generations of computer language with a fifth generation in development.
- First: Machine Language
- Second: Assembly Language
- Third: High-level Language
- Fourth: Fourth-Generation Language
- Fifth: Natural Language Processing

Assembly language, the second generation of computer languages, is a higher-level language than machine language but is also machine dependent. It uses a series of short codes, or mnemonics, to represent data or instructions.

High-level languages are machine independent and part of the third-generation of computer languages. Many languages are available, and each is designed for a specific purpose.

Fourth-generation languages (4GLs) use macro codes that can take the place of several lines of programming. The commands are powerful and easy to learn, even for people with little computer training.

Fifth-generation languages use some of the artificial intelligence technologies, such as knowledge-based systems, natural language processing (NLP), visual programming, and a graphical approach to programming. These languages are designed to facilitate natural conversations between you and the computer.

WWW.CENGAGEBRAIN.COM/LOGIN

Like 4LTR Press for MIS on Facebook
Follow 4LTRPress_MIS on Twitter
View related videos at the YouTube channel created for MIS

LEARNING OUTCOMES

6-1 Describe major applications of a data communication system.

Data communication is the electronic transfer of data from one location to another. Data communication applications enhance decision makers' efficiency and effectiveness.

6-2 Explain the major components of a data communication system.

Typical data communication systems include the following components:

- Sender and receiver devices
- Modems or routers
- Communication medium (channel) (see Exhibit 6.1)

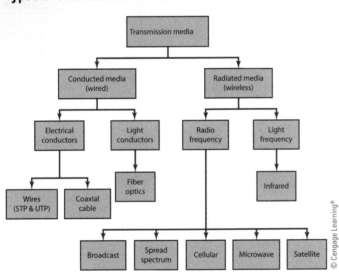

Exhibit 6.1
Types of communication media

© Cengage Learning®

6-3 Describe the major types of processing configurations.

Data communication systems can be used in several different configurations, including centralized, decentralized, and distributed processing systems.

6-4 Explain the three types of networks.

The three major types of networks are local area networks, wide area networks, and metropolitan area networks.

6-5 Describe the main network topologies.

The five common topologies are star, ring, bus, hierarchical, and mesh.

6-6 Explain important networking concepts, such as bandwidth, routing, routers, and the client/server model.

Protocols are agreed-on methods and rules that electronic devices use to exchange information. Transmission Control Protocol/Internet Protocol (TCP/IP) is an industry-standard suite of communication protocols.

The process of deciding which path data takes is called routing.

In the client/server model, software runs on the local computer (the client) and communicates with the remote server to request information or services.

KEY TERMS

6-1

Data communication is the electronic transfer of data from one location to another.

6-2

Bandwidth is the amount of data that can be transferred from one point to another in a certain time period, usually one second.

Attenuation is the loss of power in a signal as it travels from the sending device to the receiving device.

In **broadband** data transmission, multiple pieces of data are sent simultaneously to increase the transmission rate.

Narrowband is a voice-grade transmission channel capable of transmitting a maximum of 56,000 bps, so only a limited amount of information can be transferred in a specific period of time.

Protocols are rules that govern data communication, including error detection, message length, and transmission speed.

A **modem** (short for "modulator-demodulator") is a device that connects a user to the Internet.

Digital subscriber line (DSL) a common carrier service, is a high-speed service that uses ordinary phone lines.

Communication media or channels, connect sender and receiver devices. They can be conducted or radiated.

Conducted media provide a physical path along which signals are transmitted, including twisted pair copper cable, coaxial cable, and fiber optics.

Radiated media use an antenna for transmitting data through air or water.

6-3

In a **centralized processing** system, all processing is done at one central computer.

In **decentralized processing**, each user, department, or division has its own computer (sometimes called an "organizational unit") for performing processing tasks.

Distributed processing maintains centralized control and decentralized operations. Processing power is distributed among several locations.

The **Open Systems Interconnection (OSI) model** is a seven-layer architecture for defining how data is transmitted from computer to computer in a network, from the physical connection to the network to the applications that users run. It also standardizes interactions between network computers exchanging information.

6-4

A **network interface card (NIC)** is a hardware component that enables computers to communicate over a network.

A **local area network (LAN)** connects workstations and peripheral devices that are in close proximity.

A **wide area network (WAN)** can span several cities, states, or even countries, and it is usually owned by several different parties.

A **metropolitan area network (MAN)** is designed to handle data communication for multiple organizations in a city and sometimes nearby cities as well.

6-5

A **network topology** represents a network's physical layout, including the arrangement of computers and cables.

The **star topology** usually consists of a central computer (host computer, often a server) and a series of nodes (typically, workstations or peripheral devices).

In a **ring topology**, no host computer is required because each computer manages its own connectivity.

The **bus topology** (also called "linear bus") connects nodes along a network segment, but the ends of the cable are not connected, as they are in a ring topology.

A **hierarchical topology** (also called a "tree") combines computers with different processing strengths in different organizational levels.

6-6

A **controller** is a hardware and software device that controls data transfer from a computer to a peripheral device (examples are a monitor, a printer, or a keyboard) and vice versa.

A **multiplexer** is a hardware device that allows several nodes to share one communication channel.

In a **mesh topology** (also called "plex" or "interconnected"), every node (which can differ in size and configuration from the others) is connected to every other node.

Transmission Control Protocol/ Internet Protocol (TCP/IP) is an industry-standard suite of communication protocols that enables interoperability.

6-7 Describe wireless and mobile technologies and networks.

A wireless network is a network that uses wireless instead of wired technology. A mobile network is a network operating on a radio frequency (RF). See Exhibit 6.10.

6-8 Discuss the importance of wireless security and the techniques used.

Wireless security is especially important since anyone within walking or driving distance of an access point can gain use of a wireless network.

There are several techniques for improving the security of wireless networks:

- Service set identifier (SSI)
- Wired equivalent privacy (WEP)
- Extensible authentication protocol (EAP)
- Wi-Fi protected access (WPA)
- WPA2 or 802.11i

6-9 Summarize the convergence phenomenon and its applications for business and personal use.

In data communication, convergence refers to integrating voice, video, and data so that multimedia information can be used for decision making.

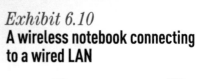

Exhibit 6.10
A wireless notebook connecting to a wired LAN

© Cengage Learning®

A **packet** is a collection of binary digits—including message data and control characters for formatting and transmitting—sent from computer to computer over a network.

Routing is the process of deciding which path to take on a network. This is determined by the type of network and the software used to transmit data.

A **routing table**, generated automatically by software, is used to determine the best possible route for a packet.

In **centralized routing**, one node is in charge of selecting the path for all packets. This node, considered the network routing manager, stores the routing table, and any changes to a route must be made at this node.

Distributed routing relies on each node to calculate its own best possible route. Each node contains its own routing table with current information on the status of adjacent nodes so packets can follow the best possible route.

A **router** is a network connection device containing software that connects network systems and controls traffic flow between them.

A **static router** requires the network routing manager to give it information about which addresses are on which network.

A **dynamic router** can build tables that identify addresses on each network.

In the **client/server model**, software runs on the local computer (the client) and communicates with the remote server to request information or services. A server is a remote computer on the network that provides information or services in response to client requests.

In the **two-tier architecture** (the most common type), a client (tier one) communicates directly with the server (tier two).

An **n-tier architecture** attempts to balance the workload between client and server by removing application processing from both the client and server and placing it on a middle-tier server.

6-7

A **wireless network** is a network that uses wireless instead of wired technology.

A **mobile network** (also called a cellular network) is a network operating on a radio frequency (RF), consisting of radio cells, each served by a fixed transmitter, known as a cell site or base station.

Throughput is similar to bandwidth. It is the amount of data transferred or processed in a specified time, usually one second.

To improve the efficiency and quality of digital communications, **Time Division Multiple Access (TDMA)** divides each channel into six time slots. Each user is allocated two slots: one for transmission and one for reception. This method increases efficiency by 300 percent, as it allows carrying three calls on one channel.

To improve the efficiency and quality of digital communications, **Code Division Multiple Access (CDMA)** transmits multiple encoded messages over a wide frequency and then decodes them at the receiving end.

6-8

In data communication, **convergence** refers to integrating voice, video, and data so that multimedia information can be used for decision making.

WWW.CENGAGEBRAIN.COM/LOGIN

Like 4LTR Press for MIS on Facebook
Follow 4LTRPress_MIS on Twitter
View related videos at the YouTube channel created for MIS

LEARNING OUTCOMES

7-1 Describe the makeup of the Internet and the World Wide Web.

The Internet backbone is a foundation network linked with fiber-optic cables that can support very high bandwidth.

The World Wide Web changed the Internet by introducing a graphical interface to the largely text-based Internet in 1989.

When information is transferred from one network to another, domain names are converted to IP addresses by the protocol Domain Name System (DNS). You see domain names used in uniform resource locators (URLs), also called "universal resource locators," to identify a Web page.

There are several methods for connecting to a network, including the Internet. These methods include dial-up and cable modems as well as Digital Subscriber Lines (DSLs).

7-2 Discuss navigational tools, search engines, and directories.

Navigational tools are used to travel from site-to-site or to "surf" the Internet.

Search engines give you an easy way to look up information and resources on the Internet by entering key words related to your topic of interest.

Directories are indexes of information based on keywords in documents and make it possible for search engines to find what you're looking for. Some Web sites, such as Yahoo!, also use directories to organize content into categories.

7-3 Describe common Internet services.

Many services are available via the Internet, and most are made possible by the TCP suite of protocols in the Application layer. These include:

- E-mail
- Newsgroups and discussion groups
- Internet Relay Chat (IRC) and instant messaging
- Internet telephony

7-4 Summarize widely used Web applications.

Several service industries use the Internet and its supporting technologies to offer services and products to a wide range of customers at more competitive prices and with increased convenience.

7-5 Explain the purpose of intranets.

An intranet is a network within an organization that uses Internet protocols and technologies for collecting, storing, and disseminating useful information that supports business activities. See Tables 7.2 and 7.3.

TABLE 7.2 THE INTERNET vs. INTRANETS

Key Feature	Internet	Intranet
User	Anybody	Approved users only
Geographical scope	Unlimited	Limited or unlimited
Speed	Slower than an intranet	Faster than the Internet
Security	Less than an intranet's	More than the Internet's; user access more restricted

© Cengage Learning®

TABLE 7.3 COMPARISON OF THE INTERNET, INTRANETS, AND EXTRANETS

	Internet	Intranet	Extranet
Access	Public	Private	Private
Information	General	Typically confidential	Typically confidential
Users	Everybody	Members of an organization	Groups of closely related companies, users, or organizations

© Cengage Learning®

KEY TERMS

7-1

The **Internet** is a worldwide collection of millions of computers and networks of all sizes. It is a network of networks.

The **Advanced Research Projects Agency Network (ARPANET)**, a project started in 1969 by the U.S. Department of Defense, was the beginning of the Internet.

The **Internet backbone** is a foundation network linked with fiber-optic cables that can support very high bandwidth. It is made up of many interconnected government, academic, commercial, and other high-capacity data routers.

With **hypermedia**, documents include embedded references to audio, text, images, video, and other documents.

The embedded references in hypermedia documents are called **hypertext**; they consist of links users can click to follow a particular thread (topic).

When information is transferred from one network to another, domain names are converted to IP addresses by the **Domain Name System (DNS)** protocol. Servers using this protocol (called DNS servers) maintain lists of computers' and Web sites' addresses and their associated IP addresses.

Uniform resource locators (URLs), also called *universal resource locators*, identify a Web page. A URL is the address of a document or site on the Internet.

7-2

Hypertext Markup Language (HTML) is the language used to create Web pages. It defines a page's layout and appearance by using tags and attributes. A tag delineates a section of the page, such as the header or body; an attribute specifies a value, such as a font color, for a page component.

Navigational tools are used to travel from Web site to Web site—as in "surf" the Internet.

A **search engine**, such as Google.com or Ask.com, is an information system that enables users to retrieve data from the Web by using search terms.

Directories are indexes of information based on keywords embedded in documents, which make it possible for search engines to find what you are looking for.

7-3

Discussion groups are usually for exchanging opinions and ideas on a specific topic, usually of a technical or scholarly nature. Group members post messages or articles that others in the group can read.

Newsgroups are typically more general in nature and can cover any topic; they allow people to get together for fun or for business purposes.

Internet Relay Chat (IRC) enables users in chat rooms to exchange text messages with people in other locations in real time.

Instant messaging (IM) is a service for communicating with others via a private "chat room" on the Internet.

Internet telephony is using the Internet rather than the telephone network to exchange spoken conversations.

Voice over Internet Protocol (VoIP) is the protocol used for Internet telephony.

7-5

An **intranet** is a network within an organization that uses Internet protocols and technologies (e.g., TCP/IP, which includes File Transfer Protocol [FTP], SMTP, and others) for collecting, storing, and disseminating useful information that supports business activities, such as sales, customer service, human resources, and marketing.

7-6

An **extranet** is a secure network that uses the Internet and Web technologies to connect intranets of business partners so communication between organizations or between consumers is possible.

7-7

Web 2.0 refers to the trend toward Web applications that are more interactive than traditional Web applications. Collaboration or e-collaboration is one of its key components.

A **blog** (short for *Weblog*) is a journal or newsletter that is updated frequently and intended for the general public. Blogs reflect their authors' personalities and often include philosophical reflections and opinions on social or political issues.

A **wiki** is a type of Web site that allows users to add, delete, and sometimes modify content.

Social networking refers to a broad class of Web sites and services that

7-6 Explain the purpose of extranets.

An extranet is a secure network that uses the Internet and Web technologies to connect intranets of business partners, so communication between organizations or between consumers is possible.

7-7 Summarize the trends of the Web 2.0 and Web 3.0 eras and Internet2.

Web 2.0 describes the trend of Web applications that are more interactive than traditional Web applications. Some of these applications include blogs, wikis, social networking sites, RSS, and podcasts. See Table 7.4.

The Internet2 (I2) is a collaborative effort involving more than 200 U.S. universities and corporations to develop advanced Internet technologies and applications for higher education and academic research.

Most experts agree that Web 3.0, or the Semantic Web, provides personalization that allows users to access the Web more intelligently.

7-8 Describe the Internet of Everything.

The Internet of Everything (IOE) refers to a Web-based development in which people, processes, data, and things are interconnected via the Internet using various means, such as RFID devices, barcodes, wireless systems (using Bluetooth and WI-FI), and QR codes. And the Internet of Things (IoT) refers to the physical objects that are connected to the Internet and, therefore, to all the other physical objects.

TABLE 7.4 WEB 1.0 VS. WEB 2.0	
Web 1.0	**Web 2.0**
DoubleClick (used for online marketing)	Google AdSense
Ofoto (sharing digital photos)	Flickr
Akamai (streaming media services)	BitTorrent
mp3.com	iTunes
Britannica Online	Wikipedia
Personal Web sites	Blogging
eVite (type of wiki for event planning)	Upcoming.org and Events and Venues Database (EVBD)
Domain name speculation	Search engine optimization
Page views	Cost per click
Content management systems	Wikis
ERoom and Groove (collaboration software)	Collaboration portals, such as IBM Quickr and Microsoft Sharepoint
Posting a movie file on a personal Web page	YouTube

© Cengage Learning®

allows users to connect with friends, family, and colleagues online as well as meet people with similar interests or hobbies.

RSS (Really Simple Syndication) feeds are a fast, easy way to distribute Web content in Extensible Markup Language (XML) format. It is a subscription service, and new content from Web sites you have selected is delivered via a feed reader to one convenient spot.

A **podcast** is an electronic audio file, such as an MP3 file, that is posted on the Web for users to download to their mobile devices—iPhones, iPods, and iPads, for example—or even their computers.

Internet2 (I2) is a collaborative effort involving more than 200 U.S. universities and corporations to develop advanced Internet

technologies and applications for higher education and academic research.

A **gigapop** is a local connection point-of-presence that connects a variety of high-performance networks, and its main function is the exchange of I2 traffic with a specified bandwidth.

7-8

The **Internet of Everything (IoE)** refers to a Web-based development in which people, processes, data, and things are interconnected via the Internet using various means, such as RFID devices, barcodes, wireless systems (using Bluetooth and WI-FI), and QR codes.

The **Internet of things (IoT)** refers to the physical objects that are connected to the Internet and, therefore, to all the other physical objects.

WWW.CENGAGEBRAIN.COM/LOGIN

Like 4LTR Press for MIS on Facebook
Follow 4LTRPress_MIS on Twitter
View related videos at the YouTube channel created for MIS

LEARNING OUTCOMES

9-1 Discuss reasons for globalization and using global information systems, including e-business and Internet growth.

The global economy is creating customers who demand integrated worldwide services, and the expansion of global markets is a major factor in developing global information systems to handle these integrated services. Many companies have become international. In 2010, for example, the Coca-Cola company generated more than 75% of its revenue from outside the United States. Airline reservation systems are considered the first large-scale interactive global system, and hotels, rental car companies, and credit card services also require worldwide databases now to serve their customers more efficiently and effectively.

E-business is a major factor in the widespread use of global information systems. The Internet can simplify communication, change business relationships, and offer new opportunities to both consumers and businesses.

Today, the Internet is a part of daily life in most parts of the world. Growth has been highest in the Middle East and lowest in North America. The number of Internet users worldwide tops 2.27 billion, with Asia having the most.

9-2 Describe global information systems and their requirements and components.

A global information system (GIS) is an information system that works across national borders, facilitates communication between headquarters and subsidiaries in other countries, and incorporates all the technologies and applications found in a typical information system to store, manipulate, and transmit data across cultural and geographic boundaries.

Although a GIS can vary quite a bit depending on a company's size and business needs, most GISs have these basic components:
- A global database
- Information-sharing technologies

A GIS must be capable of supporting complex global decisions. This complexity stems from the global environment in which multinational corporations (MNCs) operate. A GIS, like any information system, can be classified by levels of management affecting different functions: operational, tactical, and strategic. The complexities of global decision making mean that a GIS has some functional requirements that differ from a domestic information system's requirements.

GISs can be categorized in different ways, depending on their function or application. Global marketing information systems, strategic intelligent systems, transnational management support systems, and global competitive intelligent systems are some different names for GISs.

9-3 Explain the types of organizational structures used with global information systems.

The most important factor for effective operation of an MNC is coordination, and a global information system can provide essential information for this task. There are four commonly accepted types of global organizations:
- Multinational
- Global
- International
- Transnational

In a multinational structure production, sales, and marketing are decentralized, and financial management remains the parent's responsibility. An organization following a global structure manages highly centralized information systems. Subsidiaries have little autonomy and rely on headquarters for all process and control decisions as well as system design and implementation. An organization using an international structure operates much like a multinational corporation, but subsidiaries depend on headquarters more for process and production decisions. In an organization following a transnational structure, the parent and all subsidiaries work together in designing policies, procedures, and logistics for delivering products to the right market.

Offshore outsourcing is an alternative for developing information systems. With this approach, an organization chooses an outsourcing firm in another country that can provide needed services and products. See Table 9.1.

KEY TERMS

9-2
A **global information system (GIS)** is an information system that works across national borders, facilitates communication between headquarters and subsidiaries in other countries, and incorporates all the technologies and applications found in a typical information system to gather, store, manipulate, and transmit data across cultural and geographic boundaries.

Transborder data flow (TDF) restricts what type of data can be captured and transmitted in foreign countries.

A **multinational corporation (MNC)** is an organization with assets and operations in at least one country other than its home country. An MNC delivers products and services across national borders and is usually centrally managed from its headquarters.

9-3
In a **multinational structure**, production, sales, and marketing are decentralized, and financial management remains the parent's responsibility.

A **global structure** (also known as franchiser) uses highly centralized information systems. Subsidiaries have little autonomy and rely on headquarters for all process and control decisions as well as system design and implementation.

An organization with an **international structure** operates much like a multinational corporation, but subsidiaries depend on headquarters more for process and production decisions.

In an organization with a **transnational structure**, the parent and all subsidiaries work together in designing policies, procedures, and logistics for delivering products and services to the right market.

With **offshore outsourcing**, an organization chooses an outsourcing firm in another country that can provide needed services and products.

TABLE 9.1	TOP OFFSHORE LOCATIONS FOR OUTSOURCING IN 2012	
Americas	**Asia/Pacific**	**Europe, Middle East, and Africa (EMEA)**
Argentina	Bangladesh	Belarus
Brazil	China	Bulgaria
Chile	India	Czech Republic
Colombia	Indonesia	Hungary
Costa Rica	Malaysia	Mauritius
Mexico	Philippines	Morocco
Peru	Sri Lanka	Poland
Uruguay	Thailand	Romania
	Vietnam	Russia
		Slovakia
		South Africa
		Turkey
		Ukraine

© Cengage Learning®

9-4 Discuss obstacles to using global information systems.

The following factors can prevent the success of a GIS (see the adjoining information box on making a company Web site global):

- Lack of standardization (can also include differences in time zones, taxes, language, work habits, and so forth)
- Cultural differences
- Diverse regulatory practices
- Poor telecommunication infrastructures
- Lack of skilled analysts and programmers

MAKING A COMPANY WEB SITE GLOBAL

Here are some suggestions for making a company Web site global:

- *Language*—Present your Web site in one of the seven most-used languages: English, French, Italian, German, Spanish, Japanese, or Chinese.
- *Font*—Make sure the Web site's main features are readable in different languages, depending on which font is selected.
- *Cultural differences*—Keep cultural variation in mind. For example, although white is the color of purity in the United States, it is the color of mourning in Eastern cultures. Also, a woman smiling without covering her mouth would be considered sexually suggestive in Japan.
- *Currency*—Make sure the Web site includes a currency conversion feature.
- *Date format*—This varies around the world. Many countries use day/month/year instead of month/day/year.
- *Spare use of graphics and flash features*—Because of bandwidth limitations in many parts of the world, you may want to keep the Web site rather simple.
- *E-mail*—Be prepared to send and receive e-mails in foreign languages.
- *Payments*—Not everybody uses credit cards. In Germany, for example, bank transfers are popular, whereas in Japan both COD and bank transfers are popular.
- *International logistics*—To ship internationally, you must address the various regulations that pertain to each country.
- *International listing*—List your Web site with international search engines and indexes.
- *Local involvement*—Invite local people to review the Web site before you launch it to the global market.
- *International marketing*—Promote your Web site in specific languages.

WWW.CENGAGEBRAIN.COM/LOGIN

Like 4LTR Press for MIS on Facebook
Follow 4LTRPress_MIS on Twitter
View related videos at the YouTube channel created for MIS

LEARNING OUTCOMES

11-1 Explain how supply chain management is used.

A supply chain is an integrated network consisting of an organization, its suppliers, transportation companies, and brokers used to deliver goods and services to customers. Supply chains exist in both service and manufacturing organizations, although the chain's complexity can vary widely in different organizations and industries.

Supply chain management (SCM) is the process of working with suppliers and other partners in the supply chain to improve procedures for delivering products and services. An SCM system coordinates:

- Procuring materials
- Transforming materials into intermediate and finished products or services
- Distributing finished products or services

The following tools help overcome some challenges associated with SCM:

Electronic data interchange (EDI) enables business partners to send and receive information on business transactions.

An e-marketplace is a third-party exchange (a B2B business model) that provides a platform for buyers and sellers to interact with each other and trade more efficiently online.

An online auction brings traditional auctions to customers around the globe and makes it possible to sell far more goods and services than at a traditional auction.

Collaborative planning, forecasting, and replenishment (CPFR) is used to coordinate supply chain members through point-of-sale (POS) data sharing and joint planning. CPFR ensures that inventory and sales data is shared across the supply chain.

11-2 Describe customer relationship management systems.

Customer relationship management (CRM) consists of the processes a company uses to track and organize its contacts with customers. The main goal of a CRM system is to improve services offered to customers and use customer contact information for targeted marketing. CRM gives organizations more complete pictures of their customers. Typically, CRM applications are implemented with one of these approaches: on-premise CRM or Web-based CRM.

11-3 Describe knowledge management systems.

Knowledge management (KM) is a technique used to improve CRM systems (and many other systems) by identifying, storing, and disseminating "know-how"—facts about how to perform tasks.

Personalization is the process of satisfying customers' needs, building customer relationships, and increasing profits by designing goods and services that meet customers' preferences better.

Customization, which is somewhat different from personalization, allows customers to modify the standard offering, such as selecting a different home page to be displayed each time you open your Web browser.

11-4 Describe enterprise resource planning systems.

Enterprise resource planning (ERP) is an integrated system that collects and processes data and manages and coordinates resources, information, and functions throughout an organization. A well-designed ERP system offers many benefits. More than 40 vendors, such as SAP, Oracle, Sage Group, and Microsoft, offer ERP software with varying capabilities. See Table 11.1.

KEY TERMS

11-1

An **enterprise system** is an application used in all the functions of a business that supports decision making throughout the organization.

A **supply chain** is an integrated network consisting of an organization, its suppliers, transportation companies, and brokers used to deliver goods and services to customers.

Supply chain management (SCM) is the process of working with suppliers and other partners in the supply chain to improve procedures for delivering products and services.

Electronic data interchange (EDI) enables business partners to send and receive information on business transactions.

An **e-marketplace** is a third-party exchange (B2B model) that provides a platform for buyers and sellers to interact with each other and trade more efficiently online.

By using the Internet, an **online auction** brings traditional auctions to customers around the globe and makes it possible to sell far more goods and services than at a traditional auction.

A **reverse auction** invites sellers to submit bids for products and services. In other words, there is one buyer and many sellers: a one-to-many relationship. The buyer can choose the seller that offers the service or product at the lowest price.

Collaborative planning, forecasting, and replenishment (CPFR) is used to coordinate supply chain members through point-of-sale (POS) data sharing and joint planning.

11-2

Customer relationship management (CRM) consists of the processes a company uses to track and organize its contacts with customers. It improves services offered to customers and uses customer contact information for targeted marketing.

Personalization is the process of satisfying customers' needs, building customer relationships, and increasing profits by designing goods and services that meet customers' preferences better. It involves not only customers' requests, but also the interaction between customers and the company.

Customization allows customers to modify the standard offering, such as selecting a different home page to be displayed each time you open your Web browser.

Collaborative filtering (CF) is a search for specific information or patterns, using input from multiple business partners and data sources. It identifies groups of people based on common interests and recommends products or services based on what members of the group purchased or did not purchase.

11-3
Knowledge management (KM) draws on concepts of organizational learning, organizational culture, and best practices to convert tacit knowledge into explicit knowledge, create a knowledge-sharing culture in an organization, and eliminate obstacles to sharing knowledge.

11-4
Enterprise resource planning (ERP) is an integrated system that collects and processes data and manages and coordinates resources, information, and functions throughout an organization.

TABLE 11.1 ERP COMPONENTS

Component	Function
Unified database	Collects and analyzes relevant internal and external data and information needed by other functions
Inventory management	Provides inventory status and inventory forecasts
Supply chain	Provides information on supply chain members, including suppliers, manufacturing, distribution, and customers
Manufacturing	Provides information on production costs and pricing
Human resources	Provides information on assessing job candidates, scheduling and assigning employees, and predicting future personnel needs
CRM	Provides information on customers and their needs and preferences
Purchasing	Provides information related to the purchasing function, including e-procurement
Accounting	Tracks financial information, such as budget allocations and debits and credits
Vendor integration	Integrates information for vendors, such as offering automated downloads of data on product pricing, specifications, and availability
E-commerce	Provides B2C information related to order status and B2B information related to suppliers and business partners
Sales	Provides information on sales and marketing

© Cengage Learning®

WWW.CENGAGEBRAIN.COM/LOGIN

Like 4LTR Press for MIS on Facebook
Follow 4LTRPress_MIS on Twitter
View related videos at the YouTube channel created for MIS

LEARNING OUTCOMES

13-1 Define artificial intelligence, and explain how AI technologies support decision making.

Artificial intelligence (AI) consists of related technologies that try to simulate and reproduce human thought behavior, including thinking, speaking, feeling, and reasoning. The most recent developments in AI technologies promise new areas of decision-making support.

Over the years, the capabilities of these systems have improved in an attempt to close the gap between artificial intelligence and human intelligence. See Table 13.1.

KEY TERMS

13-1
Artificial intelligence (AI) consists of related technologies that try to simulate and reproduce human thought behavior, including thinking, speaking, feeling, and reasoning. AI technologies apply computers to areas that require knowledge, perception, reasoning, understanding, and cognitive abilities.

TABLE 13.1 APPLICATIONS OF AI TECHNOLOGIES

Field	Organization	Applications
Energy	Arco and Tenneco Oil Company	Neural networks used to help pinpoint oil and gas deposits
Government	Internal Revenue Service	Software used to read tax returns and spot fraud
Human services	Merced County, California	Expert systems used to decide if applicants should receive welfare benefits
Marketing	Spiegel	Neural networks used to determine most likely buyers from a long list
Telecommunications	BT Group	Heuristic search used for a scheduling application that provides work schedules for more than 20,000 engineers
Transportation	American Airlines	Expert systems used to schedule the routine maintenance of its airplanes
Inventory/forecasting	Hyundai Motors	Neural networks and expert systems used to reduce delivery time by 20 percent and increase inventory turnover from 3 to 3.4
Inventory/forecasting	SCI Systems	Neural networks and expert systems used to reduce on-hand inventory by 15 percent, resulting in $180 million in annual savings
Inventory/forecasting	Reynolds Aluminum	Neural networks and expert systems used to reduce forecasting errors by 2 percent, resulting in an inventory reduction of 1 million pounds
Inventory/forecasting	Unilever	Neural networks and expert systems used to reduce forecasting errors from 40 percent to 25 percent, resulting in a multimillion-dollar savings

© Cengage Learning®

13-2 Describe an expert system, its applications, and its components.

Expert systems mimic human expertise in a field to solve a problem in a well-defined area. A typical expert system includes:

- Knowledge acquisition facility
- Knowledge base
- Factual knowledge
- Heuristic knowledge
- Meta-knowledge
- Knowledge base management system
- User interface
- Explanation facility
- Inference engine

13-3 Describe case-based reasoning.

Case-based reasoning (CBR) is a problem-solving technique that matches a new case (problem) with a previously solved case and its solution stored in a database.

Robots are one of the most successful applications of AI. They perform well at simple, repetitive tasks and can be used to free workers from tedious or hazardous jobs.

13-2
Expert systems mimic human expertise in a particular field to solve a problem in a well-defined area.

A **knowledge acquisition facility** is a software package with manual or automated methods for acquiring and incorporating new rules and facts so the expert system is capable of growth.

A **knowledge base** is similar to a database, but in addition to storing facts and figures it keeps track of rules and explanations associated with facts.

A **knowledge base management system (KBMS)**, similar to a DBMS, is used to keep the knowledge base updated, with changes to facts, figures, and rules.

An **explanation facility** performs tasks similar to what a human expert does by explaining to end users how recommendations are derived.

An **inference engine** is similar to the model base component of a decision support system. By using different techniques, such as forward and backward chaining, it manipulates a series of rules.

In **forward chaining**, a series of "if-then-else" condition pairs is performed.

In **backward chaining**, the expert system starts with the goal—the "then" part—and backtracks to find the right solution.

13-3

Case-based reasoning (CBR) is a problem-solving technique that matches a new case (problem) with a previously solved case and its solution, both stored in a database. After searching for a match, the CBR system offers a solution; if no match is found, even after supplying more information, the human expert must solve the problem.

13-4

Intelligent agents are software capable of reasoning and following rule-based processes; they are becoming more popular, especially in e-commerce.

Shopping and information agents help users navigate through the vast resources available on the Web and provide better results in finding information. These agents can navigate the Web much faster than humans and gather more consistent, detailed information. They can serve as search engines, site reminders, or personal surfing assistants.

Personal agents perform specific tasks for a user, such as remembering information for filling out Web forms or completing e-mail addresses after the first few characters are typed.

Data-mining agents work with a data warehouse, detecting trends and discovering new information and relationships among data items that were not readily apparent.

Monitoring and surveillance agents usually track and report on computer equipment and network systems to predict when a system crash or failure might occur.

13-5

Fuzzy logic allows a smooth, gradual transition between human and computer vocabularies and deals with variations in linguistic terms by using a degree of membership.

13-6

Artificial neural networks (ANNs) are networks that learn and are capable of performing tasks that are difficult with conventional computers, such as playing chess, recognizing patterns in faces and objects, and filtering spam e-mail.

13-4 Summarize the types of intelligent agents and how they are used.

Intelligent agents, or bots, consist of software capable of reasoning and following rule-based processes. These include:
- Shopping and information agents
- Personal agents
- Data-mining agents
- Monitoring and surveillance agents

13-5 Describe fuzzy logic and its uses.

Fuzzy logic is designed to help computers simulate vagueness and uncertainty in common situations. Fuzzy logic has been used in search engines, chip design, database management systems, software development, and more.

13-6 Explain artificial neural networks.

Artificial neural networks (ANNs) are networks that learn and are capable of performing tasks that are difficult with conventional computers.

13-7 Describe how genetic algorithms are used.

Genetic algorithms (GAs) are used mostly in techniques to find solutions to optimization and search problems. Genetic algorithms can examine complex problems without any assumptions of what the correct solution should be.

13-8 Explain natural-language processing and its advantages and disadvantages.

Natural-language processing (NLP) was developed so that users could communicate with computers in their own language. The size and complexity of the human language has made developing NLP systems difficult.

13-9 Summarize the advantages of integrating AI technologies into decision support systems.

AI-related technologies add explanation capabilities (by integrating expert systems) and learning capabilities (by integrating ANNs) and create an interface that's easier to use (by integrating an NLP system).

13-10 Explain contextual computing.

Contextual computing refers to a computing environment that is always present, can feel our surroundings, and—based on who we are, where we are, and whom we are with—offer recommendations. As an example, your smartphone may soon be able to predict with 80 percent probability that you will receive a job offer if you go to a particular job fair, based on information you have included on your various social media sites.

13-7

Genetic algorithms (GAs) are search algorithms that mimic the process of natural evolution. They are used to generate solutions to optimization and search problems using such techniques as mutation, selection, crossover, and chromosome.

13-8

Natural-language processing (NLP) was developed so users could communicate with computers in human language.

13-10

Contextual computing refers to a computing environment that is always present, can feel our surroundings, and—based on who we are, where we are, and whom we are with—offer recommendations.

NOTES